Chemical Principles of
Environmental Pollution

Chemical Principles of Environmental Pollution

Second edition

B. J. Alloway
University of Reading, UK
and
D. C. Ayres
Queen Mary and Westfield College,
University of London, UK

BLACKIE ACADEMIC & PROFESSIONAL
An Imprint of Chapman & Hall

London · Weinheim · New York · Tokyo · Melbourne · Madras

**Published by Blackie Academic and Professional, an imprint of
Chapman & Hall, 2–6 Boundary Row, London SE1 8HN, UK**

Chapman & Hall, 2–6 Boundary Row, London SE1 8HN, UK

Chapman & Hall GmbH, Pappelallee 3, 69469 Weinheim, Germany

Chapman & Hall USA, 115 Fifth Avenue, New York, NY 10003, USA

Chapman & Hall Japan, ITP-Japan, Kyowa Building, 3F, 2-2-1 Hirakawacho, Chiyoda-ku, Tokyo 102, Japan

DA Book (Aust.) Pty Ltd, 648 Whitehorse Road, Mitcham 3132, Victoria, Australia

Chapman & Hall India, R. Seshadri, 32 Second Main Road, CIT East, Madras 600 035, India

First edition 1993
Reprinted 1994
Second edition 1997

© 1997 Chapman & Hall

Typeset in 10/12 pt Times by Thomson Press (I) Ltd, New Delhi
Printed in Great Britain by Clays Ltd, Bungay, Suffolk

ISBN 0 7514 0380 6

A catalogue record for this book is available from the British Library

Library of Congress Catalog Card Number: 96–85900

∞ Printed on permanent acid-free text paper, manufactured in accordance with ANSI/NISO Z39.48-1992 and ANSI/NISO Z39.48-1984 (Permanence of Paper)

Contents

Preface   xi

Glossary of acronyms and abbreviations   xiii

PART ONE   BASIC PRINCIPLES

1 Introduction   **3**

1.1	Pollution in the modern world	3
1.2	Definition of pollution	5
1.3	Air pollution	7
1.4	Environmental literature	9
1.5	Population pressures	10
1.6	Energy demand	11
1.7	Wastes	13
1.8	Concluding remarks	13
	References	14
	Further reading	15

2 Transport and behaviour of pollutants in the environment   **17**

2.1	A basic model of environmental pollution	17
2.2	Sources of pollutants	17
2.3	The pollutants	18
	2.3.1 Classification of hazardous substances in the USA	20
	2.3.2 European Community Dangerous Substances Directive	22
	2.3.3 UK priority list of pollutants	22
	2.3.4 Pesticides	24
	2.3.5 Indoor pollution	26
2.4	Physical processes of pollutant transport and dispersion	27
	2.4.1 Transport media	27
	2.4.2 Transport of pollutants in air	27
	2.4.3 Some important types of reaction which pollutants undergo in the atmosphere	34
2.5	Dispersion of pollutants in water	36
	2.5.1 Physical transport in surface waters	36
	2.5.2 Dispersion of pollutants in groundwaters	39
	2.5.3 Biochemical processes in water (involving microorganisms)	46
2.6	The behaviour of pollutants in the soil	53
	2.6.1 The composition and physico-chemical properties of soils	53
	2.6.2 Cation and anion adsorption in soils	55
	2.6.3 Adsorption and decomposition of organic pollutants	58
2.7	Overview of the fate of soil pollutants	62
	References	63
	Further reading	64

3   Toxicity and risk assessment of environmental pollutants   **65**

3.1   Basic principles of toxicology   65
3.2   Effect of pollutants on animals and plants   68
 3.2.1   Effect of pollutants on humans and other mammals   68
 3.2.2   Teratogenesis, mutagenesis, carcinogenesis and immune
   system defects   69
 3.2.3   Ecotoxicology   75
3.3   Assessment of toxicity risks   75
 3.3.1   Pollutants in contaminated land   77
 3.3.2   Pollutants in drinking water   81
 3.3.3   Toxic or explosive gases and vapours   84
References   84
Further reading   85

4   Analysis and monitoring of pollutants   **86**

4.1   Introduction   86
4.2   Chromatography   87
4.3   Thin-layer chromatography   88
 4.3.1   Separation of pesticides by HPLC   89
 4.3.2   Separation of metal cations by HPLC   89
4.4   Gas–liquid chromatography   90
 4.4.1   Detection of eluted substances   93
 4.4.2   Principal parameters   95
 4.4.3   Optimum operating conditions   97
 4.4.4   Capillary columns for GLC   98
 4.4.5   Analysis of urban air pollution   99
 4.4.6   Detection by mass spectrometry   99
 4.4.7   Analysis of common pollutants   109
4.5   High pressure liquid chromatography   113
 4.5.1   The components   114
 4.5.2   Detectors   115
 4.5.3   Analysis of polluted air   117
 4.5.4   Analysis of polluted water   120
 4.5.5   Trace enrichment followed by GLC analysis   121
4.6   Pollution by metals – atomic absorption spectroscopy   123
 4.6.1   Historical introduction   123
 4.6.2   Basic theory of atomic absorption and emission   124
 4.6.3   The Lambert–Beer law   125
 4.6.4   Instrumental details   126
 4.6.5   Interferences   127
 4.6.6   The determination of sodium in concrete   128
 4.6.7   Sample preparation   129
 4.6.8   Precision and accuracy of measurement   130
 4.6.9   Graphite furnace atomization   130
4.7   A plasma source   131
 4.7.1   Method development in ICP-OES (ICP-AES)   133
 4.7.2   Determination of metals in environmental samples   135
 4.7.3   ICP-mass spectrometry   137
4.8   Analytical quality assurance   137
4.9   Environmental monitoring   139
 4.9.1   Introduction   139
 4.9.2   Air pollution monitoring   141
 4.9.3   Water pollution monitoring   143
 4.9.4   Soil pollution monitoring   146
 4.9.5   Plant sampling and analysis   149
 4.9.6   Sampling and analysis of animals and animal tissue   150
 4.9.7   The use of biological indicators in environmental monitoring   150

	References		152
	Further reading		153

PART TWO   THE POLLUTANTS

5   Inorganic pollutants   157

5.1	Ozone		157
	5.1.1	Historical introduction	157
	5.1.2	Formation	157
	5.1.3	Physical properties and structure	157
	5.1.4	The ozone layer	158
	5.1.5	Factors which disturb the natural environment	159
	5.1.6	Chemistry of stratospheric CFC	161
	5.1.7	Control measures (*American Conference of Government Industrial Hygienists, 1990*)	162
	5.1.8	Ozone in the troposphere	162
	5.1.9	Diurnal variations of ozone levels	163
	5.1.10	Toxicity and control	164
5.2	Oxides of carbon, nitrogen and sulphur		164
	5.2.1	Carbon dioxide	165
	5.2.2	Oxides of nitrogen	174
	5.2.3	Oxides of sulphur	179
5.3	Heavy metals		190
	5.3.1	General properties	190
	5.3.2	Biochemical properties of heavy metals	191
	5.3.3	Sources of heavy metals	192
	5.3.4	Environmental media affected	199
	5.3.5	Heavy metal behaviour in the environment	200
	5.3.6	Toxic effects of heavy metals	207
	5.3.7	Analytical methods	211
	5.3.8	Examples of specific heavy metals	211
5.4	Other metals and inorganic pollutants		217
	5.4.1	Aluminium	217
	5.4.2	Beryllium	218
	5.4.3	Fluorine	219
5.5	Radionuclides		220
	5.5.1	History and nomenclature	220
	5.5.2	Types of radioactive emission	220
	5.5.3	Units of energy and measurement of toxicity	222
	5.5.4	Radioactive potassium	223
	5.5.5	Production of radionuclides by artificial means	223
	5.5.6	Nuclear fission	223
	5.5.7	Power generation in nuclear reactors	225
	5.5.8	Nuclear reactor types	226
	5.5.9	The future of nuclear power	231
	5.5.10	Observations on major accidents	233
	5.5.11	Social aspects of nuclear power generation	237
	5.5.12	Power from thermal fusion	238
	5.5.13	Cold fusion	239
5.6	Mineral fibres and particles		240
	5.6.1	General aspects	240
	5.6.2	Examples of mineral pollutants	241
	References		242
	Further reading		245

6   Organic pollutants   247

6.1	Smoke		247
6.2	Methane and other hydrocarbons – coal and oil as sources		252

	6.2.1	The formation of coal	252
	6.2.2	Petroleum	253
	6.2.3	Methane	258
	6.2.4	Higher alkanes	259
	6.2.5	Polycyclic aromatic hydrocarbons (PAHs)	267
6.3	Organic solvents	271
	6.3.1	Adhesives	272
	6.3.2	Coatings and inks	273
	6.3.3	Aerosol sprays	274
	6.3.4	Metal cleaning	275
	6.3.5	Dry cleaning of clothes	275
	6.3.6	Solvent toxicology	276
	6.3.7	Organochlorine compounds	277
	6.3.8	Detergents	279
6.4	Organohalides: pesticides, PCBs and dioxins	282
	6.4.1	Historical note	282
	6.4.2	Organochlorine production	282
	6.4.3	DDT (dichlorodiphenyl trichloroethane)	283
	6.4.4	Lindane, hexachlorocyclohexane	284
	6.4.5	Some other chlorinated pesticides	285
	6.4.6	Organochlorine herbicides	287
	6.4.7	Toxic effects of insecticides	288
	6.4.8	Control of pesticides	291
	6.4.9	Vinyl chloride and polyvinyl chloride *(Kirk-Othmer, 1996)*	292
	6.4.10	Polychlorobiphenyls	293
	6.4.11	Toxic substances in herbicides	297
	6.4.12	Metabolism of chloroaromatic compounds	305
	6.4.13	Disposal of organochlorine compounds	305
	6.4.14	Cremation or burial of wastes	307
	6.4.15	Use of biodegrading microorganisms	307
6.5	Natural, organophosphorus and carbamate pesticides	307
	6.5.1	Naturally occurring pesticides	307
	6.5.2	Organophosphorus pesticides	312
	6.5.3	Carbamate pesticides	318
6.6	Odours	322
	6.6.1	Important properties of odours	322
	6.6.2	Methods of odour control	323
	6.6.3	Methods of odour treatment	323
References		324
Further reading		327

7 Indoor pollution 329

7.1	Volatile organic compounds	329
	7.1.1	Oxidized forms of VOCs	331
7.2	Ozone	333
7.3	Oxides of nitrogen	334
7.4	Carbon monoxide	335
7.5	Other gaseous oxides	336
7.6	Cigarette smoking	336
	7.6.1	*N*-Nitrosocompounds	337
	7.6.2	Polycyclic aromatic hydrocarbons	338
	7.6.3	Other mutagens	339
7.7	Asbestos fibres	340
	7.7.1	Analysis	341
	7.7.2	Toxicity	341
	7.7.3	Other particulate pollutants	343
7.8	Lead	344
7.9	Radon	346

7.9.1	Cancer risk	346
7.9.2	Remedial measures	348
References		348
Further reading		350

PART THREE   WASTE AND OTHER MULTIPOLLUTANT SITUATIONS

8   Wastes and their disposal   353

8.1	Introduction	353
8.2	Amounts of waste produced	353
	8.2.1 Industrial wastes	354
	8.2.2 Municipal wastes	354
8.3	Methods of disposal of municipal wastes	354
	8.3.1 Landfilling	354
	8.3.2 Incineration	358
	8.3.3 Composting	359
	8.3.4 Recycling	359
8.4	Sewage treatment	360
8.5	Hazardous wastes	365
	8.5.1 The nature and amount of hazardous waste produced	365
	8.5.2 Hazardous waste management	366
	8.5.3 New technologies for waste disposal	368
8.6	Long-term pollution problems of abandoned landfills containing hazardous wastes	369
	8.6.1 Love Canal, New York, USA	369
	8.6.2 Lekkerkirk, near Rotterdam, the Netherlands	371
	8.6.3 The Chemstar fire at Carrbrook, Cheshire	371
	8.6.4 Abandoned waste tips	372
8.7	Tanker accidents and oil spillages at sea	372
8.8	Other multipollutant situations	373
	8.8.1 Pollution from warfare and military training	374
8.9	Chemical time bombs	376
References		377

Appendix	Table of units and conversions	**379**
Index		**381**

Preface

Pollution is the most serious of all environmental problems and poses a major threat to the health and well-being of millions of people and global ecosystems. Other major environmental problems are also partly caused by pollution; these include global warming, climatic change and the loss of biodiversity through the extinction of many species. In the future, environmental pressures can only increase as a result of population growth and the expectation of higher living standards.

For at least thirty years people have become increasingly aware of these issues. As a result, governments and regulatory bodies have responded by taking action against grossly polluting activities and by enforcing tighter limits on the emission of pollutants into the environment. As the level of control improves so the financial costs increase exponentially, hence an effective limit must be enforced which does not impose unacceptable burdens on industrial producers.

As a consequence of the increasing economic and legal pressures which result from pollution control, there is a growing need for greater understanding of the scientific principles underlying environmental pollution. Appropriately qualified professionals will be needed in the energy, manufacturing, service and waste disposal industries and their regulatory authorities. They will be required to enforce increasingly strict standards and to monitor the environment for accidental pollution. The rapid growth in the numbers studying for undergradute and postgraduate degrees and related qualifications in environmental science is a reflection of these requirements. Modules dealing with aspects of pollution have been introduced into many courses in the traditional disciplines of chemistry, biology, geography and civil engineering.

This book was written, in sympathy with these trends, for students and others with a basic knowledge of chemistry. It provides an introduction to the principles, and adopts a pollutant-oriented approach, rather than the more common one based on specific media such as air or water. All the main groups of substances are covered, and the principles relating to their nature, sources, transport, environmental behaviour and effects on targets. It is expected that readers will want to consult more advanced specialized texts dealing with particular topics, and an extensive bibliography has been included to further their studies.

The authors have been involved in teaching and researching this subject for many years and both have directed a BSc Environmental Science degree programme (at Westfield and Queen Mary and Westfield Colleges in the University of London). Their research interests include heavy metals and contaminated land (BJA), and organic pollutants and their analysis (DCA). Although both authors have come into environmental science from different subject backgrounds, in due course they both developed an interdisciplinary perspective. Nevertheless, it must be stressed that the investigation and management of environmental pollution requires the rigorous application of scientific principles. The theoretical basis of environmental pollution problems is intellectually demanding, and environmental scientists need to be able to understand the underlying principles in order to make effective decisions.

The advice of Dr Trevor Toube concerning mass spectrometry, of Dr Roger Brown on pollution monitoring and of Dr Peixun Zhang on emission spectroscopy, is gratefully acknowledged. We also thank Mr John Cross of the NRA and Ms Elizabeth Ayres for their help with the manuscript

BJA
DCA

Glossary of acronyms and abbreviations

AAS	Atomic absorption spectrophotometry (analysis)
ACGIH	American Conference of Govermental Industrial Hygienists
ADI	allowable daily intake (toxicology)
AEA	Atomic Energy Authority (UK)
AGR	advanced gas-cooled reactor
ALARA	as low as reasonably achievable
ALARP	as low as reasonably possible
AMU	atomic mass unit
ANOVA	analysis of variance (statistical test)
APEs	alkyl phenol ethoxylates (chemicals)
AQA	analytical quality assurance
BATNEEC	best available technology not entailing excessive cost
BBP	butyl benzyl phthalate (chemical)
BCF	biological concentration factor
BHA	butylated hydroxyanisole (chemical)
BHC	benzene hexachloride (chemical)
BHT	butylated hydroxytoluene (chemical)
BOD	biochemical oxygen demand
BPEO	best practicable environmental option
BPM	best practicable means
BSI	British Standards Institution
BTEX	benzene, toluene, ethyl benzene and xylene (suite of organic chemicals)
BWR	boiling water reactor (nuclear)
CAMP	Continuous Air Monitoring Programme (USA)
CANDU	Canadian Deuterium Uranium Reactor
CEC	cation exchange capacity (soil chemistry)
CEC	Commission of the European Communities
CERCLA	Comprehensive Environmental Response Compensation and Liability Act (US) (1980) ('Superfund') – see also SARA
CFCs	chlorofluorocarbons (chemicals)
CHCs	chlorinated hydrocarbon compounds (chemicals)
Ci	curie

CIMAH	Control of Industrial Major Accidents Regulation (1984) (UK)
CNS	central nervous system
COD	chemical oxygen demand
COPA	Control of Pollution Act (UK) (1974)
COSHH	Control of Substances Hazardous to Health (UK)
CPs	chlorophenols (chemicals)
CRM	certified reference material (analytical)
CTB	chemical time bomb
DBP	di-n-butyl phthalate (chemical)
DCE	dichloroethene (chemical)
DDE	dichlorodiphenyl dichloroethene (chemical)
DDT	dichlorodiphenyl trichloroethane (chemical)
DEHP	bis(2-ethylhexyl) phthalate (chemical)
DNA	deoxyribonucleic acid (biochemical: genetic code material)
DNAPL	dense non-aqueous-phase liquid
DOCs	dissolved organic compounds
DoE	Department of the Environment (UK)
DoE	Department of Energy (US)
DTPA	diethyltriaminepentaacetic acid (chemical reagent)
EDTA	ethylenediaminetetraacetic acid (chemical reagent)
EIA	Environmental Impact Assessment (also EA: environmental assessment)
EPA	Environmental Protection Agency (US)
EPA	Environmental Protection Act (UK) (1990)
EQO	environmental quality objective
ETA–AAS	electrothermal activation–atomic absorption spectrometry (elemental analysis)
EU	European Union (formerly European Community)
FGD	flue gas desulphurization
GC	gas chromatography (analysis)
GC–MS	GC combined with mass spectrometry (analysis)
GDP	gross domestic product
GFAAS	graphite furnace atomic absorption spectrophotometry (also referred to as ETA–AAS) (analysis)
GIS	geographical information systems
GLC	gas–liquid chromatography (analysis)
GLP	good laboratory practice
Gy	the Gray (100 rads)
HCFC	hydrogen-containing chlorofluorocarbon (chemical)
HCH	hexachlorohexane (chemical)
HMIP	Her Majesty's Inspectorate of Pollution (UK)
HPLC	high-pressure liquid chromatography (analysis) also high performance liquid chromatography

HSE	Health and Safety Executive (UK)
IAEA	International Atomic Energy Authority
IARC	International Agency for Research on Cancer
ICP–OES	inductively coupled plasma optical emission spectrometry (analysis)
ICP–MS	inductively coupled plasma mass spectrometry (analysis)
ICRP	International Commission for Radiological Protection
INES	International Nuclear Event Scale
IPC	Integrated Pollution Control (Europe)
IPCC	International Panel for Climate Change
ISO	International Standards Organization
JET	Joint European Taurus
LAAPC	Local Authority Air Pollution Control
LC_{50}	lethal concentration for 50% survival
LD_{50}	lethal dose for 50% survival
LFG	landfill gas
LMFBR	liquid metal fast-breeder reactor
LNAPL	light non-aqueous-phase liquid
LOAEL	lowest observable adverse effect level (toxicology)
LOEL	lowest observable effect level (toxicology)
LPG	liquefied petroleum gas
LULU	locally unwanted land use
LUST	leaking underground storage tank
MAC	maximum allowable concentration (USA)
MAFF	Ministry of Agriculture, Fisheries and Food
MCP	monochlorophenol (chemical)
MEL	maximum exposure limit (toxicology)
MSW	municipal solid waste
NAAQS	National Air Quality Standards (USEPA)
NAPL	non-aqueous-phase liquid
NBS	National Bureau of Standards (US): supplier of CRMs
MCP	monochlorophenol (chemical)
NIMBY	'not in my back yard'
NIOSH	National Institute of Occupational Safety and Health (USA)
NOAEL	no observable adverse effects level (toxicology)
NOEL	no observable effects level (toxicology)
NPEs	nonyl phenol ethoxylates (chemicals)
NRA	National Rivers Authority (UK) (Incorporated into Environmental Agency 1996)
NRPB	National Radiological Protection Board (UK)
NTA	nitriloacetic acid (chemical)

NTP	normal temperature (273.15 K) and pressure (101/325 Pa)
ODP	ozone depletion potential
OECD	Organization for Economic Co-operation and Development
OEL	occupational exposure limit (toxicology)
OES	occupational exposure standard
OP	organophosphorus
OSHA	Occupational Safety and Health Administration
PAH	polycyclic (or polynuclear) aromatic hydrocarbon (chemical)
PAN	peroxyacetyl nitrate (chemical)
PCB	polychlorinatedbiphenyl (chemical)
PCDD	polychlorodibenzo-p-dioxin (chemical)
PCDF	polychlorodibenzofuran (chemical)
PCP	pentachlorophenol (chemical)
PEL	permissible exposure limit (toxicology)
PM_{10}	particulate matter smaller than 10 μm
PMTWI	provisional maximum tolerable weekly intake (toxicology)
POMP	persistent organic micropollutant
PRA	probabilistic risk analysis
PTWI	provisional tolerable weekly intake (toxicology)
PVC	polyvinyl chloride (chemical)
PWR	pressurized water reactor
QA	quality assurance
QC	quality control
QSAR	quantity, structural activity relationship
Rad	absorbed radiation of 0.01 J/kg
RCRA	Resource, Conservation and Recovery Act (US) (1976)
Rem	Roentgen equivalent humans
RNA	ribonucleic acid (biochemical: genetic code material)
SARA	Superfund Amendments and Reauthorization Act (US) (1986): amendment to CERCLA (1980)
SMR	standardized mortality ratio (epidemiology)
STP	standard temperature and pressure
STEL	short-term exposure limit (toxicology)
TBT	tributyltin (chemical)
TCB	trichlorobiphenyl (chemical)
TCDD	tetrachlorodibenzo dioxin (chemical)
TCE	trichloroethene (chemical)
TDI	tolerable daily intake (toxicology)
TDS	total dissolved solids
TEF	toxic equivalent factor (toxicology)

TEL	tetraethyl lead (chemical)
TEQ	toxic equivalent quantity (TEF × concentration)
THORP	thermal oxide reprocessing plant
TLV	threshold limit value
TOMP	toxic organic micropollutant
TSP	total suspended particles
TWA	time-weighted average
UARG	Urban Air Review Group
VCM	vinyl chloride monomer (chemical)
VOC	volatile organic compound
Waldsterben	forest death
WDA	Waste Disposal Authority (UK)
WHO	World Health Organization
XRD	X-diffraction (mineralogical analysis)
XRF	X-ray fluorescence (elemental analysis)

Part One
Basic Principles

Introduction 1

1.1
Pollution in the modern world

The dramatic increase in public awareness and concern about the state of the global and local environments which has occurred in recent decades has been accompanied and partly prompted by an ever growing body of evidence on the extent to which pollution has caused severe environmental degradation. The introduction of harmful substances into the environment has been shown to have many adverse effects on human health, agricultural productivity and natural ecosystems. Nevertheless, it is increasingly surprising just how resilient global environmental systems are to many of the pollutant burdens imposed upon them.

Although problems such as the destruction of valued environments, soil erosion and the extinction of species are very important for the future of mankind, it is pollution which arouses the most interest. This is because people realize that pollution impacts on them directly through effects on their health, their food supply, the degradation of buildings, and other items of cultural heritage, as well as having overt effects on forests, rivers, coastlines and ecosystems that they are familiar with. The costs of these effects in the depreciation of resources, lost productivity and in cleaning up or improving polluted environments are high and are increasingly occupying the attention of governments and politicians around the world, especially in technologically advanced countries. With increased legislation intended to reduce and prevent pollution, there is a corresponding increase in the involvement of specialist consultants, lawyers, insurers and financiers. There are many members of a wide range of professions with a growing interest in environmental pollution.

Although cases of overt pollution have occurred since the beginning of the Industrial Revolution, the far greater awareness about pollution among the populations of most advanced countries is largely owing to the mass media (especially television and radio), which enables millions of people to know when a major environmental catastrophe, for example the wreck of an oil tanker happens. The spectacular TV images and radio reports of the massive oil slick caused by the spillage of thousands of tonnes of crude oil from a floundering oil tanker and its toll of dead and dying sea birds and mammals make most people aware, for a short time at least, of the fragility of the environment (Section 8.7). However, what is less generally appreciated, but much more important overall, is that most environmental pollution is insidious and its harmful effects only become apparent after long periods of exposure. Some people are exposed to pollutants which may cause cancer 10 or 20 years later, without realizing it (Section 3.2.2). Gradual increases in atmospheric pollution can be causing chronic toxic effects in trees which do not appear for 20 or more years and are irreversible. Likewise, lakes can become increasingly polluted (Section 5.2.3) and species die out without many obvious signs, at least in the early stages. For this reason, environmental monitoring has become recognized as being vitally important in detecting where insidious pollution is occurring, the pollutants involved and the sources from which they came. Environmental monitoring has benefited from the development over recent decades of rapid and accurate methods of chemical analysis, such as gas and liquid chromatography for organic pollutants and atomic absorption spectrophotometry and inductively coupled plasma atomic emission spectrometry for metals.

It is estimated that over the whole period of human history, around 6×10^6 chemical compounds have been created, most of these within the twentieth century. Nowadays, about 1000 new compounds are being synthesized each year and between 60 000 and 95 000 chemicals are in current commercial use. Pollutant chemicals are responsible for many human illnesses: air pollution in particular has many effects on health, ranging from smoke and sulphur dioxide (SO_2) contributing to the occurrence of chronic bronchitis, and nitrogen oxides (NO_3), volatile organic compounds (VOC_3), ozone (O_3) and carbon monoxide (CO) associated with increased incidence of asthma and other disorders. In the UK, it has recently been estimated that air pollution in urban areas, largely from road transport, is having an impact on human health costing around £3.9×10^9(c $5.8 billion) (*Guardian* 14 November, 1995). Lead from vehicle emissions and coal combustion has been dispersed so widely that there are few parts of the world not affected by it, at least to a slight extent.

In fact, it can be argued that all soils in industrialized countries have been polluted with many trace substances including Cd, Pb, polycyclic aromatic hydrocarbons (PAHs), polychlorinated biphenyls (PCBs) and polychlorinated dibenzo-*p*-dioxins (PCDDs) and furans (PCDFs) by aerial and other inputs (Jones, 1991) (Sections 5.3, 6.2.5 and 6.4).

The incidence of asthma and 'sick office' syndrome and other possible respiratory allergies are considered to be linked to people's exposure to an ever increasing range of chemicals in the atmosphere, especially in urban and industrial areas.

A human health and well-being problem which is causing growing concern is decline in male fertility in technologically advanced countries over recent decades. This is thought to be caused, at least in part, by exposure to pollutant chemicals which have effects similar to female hormones (oestrogens) (pp. 72, 73). These chemicals, which include nonylphenol ethoxylates, alkylphenol ethoxylates and even some epoxy resins are widely used and so it is important to investigate promptly any causative role they may have in order that appropriate action can be taken. Unfortunately, as with so many environmental pollution issues, the time lag is considerable. Exposure to these oestrogenic chemicals probably occurred at least 20 years previously, either before, or recently after, the males were born. Therefore, the consequences of any mitigating action taken now may not become apparent for many years to come. Many toxic chemicals are used regularly in our daily lives because their particular use is not thought to constitute a hazard to health and the advantages of using them outweigh their possible risks (see the toxicity references). An example of this is the use of chlorine compounds for water disinfection and other hygiene purposes. In many Third World countries the lack of a hygienic water supply is the cause of the deaths of many young children. Safe drinking water is taken for granted in most developed countries and this usually necessitates disinfection with chlorine (Cl_2), especially when water is recycled several times. However, the formation of trace chlorinated organic compounds can occur in treated waters and there appears to be a weak correlation between chlorinated water and bladder cancer. On balance, the current view is that the dangers from non-disinfected water outweigh the slight risk of cancer (Section 8.4, p. 362).

1.2 Definition of pollution

A widely used definition of pollution is 'the introduction by man into the environment of substances or energy liable to cause hazards to human health, harm to living resources and ecological systems, damage to structures or amenity, or interference with legitimate uses of the environment' (Holdgate, 1979). Some experts make a distinction between contamination and pollution. Contamination is used for situations where a substance is present in the environment, but not causing any obvious harm, while pollution is reserved for cases where harmful effects are apparent. However, the problem with this distinction is that with improved methods of analysis and diagnosis, it may become apparent that harmful effects have been caused and so situations initially described as contamination may have really been pollution. Holdgate's definition avoids this problem.

Pollutants are basically of two types: *primary pollutants* which exert harmful effects in the form in which they enter the environment, and *secondary pollutants*, which are synthesized as a result of chemical processes, often from less harmful precursors, in the environment. Although highly toxic substances are responsible for many cases of environmental pollution, under some circumstances materials which are normally considered harmless may cause pollution if they are present in excessive quantities or in the wrong place at the wrong time, and this makes definitions difficult (BMA, 1991). For example, milk or sugar are not normally considered as environmental pollutants. However, if a lorry carrying a load of either of these important foodstuffs was to spill much of its load into a river, severe pollution caused by the high biochemical oxygen demand (BOD) of these substances would result in the death of many fish (Section 2.5.3, p. 46).

In all cases of pollution there is (i) a source of pollutants, (ii) the pollutants themselves, (iii) the transport medium (air, water or direct dumping onto land), and (iv) the target (or receptor), which includes ecosystems, individual organisms (e.g. humans) and structures. Pollution can be classified in several ways according to (i) the source (e.g. agricultural pollution), (ii) the media affected (e.g. air pollution or water pollution) or (iii) by the nature of the pollutant (e.g. heavy metal pollution).

Although undesirable and costly, pollution is an inevitable and necessary part of life for most of the world's population, especially in large communities and those relying on technology and mechanized transport. Even in primitive cultures, accumulated human excretory products and smoke from cooking fires cause pollution. Where the volumes and rates of emission and toxicities of pollutants are relatively low, environmental processes can usually degrade or assimilate these excesses to a much greater extent that the more toxic air, water and land pollutants produced in large quantities in more technologically advanced countries. Where the costs of control are considered, it has been shown that they increase exponentially with the level of control achieved (Council on Environmental Quality, 1971); Fig. 1.1 shows a typical cost *versus* level of control curve. Up to about

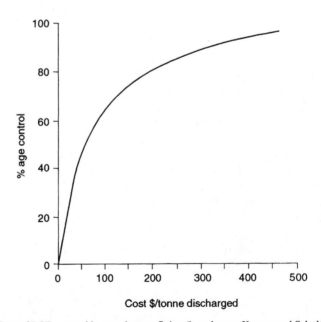

Figure 1.1   Cost of BOD control in petroleum refining (based upon Kneese and Schultz, 1975).

20% control achieved, the cost is less than \$20/unit; for 65% control, the cost rises sharply to \$110/unit. In accord with the exponential law, as control approaches completeness the cost becomes unsustainable, at 95% it is here \$500/unit.

It is significant that the above control/cost relation holds true for industries as diverse as sugar beet processing on the one hand and petrochemicals on the other. Poorer developing countries are likely to settle for a realistic level to ensure that funds are available for other high priorities, such as malaria control and the supply of good quality drinking water. Another instance of the need to balance control and cost is discussed on page 305, in relation to the stringent limit for dioxin levels set by the USEPA. In the UK, Integrated Pollution Control (IPC) has become effective through the 1990 Environmental Protection Act (p. 18).

1.3 Air pollution

Although industrial pollution in the UK in the nineteenth century was severe enough to necessitate Parliament passing the Alkali Act in 1862 and the Rivers Pollution Act in 1876, it took the notorious London smog in 1952 to bring about a major improvement in urban air quality through the introduction of the Clean Air Act of 1956. Smoke and smogs had generally been accepted as an unavoidable fact of life in towns and cities of the UK and many other industrialized countries. In the UK, a high incidence of chronic bronchitis ('the English disease') was associated with this type of air pollution. Unfortunately, the reduction of smoke and SO_2 levels which the Clean Air Act (1956) brought about were soon offset by an increase in CO, NO_x, PAH, O_3, PbBrCl and peroxyacetyl nitrate (PAN) concentrations caused by increasing numbers of motor vehicles on the roads and the replacement of coal by oil for heating buildings. The pollution produced by a single car on a journey to work is of little consequence, but when thousands of cars are involved a serious pollution situation develops. In conditions of bright sunlight, such as in Los Angeles, Mexico City, Athens and many other cities, even those in temperate areas such as London in the summer, the primary pollutants of motor vehicle exhausts (smoke, PAH, CO, NO_x and PbBrCl) are supplemented by the secondary pollutants O_3, NO_2 and PAN synthesized in photochemical reactions. These give rise to a pungent photochemical smog which irritates the eyes, nose and throat and causes more severe toxic effects in humans, animals and plants than the primary pollutants (Sections 5.2, 5.3, 6.2 and 6.3).

The worsening problems of atmospheric pollution from motor traffic in many of the world's cities have led to the introduction of legislation to try and mitigate or prevent the nuisance. This has included the phasing out of Pb additives in petrol and the introduction of catalytic converters in many countries. However, air quality problems still remain serious in most large cities and there appears no way of preventing them completely unless private cars are banned from city centres and environmentally sound public

Figure 1.2   Opencast lignite mining in northern Bohemia (Czech Republic) and air pollution from lignite burning for electricity generation (photo: B. J. Alloway).

transport systems are introduced. The expected exhaustion of petroleum reserves in the middle of the twenty-first century may help to force the issue. However, if there is a move towards using oil derived from shale rocks as a substitute for petroleum, the air quality problems may be exacerbated because the shale oils contain a wider range of compounds which would create pollutants in exhaust gases. The environmental impact of the exploitation of oil shale deposits around the world, in areas such as the Green River Basin in the USA, would also be much greater than those of petroleum (Sections 5.2 and 6.2).

Pollution problems rapidly escalate in severity when the rate of pollutant emissions exceeds the capacity of the environment to assimilate them. A vivid example of this can be found in north-west Bohemia (Czech Republic) (Fig. 1.2). After the Second World War, increased exploitation of the local brown coal (lignite) deposits for electricity generation (14×10^6 t in 1945 to 100×10^6 t in 1987) resulted in a marked increase in atmospheric SO_x pollution because of the coal's high sulphur content ($< 15\%$) and the lack of pollution control. Most bituminous coals in use for electricity generation tend to contain 0.5–4% sulphur, so this Czech lignite is a very rich source of the harmful pollutant SO_2 (Section 5.2). Although the increasing air pollution rapidly affected human health, some of its effects on the local environment took longer to become apparent. It took nearly 20 years before large numbers of trees in the forest on the tops of the nearby mountains in the Ertzgebirge range started to 'die-back'. This death of trees ('waldsterben' in German) continued to increase in severity and today most of the mountain tops in this part of the country are without trees and it is estimated that around 100 000 ha of forest are affected in the whole of the Czech Republic. The rate of sulphur deposition in the mountains most affected by the forest die-back reached $15 \, g \, S/m^2$ per year, which rates it as one of the most severely polluted places in the world (Moldan and Schnoor, 1992). The phenomenon of die-back in trees caused by atmospheric pollution is becoming a major cause for concern in many parts of Europe and North America. Problems with the acidification of lakes and the death of many species of aquatic organisms also occur in many areas affected by die-back in trees.

1.4 Environmental literature

One of the factors which led to an increase in environmental awareness and concern about pollution around 30 years ago was the publication of certain inspiring books. One of the first was *Silent Spring* by Rachel Carson (1962), which focused attention on insidious pesticide pollution. The title dramatically refers to a scenario of a spring without birdsong because most birds had been killed by pesticides or their residues. At the time this book was published, developments in plant breeding and pest control had boosted cornucopian optimism in advanced countries. Food production was increasing rapidly along with an exponential rise in the use of pesticides and fertilizers (Section 6.4) and little attention was given to the consequences of their accumulation in the environment, or of the toxicity of their degradation products (Sections 3.2 and 6.4). Carson (1962) was instrumental in drawing attention to the risk of cancer in people exposed to many pesticides, especially as consumers. Some authors consider this to be a serious misconception and have drawn attention to the fact that many normal constituents of the diet pose a greater risk of cancer than very small traces of pesticide residues in foods (Ames and Gold, 1990). Nevertheless, the fact remains that the smoking of cigarettes is by far the greatest risk of developing cancer.

Silent Spring was soon followed by many books on environmental themes, but one book which aroused much interest and controversy was *Limits to Growth* by Meadows *et al.* (1972). It is particularly noteworthy because it was one of the first books to be based on predictions from a mathematical simulation model (World 3). It predicted that exponential increases in human population, in the consumption of finite resources and in pollution would lead to a drastic reduction in the population which the earth would be able to sustain over a 100 year period. One very important prediction from the model was that food production would decrease, partly as a consequence of environmental pollution. Nine million copies of the book were sold and, although it attracted much criticism, it focused attention on the exponential rate of increase in the severity of global environmental problems and on the need for concerted international action to try and slow them down.

In the same year that *Limits to Growth* was published (Meadows *et al.*, 1972), the United Nations organized in Stockholm its first international conference on the environment and also set up the United Nations Environmental Programme (UNEP) with its headquarters in Nairobi. The USA had already established its own Environmental Protection Agency in 1970 and nine other countries had also created agencies or government departments concerned with environmental matters. The European Environmental Agency was established in 1991 in Copenhagen. By the time of the second United Nations Conference on the Environment and Development (the so-called 'Earth Summit') in Rio de Janeiro 20 years later (1992), more than 100 countries had their own environmental agencies or ministries.

1.5 Population pressures

In general, it is possible to summarize the main factors responsible for pollution and other types of environmental deterioration in any community or society as being caused by the combined effects of population, affluence and technology (Meadows *et al.*, 1992).

$$\text{Impact on environment} = \text{Population} \times \text{Affluence} \times \text{Technology}$$

Basically, the larger the population, the greater the extent of environmental deterioration resulting from related needs for food production, living space, waste disposal, communications and so on. The world's population rapidly increased after the Second World War. It more than doubled in the 40 years between 1950 (2516.44 million) and 1990 (5292.20 million). It was expected to have risen to 5759.28 million by 1995 and the predicted population by the year 2025 will be 8472.4 million. The average percentage change between 1990 and 1995 was 1.68% and is predicted to be 1.43% between AD 2000 and 2025 (World Resources Institute, 1994). The crude birth rates (births/1000 population) for the world were 31.5 in the period 1970–5 and 26.0 between 1990 and 1995. This downward trend is followed throughout the world, but there are marked variations in the birth rates in the different continents/regions: Africa 46.8 (1970–5) and 43.0 (1990–5), Asia 34.9 (1970–5) and 26.3 (1990–5), North and Central America 29.3 (1970–5) and 22.8 (1990–5), South America 33.0 (1970–5) and 24.2 (1990–5), Europe 15.7 (1970–5) and 12.7 (1990–5), former USSR 18.1 (1970–5) and 16.5 (1990–5) and Oceania 23.9 (1970–5) and 19.3 (1990–5) (World Resources Institute, 1990, cf. Fig. 5.4, p. 168). The trend of decreasing birth rate is accompanied by a trend towards longer life expectancy (for the world in 1970–5 this was 58.5 years, in 1990–5 it was 65.5 years) with a larger number of older people in the population. Even though some European countries, such as the UK,

Netherlands and Belgium, had relatively low rates of population increase in the period 1985–90 (0.22%, 0.63% and 0.03%, respectively), the UK has a population density of more than 200 persons/km^2 and both Belgium and the Netherlands have densities in excess of 300 persons/km^2. This high density of population creates severe environmental problems with the conflicting demands for land for living space, food production, waste disposal and the other many essential uses of land, and the pollution created by people living and working in relatively close proximity. In contrast, France, Greece, the Republic of Ireland, Portugal and Spain have population densities of 100/km^2 or less and consequently there is less pressure from competing uses of land.

Most of the increase in the world's population is occurring in the less developed countries (LDCs) of the Third World where levels of affluence and technology are generally low. It can be concluded that the environmental impact of these rapidly growing populations is not as great as it would be if they were more affluent and dependent on technology. Nevertheless, their dependence on technology is slowly increasing; as high yielding varieties of crops with their requirement for fertilizers, pesticides and, possibly, irrigation are introduced, the contribution of these countries to pollution and other aspects of environmental deterioration are likewise increasing. The pollution problem is compounded by the continued use in Third World countries of chemicals, including pesticides such as DDT, and Pb-containing petrol, which have either been banned or are being phased out on environmental grounds in technologically advanced countries (Sections 5.3 and 6.4).

A community's affluence (i.e. its capital stock per person) determines the amount of materials required to maintain this standard of living, and hence the greater the level of affluence, the greater the need for the exploitation of natural resources. This, in turn, causes much pollution, for example in the mining, transport and smelting of metals. Likewise, the fabrication of metal products, their corrosion in use and eventual disposal are all potentially polluting activities.

1.6 Energy demand

Technology requires energy and so the higher the level of technology in a community, the greater the need for energy and the greater the extent of pollution and environmental damage caused by energy generation. This includes the mining of coal, extraction and refining of petroleum, the mining and processing of uranium, the atmospheric and wider environmental pollution from fossil fuel combustion in electricity generation and the disposal of the ash. Nuclear power station accidents, such as that at Chernobyl, nuclear waste reprocessing and disposal are also part of the environmental price that has to be paid for a plentiful supply of convenient energy (Section 5.5). Nevertheless, the relationship between energy and technology can be modified, to a certain extent, by the use of technology to

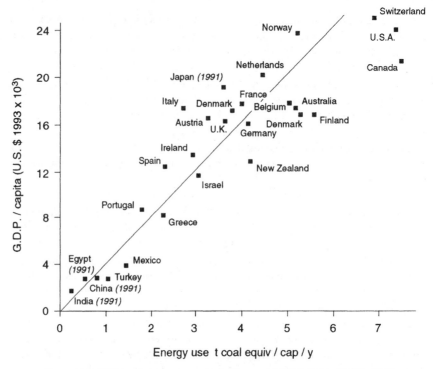

Figure 1.3 GDP/energy use per capita in the world (U.N. 1994, OECD, 1995)

control pollution (e.g. flue gas desulphurization) from energy generation (Fig. 1.3).

Energy generation is perhaps the most ubiquitous cause of pollution in the world as a whole, especially if it is taken in its widest context to include internal combustion engines in motor vehicles and fires for cooking and heating. The total world consumption of energy in 1990 was estimated to be 8013×10^6 t oil equivalent compared with 5171×10^6 oil equivalent in 1970, almost a 55% increase over 20 years (Allen, 1992). Although the burning of wood for fuel in many less developed countries will not cause so much environmental pollution as fossil fuels or nuclear power, it still has a marked environmental impact. This fuel wood is, in theory, a renewable energy resource; but in many cases the excessive and often complete removal of wood and other combustible vegetation has led to irreversible land degradation, especially through soil erosion. The use of wood for fuel ranges from 98% of the energy consumed in Nepal, 82% in Nigeria, 63% in Thailand to 33% in Brazil. Wood burning in open fires or primitive stoves results in the production of a large amount of smoke with its associated formation of PAH organo micropollutants, which can have deleterious effects on the people constantly inhaling this smoke (Sections 6.1, 6.2 and 7.6).

1.7 Wastes

In general, cities and urban areas throughout the world, with their concentrated population, high consumption of energy, transport and industrial activities, tend to have the worst environmental pollution problems. The disposal of the population's sewage, municipal and industrial wastes and the atmospheric pollution from urban sources impinge on surrounding rural areas. Urban heat-island effects tend to confine much of the dust and atmospheric pollutants within the region and this exacerbates the effects of pollution on the population.

The coastlines of densely populated regions are frequently polluted by sewage (Section 8.4). In less developed countries, untreated sewage is frequently discharged into rivers and carried into the sea, but even in more developed countries, sewage from coastal communities is piped directly into the sea. Although the sea disposal of untreated sewage should have been phased out within the European Community by 1998, many tourist beaches currently have poor quality bathing waters because of the presence of bacteria and viruses of sewage origin. Coastal fisheries and shell-fish resources have also been significantly affected. In some places, coastal vegetation has been excessively scorched by salt spray owing to the presence of detergents of sewage origin in the sea water, which have modified the effect of the salt solution droplets on previously resistant plant leaves.

Point-source pollution resulting from mining, industrial and military activities occurs all over the world, often in some very remote regions. As a result of atmospheric pollution and the use of fertilizers contaminated with potentially hazardous substances, such as Cd, most of the soils in technically advanced countries are polluted (or contaminated) to a slight extent, although their composition is much less affected than soils near to urban/industrial localities (Section 5.3.6).

1.8 Concluding remarks

In most cases, air pollution is the form of pollution which causes people the most concern. It is usually obvious by its effects on the eyes and nostrils and also causes conspicuous toxicity symptoms in vegetation. Nevertheless, some of the most harmful chemicals in the polluted air may not be detectable by smell or any of the other senses. Water pollution is the second most obvious type of pollution, especially when it affects drinking water supplies or causes the death of large numbers of fish. In contrast, soil pollution is often far less conspicuous but it is still very important. As a result of the adsorptive and buffering properties of soil, some pollutants (such as Cu, Pb and PCBs) have long half-lives in the soil, and food crops grown on these polluted soils may be affected by some of the pollutants for centuries, even millennia, because soil is difficult and expensive to clean-up

(Section 2.6). However, it is a source of comfort that ecosystem and biogeochemical processes are amazingly resilient and that the impact of people on the environment is not as severe as would be expected given the loading of pollutants emitted each year. This resilience results from the adsorption and detoxification processes operating in soils and sediments, which remove pollutants from circulation and, in many cases, either fix them more or less indefinitely or degrade them to harmless products (Section 2.6). Most of this detoxification of organic chemicals is carried out by microorganisms, which can evolve the ability to synthesize the necessary enzymes for denaturing new chemicals as a result of their rapid rates of reproduction and mutation. However, there is no room for complacency because natural processes for the assimilation and detoxification of pollutants can be overloaded if the rate of pollutant emission and transport is too great.

Given that environmental pollution poses one of the greatest threats to the health and food security of the human race, the need for a greater understanding of it becomes even more important. This book introduces the reader to the basic principles relating to the main types of environmental pollutants, their sources, chemical properties and the reactions they undergo in the air, water and soil. It is felt that the use of a pollutant-orientated approach is more logical and appropriate for an introductory text for science students than a more ecologically focused or environmental media approach. The effects of pollutants on organisms and the environment are best studied after the nature of the chemicals involved and some basics about their environmental behaviour are understood. However, the subjects of transport and behaviour of pollutants in the environment, principles of toxicology, assessment of the risks which pollutants pose and the methods of monitoring and analysis are briefly covered on a general basis at the beginning of the book and referred to later in more detail in the sections dealing with the individual groups of pollutants. The last section deals with wastes and their disposal and multipollutant situations. In a book of this type it is not possible to give an exhaustive coverage of every important pollutant, nor to discuss all relevant aspects of environmental behaviour in detail. It is expected that the reader will want to follow up this introduction by referring to more specialized texts and some suggested titles for further reading are provided. In many cases the books that give a more detailed coverage tend to focus on one particular medium, such as air or water, or a restricted group of pollutants, such as heavy metals or pesticides.

References   Allen, J. E. (1992) *Energy Resources for a Changing World.* Cambridge University Press, Cambridge.

Ames, B. N. and Gold, L. S. (1990) *Angew. Chem. Inst. Ed. Engl.,* **29**, 1197.

British Medical Association (1991) *Hazardous Waste and Human Health.* Oxford University Press, Oxford.

Carson, R. (1962) *Silent Spring*. Houghton Miflin, Boston.
Council on Environmental Quality (1971) *Second Annual Report*. Government Printing Office, Washington, DC.
Holdgate, M. W. (1979) *A Perspective of Environmental Pollution*. Cambridge University Press, Cambridge.
Jones, K. C. (1991) *Environ. Pollut.* **69**, 311.
Kneese, A. U. and Schultze, C. L. (1975) *Pollution, Prices and Public Policy*. Brookings Institute, New York.
Meadows, D. H., Meadows, D. L. and Randers, J. (1972) *Limits to Growth*. Universe Books, New York.
Meadows, D. H., Meadows, D. L. and Randers, J. (1992) *Beyond the Limits*. Earthscan Publications, London.
Moldan, B. and Schnoor, J. L. (1992) *Environ. Sci. Technol.* **26**, 14.
United Nations (1994) *Statistical Yearbook*, 39th Issue. WHO, Geneva. OECD (1995) OECD in Figures.
World Resources Institute (1994) *World Resources 1994/5*. Oxford University Press, New York.

Air pollution

Further reading

Boubel, R. W., Fox D. L., Turner, D. B. and Stern, A. C. (1994). *Fundamentals of Air Pollution*, (3rd edn.) Academic Press, San Diego.
Bridgeman, H. (1990) *Global Air Pollution*. Belhaven Press, London.
Elsom, D. (1987) *Atmospheric Pollution*. Blackwell, Oxford.

Environmental chemistry

Hemond, H. F. and Fechner, E. (1994) *Chemical Fate and Transport in the Environment*. Academic Press, San Diego.
Manahan, S. E. (1991) *Environmental Chemistry*. Lewis Publishers, Chelsea, Michigan.
O'Neill, P. (1993) *Environmental Chemistry* (2nd edn). Chapman & Hall, London.
Richardson, M. L. (1991) (ed) *Chemistry, Agriculture and the Environment*. Royal Society of Chemistry, Cambridge.
Schwarzenbach, R. P., Gschwend, P. M. and Imboden, D. M. (1993) *Environmental Organic Chemistry*, Wiley Interscience, New York.

Environmental science

Ellis, D. (1989) *Environments at Risk: case histories of impact assessment*. Springer Verlag, Berlin, Heidelberg.
Masters, G. M. (1991) *Introduction to Environmental Engineering and Science*. Prentice Hall, Englewood Cliffs, New Jersey.
Tolba, M. K., El-Kholy, O. A., El-Hinnawi, E., Holdgate, M. W. and Munn, R. E. (1992) *The World Environment 1972–1992*. UNEP/Chapman & Hall, London.

Pollution

Coughtrey, P. J., Martin, M. H. and Unsworth, M. H. (eds) (1987) *Pollutant Transport and Fate in Ecosystems*. Blackwells, Oxford.
Harrison, R. M. (ed) (1990) *Pollution: Causes, Effects and Control*. Royal Society of Chemistry, London.

Holdgate, M. W. (1979) *A Perspective of Environmental Pollution*. Cambridge University Press, Cambridge.
Murley, L. (ed) (1995) *The National Association of Clean Air Pollution Handbook*. The National Association of Clean Air, Brighton (updated annually).
Sell, N. J. (1992) *Industrial Pollution Control* (2nd edn). Van Nostrand, New York.

World resources

Brown, L. R. and Kane, H. (1995) *Full House: Reassessing the Earth's Population Carrying Capacity*, Worldwatch, Earthscan, London.
World Resources Institute (1994) *World Resources 1994/5*. Oxford University Press, New York. (Earlier editions for 1986, 1987, 1988/89, 1990/91, 1992.)

Soil science

Ross, S. (1989) *Soil Processes*. Routledge, London.
Rowell, D. L. (1994) *Soil Science: Methods and Applications*. Longman, Harlow.
White, R. E. (1987) *Introduction to the Principles and Practice of Soil Science*. Blackwells, Oxford.

Toxicity

Harte, J., Holdren, C., Schneider, R. and Shirley, C. (1991) *Toxics A to Z*. University of California Press, Berkeley and Los Angeles.
Rodricks, J. V. (1992) *Calculated Risk*. Cambridge University Press, Cambridge.
Sax, N. I. and Lewis, R. J. (eds) (1988), *Dangerous Properties of Industrial Materials*, Vol. 1–3. Van Nostrand, New York.
Various authors (1975–to date) *J. Hazardous Materials*. Elsevier, Amsterdam.

Wastes

British Medical Association (1991) *Hazardous Waste and Human Health*. Oxford University Press, Oxford.
Clark, J. H. (ed) (1995) *Chemistry of Waste Minimimisation*. Blackie, London.
La Grega, M. D., Buckingham, P. L. and Evans, J. C. (1994) *Hazardous Waste Management*. McGraw-Hill, New York.

Water pollution

Clark, R. B. (1989) *Marine Pollution*. Clarendon Press, Oxford.
Haslam, S. M. (1992) *River Pollution: An Ecological Perspective*. John Wiley, Chichester.
Langford, T. E. I. (1992) *Biology of Pollution in Major River Systems*. Chapman & Hall, London.
Laws, E. A. (1993) *Aquatic Pollution*. John Wiley, New York.

Transport and behaviour of pollutants in the environment 2

2.1
A basic model of environmental pollution

The pollution pathway concept put forward by Holdgate (1979) is a very convenient way for studying and appreciating environmental pollution. All pollution events have certain characteristics in common: all involve (i) the pollutant, (ii) the source of the pollutant, (iii) the transport medium (air, water or soil), and (iv) the target (the organisms, ecosystems or items of property affected by the pollutant). These are shown in Fig. 2.1.

Varying degrees of sophistication can be added to the simple model, including the rate of emission of the pollutant from the source, the rate of transport, chemical and physical transformations which the pollutant undergoes either during transport or after deposition at the target, amounts reaching the target, movement within the target to sensitive organs, and quantification of the effects on the target. The various components of this model are considered below.

2.2
Sources of pollutants

Sources of pollutants can either be discrete point sources or diffuse sources (non-point sources). Even at point sources, fugitive emissions may occur as a result of leakages in an apparently closed system, in addition to overt emission from chimneys or effluent discharge pipes. Some of the major sources and examples of the groups of pollutants likely to be involved are given in Table 2.1.

The following list of the major contaminating uses of land provides an indication of the wide range of possible sources: waste disposal sites, scrapyards, ship and vehicle breaking yards, gas works, petroleum refineries, petrol storage and distribution, petrol service stations, coal mining and coal storage, electricity generation, iron and steel works, metalliferous mining and smelting, metal products fabrication and metal finishing, chemical works, glass-making and ceramics, textile plants, dye works, leather tanneries, timber and timber products treatment works, manufacture of integrated circuits and semi-conductors, food processing, water treatment and sewage works, asbestos works, docks and railway land, paper and printing works, heavy engineering installations, radioactive waste processing, military bases and training areas and the burial of diseased livestock

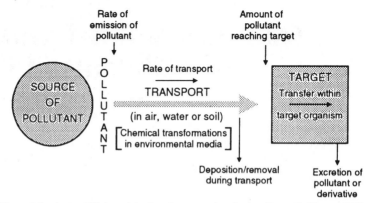

Figure 2.1   A simplified model of environmental pollution (from Holdgate, 1979).

(House of Commons Environment Committee, 1990). Although specifically related to land contamination, many of these sources also give rise to air and water pollution.

In the UK, the Environmental Protection Act 1990 embraces the concept of Integrated Pollution Control (IPC, Franklin *et al.*, 1995) and is intended to contain pollution across the air–land–water boundaries.

IPC is concerned with the most heavily polluting processes and is centrally administered by HMIP. It is based upon the Part A *Prescribed Processes Regulations*, involving some 200 process types operated by about 5000 individual companies. Industry and the regulator are committed in a high level of monitoring and surveillance.

Linked to IPC is Local Authority Air Pollution Control (LAAPC), which is concerned only with air pollutants associated with Part B of the *Prescribed Processes Regulations*. This responsibility devolves on local authorities, principally the London Borough Councils and elsewhere the District Councils.

Authorizations of processes under the 1990 Act became effective over the period April 1991 to January 1995.

2.3
The pollutants

Pollutants have certain intrinsic properties which determine the likely effect that they will have after emission or discharge into the environment. Holdgate (1979) divided these into two types: effect-generating properties, such as toxicity in living organisms or corrosion of metals, and pathway-determining properties, which determine the distance and the rate of dispersion of pollutant in the environment. These properties, which Holdgate (1979) suggested ought to be evaluated, include: (i) short- and long-term toxicity, (ii) persistence, (iii) dispersion properties, (iv) chemical reactions that the compound undergoes, including its decomposition,

Table 2.1 Major types of source of environmental pollutants and the environmental media they are transported or reside in

1. *Agricultural sources*
   Air: Pesticide aerosols, feather dusts, NH_3, H_2S, noxious odours, soil particles
   Water: Leachates from silage clamps and slurry lagoons, NO_3^-, HPO_4^{2-}, pesticide spillages, runoff, soil particles, HCs (fuel spillages)
   Soil: Fertilizers – e.g. As, Cd, Mn, U, V and Zn in some phosphatic fertilizers
              Manures – e.g. Zn, As and Cu in pig and poultry manures
              Pesticides – As, Cu, Mn, Pb, Zn, persistent organics (e.g. DDT, lindane)
              Corrosion of metals – e.g. galvanized metal objects (fencing, troughs, etc.)
              Fuel spillages – HCs
              Burial of dead livestock – pathogenic microorganisms

2. *Electricity generation*
   Air: CO_x, NO_x, SO_x, UO_x and PAHs from coal, radioisotopes from nuclear fission
   Water: Heat, biocides from cooling water, soluble B and As compounds and PAHs from ash
   Soil: Ash, fallout – Si, SO_x, NO_x, heavy metals, coal dust

3. *Derelict gas works sites*
   Air: VOCs, H_2S, NH_3
   Water: PAHs, phenols, Cu, Cd, As, CN, sulphates
   Soil: Tars (containing HCs, phenols, benzene, xylene, naphthalene and PAHs), CN, spent Fe oxides, Cd, As, Pb, Cu, sulphates, sulphides

4. *Metalliferous mining and smelting*
   Air: SO_x, Pb, Cd, As, Hg, Ni, Tl, etc. particulates/aerosols
   Water: SO_4^{2-}, CN, frothing agents, metal ions, tailings (ore minerals e.g. PbS, ZnS, $CuFeS_2$, etc.)
   Soil: Spoil and tailing heaps – wind erosion, weathering ore particles
              Fluvially dispersed tailings – deposited on soil during flooding, river dredging, etc.
              Transported ore separates – blown from conveyance etc. onto soil
              Ore processing – cyanides, range of metals
              Smelting – wind-blown dust, aerosols from smelter (range of metals)

5. *Metallurgical industries*
   Air: Particulates/aerosols: As, Cd, Cr, Cu, Mn, Ni, Pb, Sb, Tl and Zn, VOCs, acid droplets
   Water: Metal ions, acid wastes and solvents (VOCs) from metal cleaning
   Soil: Metals in wastes, solvents, acid residues, fallout of aerosols, etc. from casting and other pyrometallurgical processes

6. *Chemical and electronic industries*
   Air: VOCs, Hg, numerous volatile compounds
   Water: Waste disposal, wide range of chemicals in effluents, solvents from microelectronics
   Soil: Particulate fallout from chimneys
              Sites of effluent and storage lagoons, loading, packaging areas
              Scrap and damaged electrical components – PAHs, metals, etc.

7. *General urban/industrial sources*
   Air: VOCs, particulates, aerosols (e.g. Pb, V, Cu, Zn, Cd, PAHs, PCBs, dioxins, smoke)
              Fossil fuel combustion – CO_x, SO_x, NO_x, As, Pb, U, V, Zn, PAHs
              Bonfires – PAHs, PCDDs, PCDFs, Pb, Cd, etc.
              Cement manufacture – particulates, Ca, SO_4^{2-}, Si, etc.
   Water: Wide range of effluents, PAHs from soots, Pb, Zn, etc., waste oils – HCs, PAHs, detergents
   Soil: Pb, Zn, V, Cu, Cd, PCBs, PAHs, dioxins, HCs, dumped cars, etc. – HCs, asbestos, PAHs

(Contd.)

Table 2.1   *(Contd.)*

8. *Waste disposal*
   Air: Incineration – fumes, aerosols and particulates (Cd, Hg, Pb, CO_x, NO_x,
             PCDDs, PCDFs, PAHs)
             Landfills – CH_4, VOCs
             Livestock farming wastes: CH_4, NH_3, H_2S
             Scrapyards – combustion of plastics (PAHs, PCDDs, PCDFs)
   Water: Landfill leachates, NO_3^-, NH_4^+, Cl^-, Cd, PCBs, microorganisms
             Effluents from water treatment – organic matter, HPO_4^{2-}, NO_3, NH_4^+, Cl^-,
             SO_4^{-2}
   Soil: Sewage sludge PCDDs, chlorophenols – NH_4^+, PAHs, PCBs, metals (Cd, Cr,
             Cu, Hg, Mn, Mo, Ni, Pb, V, Zn, etc.)
             Scrapheaps – Cd, Cr, Cu, Ni, Pb, Zn, Mn, V, W, PAHs, PCBs
             Bonfires, coal ash, etc. – Cu, Pb, PAHs, B, As
             Fallout from waste incinerators – Cd, PCDFs, PCBs, PAHs
             Fly tipping of industrial wastes (wide range of substances)
             Landfill leachate – Cl^-, NO_3^-, NH_4^+, Cd, PCBs, microorganisms
9. *Transport*
   Air: Exhaust gases, aerosols and particulates (e.g. CO_x, NO_x, SO_x, smoke, PAHs,
             PAN, O_3, PbBrCl, V, Mo)
   Water: Spillages of fuels, spillages of transported loads (e.g. HC, pesticides,
             manufactured organic chemicals, wastes in transit – especially marine pollution
             from oil tanker operations and accidents), road and airport de-icers (e.g.
             ethylene glycol, various salts), deposition of fuel combustion products, smoke,
             PAHs, SO_x, NO_x, PbBrCl
   Soil: Particulates (PbBrCl, PAHs) acid deposits, de-icers, wide range of soluble/
             insoluble compounds at docks and marshalling yards and sidings, deposition of
             fuel combustion products, smoke, PAHs, SO_x, NO_x, rubber tyre particles
             (containing Zn and Cd)
10. *Incidental sources*
   Water: Leakage from underground storage tanks e.g. solvents, petrol products,
             LNAPLs, DNAPLs
   Soil: Preserved wood (e.g. PCP, creosote, As, Cr, Cu, etc.), discarded batteries (Hg,
             Cd, Ni, Zn) fishing and shooting (Pb, Sb), galvanized roofs and fences Zn, Cd
   All media: Warfare (e.g. fuels, explosives, ammunition, bullets, electrical components,
             poison gases, combustion products – PAHs), corrosion of metal objects – Cu,
             Zn, Cd, Pb
             Industrial accidents e.g. Bhopal, Séveso, Chernobyl nuclear reactor (wide range
             of pollutants)
11. *Long-range atmospheric transport* (deposition of transported pollutants)
   Water   As, Pb, Cd, Hg, UO_x Zn, SO_4^{2-}, NO_x, pesticides PAHs
   and soil: Wind-blown soil particles with adsorbed pesticides and pollutants

(v) tendency to be bioaccumulated in food-chains, and (vi) ease of
control.

2.3.1   *Classification of hazardous substances in the USA*

In the USA, hazardous substances have been defined by the Environmental
Protection Agency under the authority of the Resource, Recovery and
Conservation Act of 1976 and its Hazardous and Solid Wastes Amendments
of 1984. Hazardous substances are defined in terms of their ignitability,

Table 2.2 USA classification of hazardous substances (from Identification and Listing of Hazardous Wastes Code of Federal Regulations (1986) in Manahan (1984, 1991))

Waste type	Example
F-type (waste from non-specific sources)	
F001	Spent halogenated solvents used in degreasing; tetrachlorethane, trichloroethylene and methylene chloride.
F004	Spent non-halogenated solvents: cresols, cresylic acid and nitrobenzene, and the still bottoms from the recovery of these solvents
F007	Spent plating-bath solutions
F010	Quenching bath sludge from oil baths from metal heat treating operations
K-type (hazardous wastes from specific sources)	
K001	Bottoms sediment sludge from the treatment of wastewaters from wood-preserving processes that use creosote and/or pentachlorophenol
K002	Wastewater treatment sludge from the production of chrome yellow and orange pigments
K008	Oven residue from the production of chrome oxide green pigments
K019	Heavy ends from the distillation of ethylene dichloride
K020	Heavy ends (residue) from the distillation of vinyl chloride
K027	Centrifuge residue from toluene diisocyanate production
K043	2,6-Dichlorophenol waste from the production of the herbicide 2,4-D
K047	Pink/red water from TNT operations
K049	Slop oil emulsion solids from the petroleum refining industry
K060	Ammonia lime sludge from coking operations
K067	Electrolytic anode slimes and sludges from primary zinc production
P-type (acute hazardous wastes)	
P003	Acrolein
P013	Bariun cyanide
P024	*p*-Chloroaniline
P050	Endosulfan
P056	Fluorine
P063	Hydrocyanic acid
P065	Mercury fulminate
P081	Nitroglycerine
P095	Phosgene
P105	Sodium azide
P110	Tetraethyl lead
P115	Thallium sulphate
P122	Zinc phosphide
U-type (generally hazardous wastes)	
U001	Acetaldehyde
U021	Benzidine
U032	Calcium chromate
U071	Dichlorobenzene
U115	Ethylene oxide
U133	Hydrazine
U147	Maleic anhydride
U190	Phthalic anhydride
U220	Toluene

corrosivity, reactivity and toxicity. In addition, more than 450 wastes are listed as specific substances or types of substance. These are given one of the following letters and a three digit number (Table 2.2):

F-type   wastes from non-specific sources
K-type   wastes from specific sources
P-type   acute hazardous wastes
U-type   generally hazardous wastes

2.3.2   *European Community Dangerous Substances Directive*

Although almost any substance present in excess in the wrong place in the environment at the wrong time can cause pollution, chemicals belonging to the following groups of substances are considered to be the major priority concerns. These constitute the EC Dangerous Substances Directive List I (black list) of substances and List II (grey list) (EC, 1976; Murley, 1995). Table 2.3 lists substances already controlled by daughter directives. Table 2.4 gives List I and List II.

2.3.3   *UK priority list of pollutants*

The UK Department of the Environment has established a 'red list' of priority pollutants and the initial version of this list (1988) is presented in

Table 2.3 Directive 76/464/EEC – substances already controlled by daughter directives

	Directive number	Date of entry into force	EEC Reference[a]
Aldrin, Dieldrin, Endrin and Isodrin	88/347	1/1/89	1,71,77,130
Cadmium and its compounds	83/513[b]	1/4/86	12
Carbon tetrachloride	86/280[b]	1/1/88	13
Chloroform	88/347	1/1/90	23
DDT (all isomers)	86/280	1/1/88	46
Hexachlorobenzene	88/347	1/1/90	83
Hexachlorobutadiene	88/347	1/1/90	84
Hexachlorocyclohexane (all isomers)	84/491	1/4/86	85
Mercury and its compounds:			
Chlor-alkali	82/176	1/7/83	92
Other sectors	84/156	12/3/86	92
Pentachlorophenol	86/280	1/1/88	102
1,2-Dichloroethane	90/415	1/1/93	59
Trichloroethylene	90/415	1/1/93	121
Tetrachloroethylene	90/415	1/1/93	111
Trichlorobenzene (1,2,4 and technical mixture)	90/415	1/1/93	118,117

[a] Taken from numbers used in the original list of 129.
[b] The Standard Articles Directive.

Table 2.4 The European Commission priority candidate list (the list of 129) (Environmental Data Services, 1992)

List I (129 chemicals specified)

2	2-Amino-4-chlorophenol
3	Anthracene
5	Azinphos-ethyl
6	Azinphos-methyl
8	Benzidine
9	Benzyl chloride
10	Benzylidene chloride
11	Biphenyl
14	Chloral hydrate
16	Chloroacetic acid
17	2-Chloroaniline
18	3-Chloroaniline
19	4-Chloroaniline
21	1-Chloro-2,4-dinitrobenzene
22	2-Chloroethanol
24	4-Chloro-3-methylphenol
25	1-Chloronaphthalene
26	Chloronaphthalenes (technical mixture)
27	4-Chloro-2-nitroaniline
28	1-Chloro-2-nitrobenzene
29	1-Chloro-3-nitrobenzene
30	1-Chloro-4-nitrobenzene
31	4-Chloro-2-nitrotoluene
32	Chloronitrotoluenes (other than 4-chloro-2-nitrotoluene)
33	2-Chlorophenol
34	3-Chlorophenol
35	4-Chlorophenol
36	Chloroprene
37	3-Chloropropene
38	2-Chlorotoluene
39	3-Chlorotoluene
40	4-Chlorotoluene
41	2-Chloro-*p*-toluidine
42	Chlorotoluiines (other than 2-chloro-*p*-toluidine)
43	Coumaphos
44	Cyanuric chloride
45	2,4-D (including salts and esters)
47	Demeton (including demeton-*O*; -*S*; -*S*-methyl; -*S*-methyl-sulphone)
48	1,2-Dibromoethane
49	Dibutyltin dichloride
50	Dibutyltin oxide
51	Dibutyltin salts (other than dibutyltin chloride and dibutyltin oxide)
52	Dichloroanilines
53	1,2-Dichlorobenzene
54	1,3-Dichlorobenzene
55	1,4-Dichlorobenzene
56	Dichlorobenzidines
57	Dichlorodiisopropyl ether
58	1,1-Dichloroethane
60	1,1-Dichloroethylene
61	1,2-Dichloroethylene
62	Dichloromethane
63	Dichloronitrobenzenes
64	2,4-Dichlorophenol
65	1,2-Dichloropropane
66	1,3-Dichloropropan-2-ol
67	1,3-Dichloropropene
68	2,3-Dichloropropene
69	Dichlorprop
70	Dichlorvos
72	Diethylamine
73	Dimethoate
74	Dimethylamine
75	Disulfoton
76	Endosulfan
78	Epichlorohydrin
79	Ethylbenzene
80	Fenitrothion
81	Fenthion
86	Hexachloroethane
87	Isopropylbenzene
88	Linuron
89	Malathion
90	2-Methyl-4-chlorophenoxyacetic acid
91	2-Methyl-4-chlorophenoxypropanoic acid
93	Methamidophos
94	Mevinphos
95	Monolinuron
96	Naphthalene
97	Omethoate
98	Oxydemeton-methyl
99	PAH (with special reference to 3,4-benzo-pyrene and 3,4-benzofluoranthene)
100	Parathion (including parathion-methyl)
101	PCBs (including PCTs)
103	Phoxim
104	Propanil
105	Pyrazon
106	Simazine
107	2,4,5-T (including salts and esters)
108	Tetrabutyltin
109	1,2,4,5-Tetrachlorobenzene
110	1,1,2,2-Tetrachloroethane
112	Toluene
113	Triazophos
114	Tributyl phosphate
115	Tributyltin oxide
116	Trichlorofon
119	1,1,1-Trichloroethane
120	1,1,2-Trichloroethane
122	Trichlorophenols
123	1,1,2-Trichlorotrifluoroethane
124	Trifluralin
125	Triphenyltin acetate
126	Triphenyltin chloride
127	Triphenyltin hydroxide
128	Vinyl chloride
129	Xylenes (technical mixture of isomers)
131	Atrazine
132	Bentazone

(*Contd.*)

Table 2.4   (*Contd.*)

List II
Includes certain substances belonging to the families and groups in List I which do not have
    specified limit values, plus:
    Zn, Cu, Ni, Cr, Pb, Se, As, Sb, Mo, Ti, Sn, Ba, Be, B, U, V, Co, Tl, Te, Ag and their
        compounds
    Biocides and their derivatives not in List I
    Organic silicon compounds
    Inorganic phosphorus compounds
    Non-persistent mineral oils and hydrocarbons of petroleum origin
    Cyanides and fluorides
    NH_3 and NO_3

Table 2.5   The UK priority red list of pollutants (DOE, 1988)

Aldrin
Atrazine
Azinphos-methyl
Cadmium and its compounds
DDT (including metabolites DDD and DDE)
1,2-Dichloroethane
Dichlorvos
Dieldrin
Endosulfan
Endrin
Fenitrothion
Hexachlorobenzene
Hexachlorobutadiene
Gamma-hexachlorocyclohexane
Malathion
Mercury and its compounds
PCBs (polychlorinated biphenyls)
Pentachlorophenol
Simazine
Trichlorobenzene (all isomers)
Trifluralin
Triorganotin compounds

Table 2.5. It is expected that this list will be added to in due course. The first
priority candidate red list is given in Table 2.6.

2.3.4   Pesticides

Pesticides are used widely in agriculture and for many other purposes
all over the world with more than 10 000 commercial formulations of
around 450 pesticidal compounds currently in use. Their use and dispersion
in the environment has mainly occurred since the late 1940s and they

Table 2.6   The first priority candidate red list

2-Amino-4-chlorophenol
Anthracene
Azinphos-ethyl
Biphenyl
Chloroacetic acid
2-Chloroethanol
4-Chloro-2-nitrotoluene
Cyanuric chloride
2,4-D (including salts and esters)
Demeton-O
1,4-Dichlorobenzene
1,1-Dichloroethylene
1,3-Dichloropropan-2-ol
1,3-Dichloropropene
Dimethoate
Ethylbenzene
Fenthion
Hexachloroethane
Linuron
Mevinphos
Parathion (including parathion-methyl)
Pyrazon
1,1,1-Trichloroethane

Table 2.7   The main types of compound used as pesticides (from Ross (1989) and Manahan (1991))

Insecticides
　　Organochlorines (e.g. DDT, Lindane, Aldrin and Heptachlor)
　　Organophosphates (e.g. Parathion, Malathion)
　　Carbamates (e.g. carbaryl, carbofuran)

Herbicides
　　Phenoxyacetic acids (e.g. 2,4-D, 2,4,5-T, MCPA)
　　Toluidines (e.g. Trifluralin)
　　Triazines (e.g. Simazine, Atrazine)
　　Phenylureas (e.g. Fenuron, Isoproturon)
　　Bipyridyls (e.g. Diquat, Paraquat)
　　Glycines (e.g. Glyphosphate 'Tumbleweed')
　　Phenoxypropionates (e.g. 'Mecoprop')
　　Translocated carbamates (e.g. Barban, Asulam)
　　Hydroxyarylnitriles (e.g. Ioxynil, Bromoxydynil)

Fungicides
　　Non-systemic fungicides
　　　　Inorganic and heavy metal compounds (e.g. Bordeaux Mixture–Cu)
　　　　Dithiocarbamates (e.g. Maneb, Zineb, Manozeb)
　　　　Pthalimides (e.g. Captan, Captafol, Dichofluanid)
　　Systemic fungicides
　　　　Antibiotics (e.g. cycloheximide, blasticidin S, kasugamycin)
　　　　Benzimidazoles (e.g. carbendazim benomyl, thiabendazole)
　　　　Pyrimidines (e.g. ethirimol, triforine)

Table 2.8 Common indoor pollutants and their sources (Adapted from Tolba *et al.* (1992) and Masters (1991))

Pollutant	Source
Formaldehyde	Particle board, plywood, foam insulation, smoking[a]
NO_2	Gas stoves, kerosene heaters
CO	Kerosene heaters, wood stoves, smoking, vehicles in garages
PAHs	Burning wood, coal or dung, industrial solvents, wood stoves
SO_2	Kerosene heaters
Cl_2	Household bleaches and toilet cleaners
O_3	Photocopiers, laser printers, electrostatic air cleaners
VOCs	Cooking, room deodorizers, cleaning sprays, paints, varnishes, solvents, carpets, furniture
Smoke and other particulates	Smoking, cooking, aerosol sprays, rubber-backed carpets, wood stoves, Pb etc. from erosion of paint emulsions, asbestos and fibrous insulation, decoration, flooring and cement products
Moulds, fungi Viruses	Dampness (humid, cold and poorly ventilated rooms)
Radon[b]	Rock, soil, concrete

[a] Smoking, tobacco/cigarette smoking.
[b] Radon often naturally occurring.

have become relatively ubiquitous pollutants in human and animal tissues, in soils and crops, and in groundwater, rivers and lakes, especially in technologically advanced countries. However, transport in the atmosphere, the oceans and oceanic food chains has resulted in their wider global distribution. Traces of several pesticides can be found in the Arctic snows (Welch *et al.*, 1991) and in Antarctic penguins (Holdgate, 1979).

Pesticides mainly comprise insecticides, herbicides (weed killers) and fungicides and some of the main types of compound currently in use are given in Table 2.7.

2.3.5   Indoor pollution (see also Chapter 7)

Indoor air pollution in domestic houses is particularly important in considerations of health because people spend a high proportion of their life breathing the air in this environment, especially in their bedrooms. Although many types of pollutant could be present in houses near to sources of industrial pollution, the most common pollutants which are likely to be found in the air of most homes are given in Table 2.8. These substances are dealt with in more detail in various places throughout this book.

2.4.1   Transport media

Pollutants emitted from a source are dispersed in the environment in air (e.g. smoke, SO_x, NO_x, and $PbClBr$), in water (e.g. industrial effluents, pesticides, phenols, landfill leachate, NO_3^-) and in soil. Pollutant movement through the soil by pedological processes is often much slower than in other transport media, but rapid secondary transport of the pollutant while adsorbed on soil particles can occur when these are carried by the wind or running water. In the context of toxicology, humans and other animals can be exposed to pollutants in the following media: diet (often involving soil–plant pathways), air, water and dust. The transport mechanisms include movement in wind and water, gravity (e.g. movement of particles down the sides of waste heaps into rivers or onto soil), and arthropogenic transport (by land, air or sea) and placement (direct tipping of material).

2.4.2   Transport of pollutants in air

Most air pollutants are discharged into the boundary layer, which is the thin layer of the atmosphere in contact with the earth's surface where the airflow is frequently turbulent because of surface roughness. Many pollutants, especially the larger particulates ($1-0\,\mu m$) remain in the boundary layer but gases and smaller aerosol particles ($< 5\,\mu m$) are transferred into the troposphere zone above by vertical movements in thermal plumes, storms and flow over mountains. The troposphere lies immediately above the boundary layer and the air gets cooler with height up to the tropopause, which occurs at around 10 km at the poles and 16 km at the equator. Above the tropopause is the stratosphere where there is a reversal in the temperature gradient up to around 50 km. The mesosphere lies above the stratosphere but pollutants do not normally enter this zone (Section 5.1).

The transport of atmospheric pollutants depends on the height they reach in the atmosphere, their particle size and climatic factors. Little transfer of air and pollutants occurs between the northern and southern hemispheres in the troposphere, but some transfer occurs between the troposphere and the stratosphere near the equator and in other places. Explosions, volcanic eruptions and high-flying aircraft inject some pollutants directly into the stratosphere. Gaseous pollutants tend to remain in the stratosphere for a long time because of the lack of washout. However, some pollutant molecules may be degraded by the intense radiation which occurs above the ozone layer (at around 20 km altitude) (Holdgate, 1979).

Air pollutants tend to be transported by the wind and mixed with the surrounding air until their concentration in the turbulent boundary layer is relatively uniform. Dispersion in the horizontal plane is generally unrestricted and occurs more rapidly than vertical mixing. The extent to which air pollutants become diluted after emission is largely controlled by the factors determining the degree of turbulence in the boundary layer and these include: incoming solar radiation, wind speed, cloud cover and land surface roughness (Masters, 1991). The density of air is inversely proportional to its temperature and, therefore, warm air is less dense and rises, while cooling air becomes denser and descends. However, in the troposphere temperature decreases with increasing height; therefore, the warm rising air gradually cools and descends again. The temperature of the air also determines how much water vapour it can hold. Descending cool air can become saturated with water vapour and form clouds or fog (Dix, 1981).

The adiabatic lapse rate of the cooling of air masses plays a major part in determining the stability of the air into which pollutants are discharged. Upward movement of air in the troposphere results in expansion and cooling whereas downward movement leads to compression and warming (Masters, 1991). The adiabatic lapse rate of dry air is $9.76°C/km$ (or about $1°C/100\,m$). When the temperature of rising air decreases faster than the adiabatic lapse rate, the air mass becomes unstable and rapid mixing and dilution of pollutants occurs. However, if the temperature decreases more slowly than the adiabatic lapse rate, the air will remain stable and the pollutants will concentrate. If there is sufficient moisture in the air for condensation to occur, latent heat will be released and the air mass will cool more slowly. On average, the wet adiabatic lapse rate is about $6°C/km$ but will be higher at the poles and lower in the tropics (Masters, 1991).

Radiation inversions occur on cold nights (especially in winter) when the temperature of the air above the ground is colder than the ground temperature. They start at ground level around dusk and can move upwards to an altitude of several hundred metres before dispersing the following morning. Since these inversions occur in darkness, no photochemical reactions occur and so primary pollutant gases (such as CO and SO_2) will accumulate and not secondary pollutants (such as O_3, NO_2 and PAN) (Section 6.2).

Subsidence inversions differ markedly from radiation and are associated with high-pressure weather systems (anticyclones). At higher elevations they can last for months at a time and are more common in summer than winter. In a high-pressure zone, air in the middle descends and moves outwards near the ground, while air at the edges rises. The descending air in the middle of the high-pressure system undergoes adiabatic heating to a temperature above that of the air near the ground and results in an inversion of up to several thousand metres' vertical extent that persists for as long as the high pressure system lasts. These inversions are responsible for many regional air pollution problems, such as those of the Los Angeles and San Francisco areas of the south-west USA. A large high-pressure zone off the Californian coast remains in position from spring to autumn resulting in clear sunny skies which are conducive to photochemical reactions, causing the formation of smogs. There is virtually no washing-out of pollutants because of the lack of rain. The topography in the Los Angeles area further exacerbates the problem because the mountains inland prevent winds from dispersing pollutants away from the area (Masters, 1991).

The amount of air available for the dilution of atmospheric pollutants is dependent on the wind speed and the extent to which the emission can rise into the atmosphere. The mixing depth of the atmosphere is determined by the altitude at which rising adiabatically cooling air would reach the same temperature as the surrounding air. The product of the maximum mixing depth and the average wind speed within this depth is known as the 'ventilation coefficient' (m^2/s) and values of this coefficient of $< 6000\,m^2/s$ are regarded as indicative of high potential for pollution (Masters, 1991). Wind speed increases with height above the ground and the more stable the atmosphere, the higher the wind speed becomes with increasing altitude. The differential between the wind speed at ground level and at higher altitude is exacerbated by the roughness of the land surface (owing to surface drag).

Transport of pollutants in chimney plumes (point sources of pollution)

Since many atmospheric pollutants are emitted from chimneys (or stacks) (Fig. 2.2), it is convenient to use an example of the dispersion of a plume of smoke from a single chimney to illustrate the principal factors involved. As in all other considerations of air pollutant dispersion, the stability of the atmosphere and the speed of the wind are very important, as shown in Fig. 2.3, which shows the typical range of atmospheric conditions over 24 hours (Masters, 1991; Briggs, 1969; LaGrega *et al.*, 1994).

Figure 2.2 Plume of steam from an incinerator (photo: B. J. Alloway).

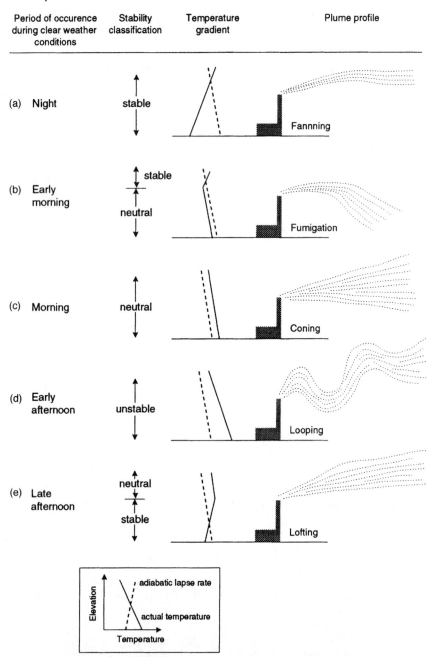

Figure 2.3   Plume profile for various atmospheric stability classifications (after Briggs, 1969 in La Grega *et al.*, 1994 reproduced by permission of the publishers).

From Fig. 2.3, starting with night time (a) when conditions are often relatively stable, the vertical dispersion of the plume is restricted but horizontal dispersion will be unlimited and so *fanning* will take place (i.e. the plume assumes a fanshape). In the early morning (b) conditions will be neutral near the ground below a temperature inversion and stable above and so the chimney emissions will be held down near the ground and result in increased concentrations of the pollutants down-wind from, but close to, the source; this is called *fumigation*. Later in the morning (c) when the chimney is discharging into a neutrally stable atmosphere (where the adiabatic and ambient lapse rates are similar), a symmetrical cone-shaped plume will form (*coning*) because of the relatively unrestricted dispersion in both vertical and horizontal directions. In contrast, in the early afternoon (d) or when the atmosphere is unstable because the ambient temperature decrease is greater than the adiabatic lapse rate, there is rapid vertical air movement, both up and down, giving a *looping* plume. In the late afternoon (e) or when there is an inversion near the ground, if the top of the chimney is above the inversion, vertical mixing is not restricted but downward mixing is limited by the inversion's stable air. This gives rise to *lofting*, which assists in the dilute and disperse approach to pollution abatement. Over recent decades, high chimneys have been built at many major sources of atmospheric pollution, such as coal-burning power stations and smelting works (e.g. 300 m high stack at a Cu–Co smelter at Sudbury, Ontario, Canada), to discharge the emission well above inversion layers. Areas immediately downwind from these high chimneys will generally be less troubled by atmospheric pollution but long-distance tropospheric transport of the pollutants may result in more chronic effects on ecosystems over larger areas, such as the acidification of soils and lakes and die-back of trees (Section 5.2).

It is important to realize that the presence of tall buildings, other chimneys and many local climatic features of the urban environment can greatly complicate the pattern of dispersion of atmospheric pollutants from a single chimmey. Down-draughts, 'street canyon' phenomena and concentration in urban heat islands can all have marked effects.

Point-source Gaussian model for pollutant transport. Most models of time-averaged concentrations of pollutants down-wind from a point source, such as a chimney, are based on a normal (or Gaussian) distribution curve of the pollutants (see Fig. 2.4). The main variables which have to be considered include the rate of emission from the source (assumed to be constant in simple models) and the wind speed. It is also assumed that the pollutant is conservative (i.e. not lost by decay or deposition) and that the terrain is relatively flat. Masters (1991) quotes a model:

$$C(x,y) = \frac{Q}{\pi\mu\sigma_y\sigma_z}\left(\exp\frac{-H^2}{2\sigma_z^2}\right)\left(\exp\frac{-y^2}{2\sigma_y^2}\right) \qquad (2.1)$$

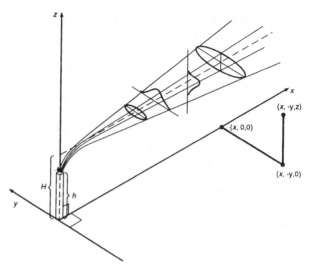

Figure 2.4 A Gaussian model for the dispersion of a plume in the vertical and horizontal directions (reproduced from Turner (1970) in Masters, 1991).

where $C(x, y) =$ concentration at ground level at point x, y (μg/m); $x =$ distance downwind (m); $y =$ horizontal distance from the plume centre-line (m); $Q =$ emission rate of pollutants (μg/s); $H =$ effective stack height (m) ($H = h + \Delta h$, where $h =$ chimney height and $\Delta h =$ plume rise); $\mu =$ average wind speed at the effective height of the stack (m/s), $\sigma_y =$ horizontal dispersion coefficient (standard deviation) (m); and $\sigma_z =$ vertical dispersion coefficient (standard deviation) (m).

The concentration of pollutants at ground level is directly proportional to the source strength Q. The ground-level concentrations decrease for higher chimneys (but the relationship is not linear). Although the down-wind pollutant concentration appears approximately inversely proportional to wind speed, the relationship is slightly affected by the wind speed, which influences the rise of the plume. Higher wind speeds reduce the effective height of the chimney (H) and so the ground-level pollution does not decrease as much as would be expected from the inverse relationship (Masters, 1991). The dispersion coefficients σ_y and σ_z are the standard deviations of the vertical and horizontal Gaussian distributions and are functions of down-wind distance and atmospheric stability, and they will increase with distance down-wind. Even though this simple model is probably only reliable to $\pm 50\%$ in most cases, it is still useful for predicting the fate of pollutant smoke emitted from chimneys.

The rise of the plume after emission is an important factor and is dependent on the buoyancy and momentum of the gaseous emissions and on the stability of the atmosphere. The equation for buoyancy flux (F) is:

$$F = gr^2 v_s (1 - T_a/T_s) \tag{2.2}$$

where F = buoyancy flux (m^4/s^3); g = gravitational acceleration (9.8 m/s^2); r = inside radius of chimney; v_s = chimney gas exit velocity (m/s); T_s = chimney gas temperature (K); and T_a = ambient temperature (K) (Masters, 1991).

In a temperature inversion, the pollutants will reflect off the inversion layer and if the inversion is above chimney height then the basic Gaussian equation must be modified to take account of the restricted vertical dispersion.

Transport of pollutants from non-point sources

(a) Linear sources. Dispersion from a linear source, such as a major road or a burning of stubble in a farm field, can be described by a simple equation, assuming an infinitely long source and that the wind blows perpendicular to the line. The concentration at a point x metres away from the line of emission is given by:

$$C(x) = \frac{2q}{\sqrt{2\pi}\,\sigma_z u} \tag{2.3}$$

where u = wind velocity (m/s) and q = emission rate per unit of distance along an infinite line source (g/m per s) (Masters, 1991).

(b) Area sources. The simplest way to deal with an area source, such as a city, is to use a box to represent the airshed over the area:

$$\begin{pmatrix} \text{rate of change of} \\ \text{pollution in box} \end{pmatrix} = \begin{pmatrix} \text{rate of pollution} \\ \text{entering box} \end{pmatrix} + \begin{pmatrix} \text{rate of pollution} \\ \text{leaving box} \end{pmatrix}$$

$$LWH\,\frac{dC}{dt} = q_s\,LW + WHu\,C_{in} - WHu\,C \tag{2.4}$$

where C = concentration of pollutant in the airshed; C_{in} = concentration in incoming air; L = length of airshed; q_s = emission rate per unit area; W = width of airshed; H = mixing heights and u = average windspeed against the side of the box (Masters, 1991).

(c) Indoor air pollution. Models of indoor air pollution can be modified to take account of the decay of the pollutant:

$$\begin{pmatrix} \text{Rate of} \\ \text{increase in} \\ \text{the box} \end{pmatrix} = \begin{pmatrix} \text{Rate of} \\ \text{pollution} \\ \text{entering the} \\ \text{box} \end{pmatrix} - \begin{pmatrix} \text{Rate of} \\ \text{pollution} \\ \text{leaving the} \\ \text{box} \end{pmatrix} - \begin{pmatrix} \text{Rate of} \\ \text{decay in the} \\ \text{box} \end{pmatrix}$$

$$V\frac{dC}{dt} = S + C_a IV - CIV - KCV \tag{2.5}$$

where V = volume of conditional space in building (m^3/air change); I = air exchange rate (air changes/h); S = pollutant source strength (mg/h); C = indoor concentration (mg/m^3); C_a = ambient concentration (mg/m^3); and K = pollutant decay rate or reactivity (l/h).

A steady-state solution is found by setting $dC/dt = 0$

$$C = \frac{S/V + C_a I}{I + K}$$

A general solution is:

$$C(t) = \frac{S/V + C_a I}{I + K}(I - e) + C(0)\,e \tag{2.6}$$

where $C(0)$ = the initial concentration in the building.

In the case of conservative pollutants, such as CO and NO, where $K = 0$, if the ambient concentration is negligible and the initial concentration is 0, the equation will be:

$$C(t) = \frac{S}{IV}(1 - e^{-It}) \tag{2.7}$$

With the exception of CO_2, few other pollutants are accumulated in air; most are transported and then deposited or removed from the air by various mechanisms (Masters, 1991).

2.4.3 Some important types of reaction which pollutants undergo in the atmosphere

(a) Oxidation

An example is the conversion of CO into CO_2: This occurs as a result of CO reacting with the hydroxyl radical HO$^•$.

$$CO + HO^• \longrightarrow CO_2 + H^•$$

which leads to the hydroperoxide radical

$$O_2 + H^• + M \longrightarrow HOO^• + M$$

The HO$^•$ radical is regenerated by

$$HOO^• + NO \longrightarrow NO_2 + HO^•$$

and by

$$HOO^• + HOO^• \underset{-O_2}{\longrightarrow} H_2O_2 \longrightarrow 2\,HO^•$$

Other examples of radical reactions are:

(i) abstraction of H atom from a hydrocarbon and capture of oxygen

$$HO^\bullet + CH_4 \longrightarrow CH_3{}^\bullet + H_2O$$
$$\text{(RH)}$$
$$\downarrow O_2 + M$$
$$CH_3OO^\bullet + M$$

With subsequent reactions of the alkyl peroxide so formed

$$CH_3OO^\bullet \xrightarrow[HOO^\bullet]{} CH_3OOH + O_2$$
methyl hydroperoxide

$$NO \downarrow \qquad\qquad \downarrow$$

$$CH_3O^\bullet + NO_2 \qquad\qquad CH_3O^\bullet + HO^\bullet$$

$$O_2 \downarrow \qquad\qquad \downarrow O_2$$

$$H-\overset{\displaystyle |}{\underset{\displaystyle H}{C}}{=}O + HOO^\bullet \qquad H-\overset{\displaystyle |}{\underset{\displaystyle H}{C}}{=}O + HOO^\bullet$$

(ii) Photochemical reactions

$$NO_2 + h\nu \xrightarrow[400\,nm]{} NO + O$$

$$O_3 \xrightarrow[c.\,275\,nm]{} O_2 + O$$

$$O + H_2O \longrightarrow {}^\bullet OH + {}^\bullet OH$$

$$O + CH_4 \longrightarrow {}^\bullet OH + {}^\bullet CH_3$$

$$O + N_2O \longrightarrow NO + NO$$

(iii) Formation of PAN (Section 6.2.4)

Both the atmosphere and water are fluid media and have many properties in common. The transport and dispersion of pollutants in the aquatic environment is controlled by advection (mass movement) and mixing or diffusion (without net movement of water) (Hewitt and Harrison, 1986). Also, in common with the situation in the atmosphere, the vertical movement of water is often restricted. In large water bodies this is because of stratification, caused by differences in temperature and density or salinity, but in rivers this results from limited depth. However, unlike the atmosphere, pollutants tend to accumulate in lakes and seas and in the sediments at the bottom of these bodies of water.

2.5.1   Physical transport in surface waters

Rivers

Rivers are characterized by water flow downstream under the influence of gravity (gravity advection), although in tidal sections this is modified by the state of ocean tides. Two equations are used to model the velocity of water in uniform stream channels, these are the Chezy and the Manning equations. The Chezy equation is

$$V = C\sqrt{RS} \tag{2.8}$$

where V = velocity (L/T); C = the Chezy coefficient $[L^{1/2}/T]$, R = the hydraulic radius (L) and S the slope. The Manning equation is

$$V = \frac{1.486R^{2/3}S^{1/2}}{n} \tag{2.9}$$

where n = Manning's coefficient describing river channel roughness $[T/L^{1/3}]$. The coefficients C and n are determined experimentally (Hemond and Fechner, 1994).

It is important to be able to estimate the travel time of a chemical pollutant in a river, especially in cases of accidental spillages of hazardous chemicals, in order that precautions can be taken to protect supplies of potable water. For example, the fire at the Sandoz works in Basle, Switzerland in November 1986 resulted in 20–30 t of pesticides being washed into the Rhine. Benthic (bottom-living) organisms and eels were completely eradicated over a distance of 400 km downstream and the damage caused has been estimated at 50×10^6. Most of the chemicals had been purged from the river within months and after 1 year most aquatic life had returned to the river. However, the groundwaters in the Rhine alluvium remained seriously contaminated (Tolba *et al.*, 1992).

Hemond and Fechner (1994) state that when a mass of chemical is released into a river, the centre of the chemical's mass will move down the river at the average velocity of the river. Under constant velocity conditions, the travel time (r) for movement down a section of river of length L flowing at velocity V is given by

$$r = L/V \qquad (2.10)$$

where the velocity is not constant, the travel time is given by an integral

$$r = \int \frac{1 \mathrm{d}x}{V(x)}, \qquad (2.11)$$

where x = the distance along the river (L), x_1 and x_2 are at each end of the reach being considered, and $V(x)$ = the magnitude of the velocity of the river any given point $x(L/T)$. The discharge of a river (Q) is the volume of water passing a given point per unit of time and is the product of the cross-sectional area of the river (A) and the average velocity (V). The mass of chemical (M) transported past a given point per unit time (the total flux) is the product of discharge (L^3/T) and the average concentration of the chemical (M/L^3) and has the units of (M/T) (Hemond and Fechner, 1994). The chemical will disperse within the body of the river as it moves downstream because of both turbulent diffusion and non-uniform velocity across the river. The velocity of water in a river is usually at a maximum near the centre and below the surface because the water near the bed and banks of the river will be slowed down by friction. As in air, the more turbulent the water, the greater the degree of mixing. The concentration of a chemical at time t after injection into the river at distance x downstream is given by

$$C(x,t) = \frac{M}{\sqrt{4\pi DE}} \, e^{-Kt} \qquad (2.12)$$

where C = concentration of the chemical (M/L^3), M = the mass of the chemical injected per cross-sectional area of the river (M/L^2), x is the distance downstream (L), and D is the Fickian mixing coefficient owing to the effects of both turbulent diffusion and dispersion;

$$D = \sigma^2/2r \qquad (2.13)$$

where σ^2 = the spatial variance (square of standard deviation) of the distribution of the chemical (L^2), and r is the travel time from the point of injection. The part of the river lying within one standard deviation on either side of the maximum concentration contains 68% of the mass of the chemical.

Lakes

Lakes are bodies of water which do not have the downstream flow that is characteristic of rivers; instead the water tends to move around as a result of the influence of the wind. Wind-generated water currents in lakes are responsible for chemical transport by turbulent diffusion. The down-wind surface current, called the wind drift, tends to move at about 2–3% of the velocity of the average wind speed. At the end of the lake, the moving water moves downwards and returns in the up-wind direction (called the return current) at depth. The movement of water in a lake is influenced by the shape of the lake basin, variations in the density of the water, inflowing streams and, in large lakes, the Coriolis effect, which is caused by the rotation of the earth. Although thorough mixing can occur within the upper water layer, in some lakes stratification occurs because of differences in the density of the water. In most cases, this stratification is caused by temperature differences in the water, when the lake is described as 'thermally stratified'. The upper (well-mixed) water layer is called the *epilimnion*, the lower layer is the *hypolimnion* and the zone dividing them is known as the *thermocline*.

Thermal stratification is most likely to occur in regions with warm and cold seasons. The epilimnion is thickest in the summer as a result of warmer climatic conditions, but decreasing autumn (Fall) temperatures result in the temperature and density of the epilimnion and the hypolimnion becoming similar, and the lake is then described as isothermal. In regions with very cold winters, reverse stratification will occur because the surface layer is colder than the lower layer; ice will form and wind mixing of the lake water will not occur until after the ice melts. Some of the world's largest lakes, such as the Great Lakes in North America and Lake Biakal in Siberia (Russia), are affected by this phenomenon. Stratification can have very important effects on the behaviour of pollutants. In the warm season, the temperature of the epilimnion water may be 15–20°C but that of the hypolimnion may be much lower at depth. The difference in temperatures results in marked variations in the rate of chemical reactions, including the decomposition of pollutants. The isolation of the bottom waters from the atmosphere prevents the replacement of oxygen consumed by organisms and anoxic conditions may arise. The behaviour of pollutant chemicals in the cold anoxic water will differ considerably from that in the oxygenated epilimnion (Hemond and Fechter, 1994).

Estuaries

With regard to the behaviour of pollutant chemicals, an estuary is the zone where river water and sea water become mixed and it is characterized by steep gradients in ionic strength and chemical composition (De Mora, 1992). The flow of water in estuaries is influenced by ocean tides, the inflow of rivers and by the difference in density between the more dense salt water ($1.03\,g/cm^3$) and fresh water ($1.0\,g/cm^3$). The higher density of salt water causes marked stratification, but the ebb and flow of the tides encourages mixing (Hemond and Fechter, 1994). River flow into estuaries can show marked seasonal variations, such as in a dry season or in a region with a pronounced freeze–thaw cycle.

River waters tend to be slightly acidic and have a low ionic strength; the predominant salt usually being calcium bicarbonate ($Ca(HCO_3)_2$). In contrast, sea water has an alkaline pH of around 8.0 and a high ionic strength dominated by NaCl.

Estuaries are highly vulnerable to pollution owing to the deposition of sediment particles of terrestrial origin with pollutants already sorbed into them and their acting as a sink for various pollutants washed down rivers in solution. The coastal region is, in general, the part of the oceanic environment which is most prone to pollution and estuaries are the part of the coastal zone most affected by pollution. Industrial and sewage discharges (the latter only permitted in the European Union until the end of 1998), runoff from urban and agricultural areas and the release of biocidal compounds from marine antifouling paints can all add to the load of pollutants in estuarine sediments.

Sorption of pollutant metals occurs in sediments in the same way that it does in soils (Section 2.6.2) except that, with the exception of a thin surface layer, all the body of sediment usually has anoxic conditions and so chemical reduction and the activities of anaerobic organisms predominate. Organic pollutants will tend to be sorbed by organic matter in estuarine sediments. Estuaries of most major rivers, such as the Rhine, Danube, Rhone, Po, Mississippi, Chang Jiang (Yangtze), and sediments dispersed from them along adjacent coastlines are all polluted by a wide range of inorganic and organic pollutant chemicals.

2.5.2 Dispersion of pollutants in groundwaters

The porous rock material at the surface of the earth (under the thin cover of biologically active soil) can contain groundwater. This is the term given to the water occupying the pore spaces in the permeable rock and unconsolidated rock material; the water-bearing geological material is referred to as an aquifer (Fig. 2.4). The upper surface of the saturated zone is called the water table and springs form where the saturated zone reaches the land surface; these springs contribute to the flow of river systems. Aquifers can be either confined, where the water is trapped and cannot flow out but when wells are drilled into them the water will often rise under pressure (Artesian wells), or unconfined, where water flows both out of and into the aquifer. Groundwater is both an important direct source of potable water from wells sunk down into the aquifer, and an indirect source through its contribution to surface waters. Therefore, polluted ground-water can have a marked impact both on water supplies and aquatic ecosystems.

Groundwater in aquifers can be polluted from many sources, but some of the most important are: (i) leaking underground pipes, (ii) leaking underground storage tanks (LUSTs), (iii) leachate from landfills, and (iv) chemicals (including fertilizer elements such as nitrates, agrochemicals such as the herbicide atrazine, and pollutants in sewage sludge or materials discharged onto soil on industrially contaminated sites) which infiltrate down to the aquifer from the soil surface.

The dispersion of pollutants in aquifers is dependent upon: (i) the physics of groundwater movement, (ii) the solubility of the pollutant, and (iii) retardation owing to the pollutants becoming sorbed on the solid surfaces.

Physics of groundwater movement

In a saturated aquifer, water flows under a hydraulic gradient from zones of high hydraulic head to zones of lower hydraulic head (called a hydraulic gradient). Darcy's law relates groundwater flow, hydraulic head gradient and hydraulic conductivity, which is a measure of the ease of flow though the porous medium. Darcy's law is expressed as:

$$q = -K \cdot \mathrm{d}h/\mathrm{d}x \qquad (2.14)$$

where q = the specific discharge (amount of water flowing across unit area of the porous medium perpendicular to flow over unit time), K = the hydraulic conductivity, and $\mathrm{d}h/\mathrm{d}x$ = the hydraulic head gradient. The hydraulic conductivity can vary widely depending on the size and shape of the voids in the aquifer. For example, marine clays may have values of K (cm/s) from 10^{-10} to 10^{-7}, silt can have values 10^{-7} to 10^{-2}, clean sand 10^{-4} to 1 and gravel from 10^{-1} to 10^{2}. The porosity of an aquifer (n) is the volume of the voids divided by the total volume of the porous medium, and typical values are between 0.2 and 0.4.

The rate at which non-sorbing chemicals move in the groundwater is given by

$$v = q/n \qquad (2.15)$$

where v = the seepage velocity, q = the specific discharge as in Equation 2.14 and n = the porosity. Seepage velocity is greater than the specific discharge because flow is confined to only a fraction of the cross-sectional area.

Groundwater flow is usually plotted as a *flow net*, in which the streamline crosses the isopotentials at right angles. Space does not allow this topic to be discussed, but the reader is referred to a specialized text, such as Hemond and Fechner (1994), LaGrega *et al.* (1994), or any other book concerned with groundwater for details of how the parameters related to groundwater flow are measured and used in models.

Wells in aquifers have the effect of drawing down the water table (relative to the ground surface) in their immediate vicinity and giving rise to a *cone of depression*. The water removed from an unconfined aquifer is usually replaced by infiltration of rain water or inflow from a river. In the latter case, polluted rivers can cause serious pollution of potable groundwater supplies, as in the case of the pollution of the Rhine with pesticides after the fire at the Sandoz manufacturing plant in Basle. It is important to be able to predict the amount of drawdown with regard to both potential pollution and remediation of groundwater. The rate at which water is pumped from a well is given by

$$Q_w = -K(dh/dr) \cdot 2\pi rb \qquad (2.16)$$

where $Q_w =$ the rate at which water is pumped from a well (L^3/T), $K =$ the hydraulic conductivity, $dh/dr =$ the hydraulic gradient in the radial direction, and $b =$ the thickness of the aquifer.

The drawdown (s) is the reduction in height of the water table below that which it had before abstraction of water from the well; therefore, the hydraulic gradient can be replaced with ds/dr:

$$Q_w = K(ds/dr) \cdot 2\pi rb$$

or

$$ds/dr = Q_w/2\pi Kbr \qquad (2.17)$$

From this, it can be seen that the slope of the water table in the radial direction (ds/dr) is proportional to the rate at which the well is pumped and is inversely proportional to the hydraulic conductivity, distance from the well and thickness of the aquifer.

The radius of influence of a well (R), which is the horizontal distance away from a well beyond which pumping water from the well will have little effect on the aquifer, can be given by:

$$R = b\sqrt{K/2N} \qquad (2.18)$$

where $N =$ the annual recharge by precipitation. From this it is possible to determine the drawdown by the following equation:

$$s = (Q_w/2\pi Kb) \cdot \ln(R/r) \qquad (2.19)$$

where $s =$ the drawdown at radius r, $Q_w =$ the rate at which water is pumped from a well, $K =$ the hydraulic conductivity, $b =$ the aquifer thickness, $R =$ the radius of influence, and $r =$ the radius from the well.

Chemical dispersion in aquifers

Soluble chemicals disperse laterally in moving groundwater just as they do in surface waters and this mixing effects dilution into a larger mass of water. However, unlike surface waters, the generally slower flow in aquifers is not turbulent, and so mixing and dispersion are slower mainly owing to the tortuosity of the paths the water takes moving through the voids in the aquifer. Fick's first law can be applied to this dispersion in one dimension:

$$D = \alpha \cdot v \tag{2.20}$$

where $D =$ the mechanical dispersion coefficient, $\alpha =$ the dispersivity of the aquifer, and $v =$ the seepage velocity. In reality, two- and three-dimensional dispersion occurs not only along the axis of flow but also perpendicular to it, although dispersion along the direction of seepage flow is usually the most pronounced.

Flow of non-aqueous phase liquids (NAPLs)

Retardation can be brought about by sorption of the chemical on solid surfaces. The extent to which sorption takes place can be expressed by the distribution coefficient K_d which is:

$$K_d = \frac{\text{Concentration sorbed on solid surface}}{\text{Equilibrium concentration in solution}}$$

The retardation factor, R, is

$$R = 1 + \frac{\text{mobile chemical conc.} + \text{sorbed chemical conc.}}{\text{mobile chemical conc.}}$$

Therefore, Fickian dispersion must be modified by including the R value. On the basis of aquifer volume, the total mobile concentration of a chemical in an aquifer is given by:

$$R = (C_{aq} \cdot n) + (C_s \cdot \rho_b) \tag{2.21}$$

where $C_{aq} =$ the aqueous concentration, $C_s =$ the sorbed concentration, $n =$ the water-filled porosity, and ρ_b the bulk density of the particles in the aquifer. However, since $K_d = C_s/C_{aq}$ the following equation can be used:

$$R = 1 + K_d \cdot \rho_b/n \tag{2.22}$$

Contamination of aquifers

Contamination of aquifers is a major environmental pollution problem in most technologically advanced countries and is especially serious where the groundwater is used as a supply of drinking water. Figure 2.5 shows a typical situation that might occur with non-aqueous phase liquids (DNAPLs and LNAPLs: dense and light non-aqueous phase liquids) and water-soluble constituents in a landfill leachate.

The DNAPL seeping from the leaking storage tank (A) shown in Fig. 2.5 can be seen to pass vertically down (though macropores and fissures)

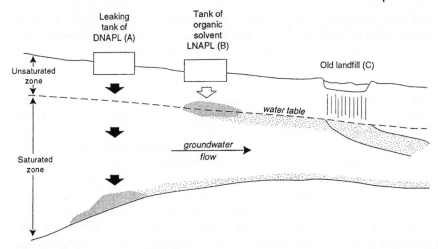

Figure 2.5 Typical patterns of dispersion in an aquifer of pollutants differing in solubility and density (adapted from Miller, 1985)

through the unsaturated zone in the subsoil; even after entering the saturated zone, it continues to move downwards under gravity until it reaches the base of the aquifer formed by the underlying impermeable stratum. However, as it accumulates at the base of the aquifer, it will be subject to the mass movement of the flow of groundwater and will gradually be dispersed along the base of the aquifer. There is also a strong possibility that some of the constituents of the DNAPL may be slightly soluble and these will tend to be dispersed more rapidly within the body of the aquifer.

In the case of the low-density pollutant (LNAPL) (B), such as an organic solvent, the compound will seep down through the unsaturated zone to the water table (top of the saturated zone). Then, owing to its low density, it will accumulate at this point and will tend to be dispersed in the direction of the groundwater flow. As with the DNAPL, any water-soluble constituents will start to dissolve in the groundwater and disperse within the body of the aquifer. As discussed in Chapter 8, leakages of aviation fuel at airports and military air bases can give rise to major groundwater pollution problems. Although most of the hydrocarbon content is immiscible with water, there are still significant amounts of water-soluble substances which will dissolve and migrate for much greater distances than the insoluble constituents.

The old landfill (C) will have given rise to leachates containing a wide range of substances (Section 8.3), many of which, such as chlorides and metal ions, are water soluble. These will seep vertically down to the water table and form a plume which moves in the direction of the groundwater flow. The size and shape of the plume will depend on the hydraulic conductivity of the aquifer, the flow of water in it, the amounts of soluble pollutants entering and the extent of retardation owing to sorption on the aquifer mineral surfaces. In general, for pollutants which are not sorbed to a large extent, the shape of the plume tends to be long and narrow in highly permeable aquifer materials and shorter but dispersed more horizontally in less permeable media. This subject is covered in considerable detail in specialist texts, such as Freeze and Cherry (1979) and Ward et al. (1985), and in environmental pollution books such as Hemond and Fechner (1994) and LaGrega et al. (1994).

Aquifers which are particularly at risk are limestones (including chalk), sandstones and glacial sands and gravels because they underlie large areas with a diversity of land uses, many of which are potentially polluting. In the UK, the chalk (limestone) and the Permo–Triassic sandstone are large aquifers which are used for potable water supplies (Mather, 1993). Large areas of Denmark and the Netherlands are underlain by aquifers which also form the potable water supply. Apart from large industrial sites, leakages from petrol/diesel filling stations and farm waste disposal can cause significant pollution problems. In countries or regions where houses are widely dispersed in the landscape, it is not feasible to connect all properties to the main drainage sewers and so most houses will have a septic tank system for dealing with wastewater and human excrement. These septic tanks function as bioreactors which gradually discharge into the unsaturated zone and, ultimately, the saturated zone, where dilution will occur. Nevertheless, septic tanks can be regarded as small, localized sources of groundwater pollution. The main danger is that a borehole for drinking water supply for a house downslope may intercept the plume of septic tank discharge from a property unslope.

Mather (1993) quotes several cases of pollutants migrating considerable distances in aquifers in the UK. The Permo–Triassic sandstone underlying part of the Birmingham industrial conurbation has become significantly contaminated with organic solvents including trichloroethylene (TCE), 1,1,1-trichloroethane (TCA) and perchloroethylene (PCE). Similarly, the chalk aquifer underlying the industrial towns of Luton and Dunstable has become contaminated with organic solvents from engineering and printing industries. Chlorinated solvents, including chloroform, trichloroethylene and carbon tetrachloride were found to have migrated from a waste-disposal facility at the Harwell Laboratory in Oxfordshire to a public water supply borehole 6 km away. A spillage of diesel fuel onto topsoil over the chalk aquifer moved down to the water table and migrated to a borehole 665 m away. Saline water from a coal mine in Kent has contaminated an area of 27 km^2 of the chalk aquifer in that area. t-Butyl methyl ether (TBME) in unleaded petrol leaking from a petrol filling station contaminated boreholes 500 m away.

It is important to recognize that some of the most serious pollution problems in fresh waters involve substances which are not considered to pose a high toxicity risk. Firstly, organic matter which can be easily oxidized biochemically by microorganisms is a major water pollution problem because it can lead to the removal of most of the dissolved oxygen in the water by microorganisms leaving insufficient for fish and other aquatic animals (as discussed in Section 2.5.1). The second is nitrate (NO_3^-) pollution. Nitrates occur naturally in soils and waters as a product of the microbial mineralization of dead plant and animal tissues in the soil and forms a very important stage in the nitrogen cycle because NO_3^- is the main form of nitrogen taken up by the plants. Ammonium ions from the mineralization process, or from fertilizers such as NH_4NO_3, tend to be rapidly nitrified to NO_3^-. There has been a marked increase in the NO_3^- content of both surface waters and groundwaters since the 1940s and this is the result of the increased use of the nitrogen fertilizers and the degradation of a greater mass of roots and litter. Owing to beneficial conditions under the EC Common Agricultural Policy (CAP), it has been economic for farmers to apply up to 300 kg N/ha or more to high-yielding crops. Since NO_3^- is not adsorbed in soils, that not taken up by crop roots is leached down the soil profile, so a considerable proportion of the nitrogen applied in fertilizers is leached into groundwaters or surface waters. The effects of excess NO_3^- on human health are not fully understood, but the possibility of them contributing to the synthesis of carcinogenic nitrosamines has been considered. Bacteria in the mouth and gut can reduce NO_3^- to NO_2^- and $NaNO_2^-$ is itself present in foodstuffs. Contact with amines can then lead to nitrosoamine synthesis. This risk is not highly rated in the UK because, in areas where gastric cancer is relatively low, NO_3^- levels in water are relatively high. However, there is a further risk from nitrosoamine synthesis from NO_2^- in tobacco (Section 7.6).

Nitrate in drinking water is a matter for concern because it may include a condition known as methaemaglobinaemia, or the 'blue-baby' syndrome. This can arise in infants in the first 6 months of life and is caused by defective carriage of oxygen. For haemoglobin to carry oxygen in the blood stream, the iron must be present as Fe(II), NO_3^- will oxidize it to Fe(III). When oxygen transport fails and cyanosis, a bluing of the lips appears. Older children and adults are protected from oxidative stress of this kind by an enzyme – methaemoglobin reductase – which develops gradually in the body after birth. The maximum allowable NO_3^- concentration in potable (drinking) waters is 50 mg NO_3/l in the UK, but some water supplies in intensive arable regions exceed this value. Peak winter values can frequently exceed 100 mg NO_3/l in rivers draining intensively farmed arable areas, such as much of East Anglia in the UK, and some mathematical models have predicted levels of < 150 mg NO_3/l (Crathorne and Dobbs, 1990). However, recent policies to protect aquifer recharge areas, campaigns to encourage farmers to use nitrogen fertilizers more efficiently, and falling prices for crop products have helped to slow down the NO_3^- increase in waters in many parts of the EC.

2.5.3   Biochemical processes in water (involving microorganisms)

Figure 2.6 gives examples of the biodegradation of organic compounds.

Oxidation

Oxidation of organic carbon compounds

$$\{CH_2O\} + O_2 \longrightarrow CO_2 + H_2O$$

Oxidation of ammonia

$$NH_4^+ + 2O_2 \longrightarrow 2H^+ + NO_3^- + H_2O$$

The first of the two oxidation reactions requires the presence of dissolved oxygen and temperatures above 0°C for oxidation of carbon compounds and >4°C for oxidation of NH_3 (Fish, 1992). This requirement for oxygen is called the *biochemical oxygen demand* (or biological oxygen demand), usually abbreviated to BOD. If there is an excess of oxidizable organic matter in a river or pond, arising from a discharge of an effluent such as liquid manure slurry from a farm, the bacteria carrying out the oxidation may utilize all the available dissolved oxygen causing an acute shortage of oxygen for fish, which then die from asphyxiation. The BOD of a stream or river is measured by determining the quantity of oxygen utilized by aquatic microorganisms over a 5-day period. Typical values of BOD are <3 mg/l for Class 1A rivers in the UK (the least polluted class), <5 mg/l for Class 1B, 9 mg/l for Class 2 (more polluted and only suitable for potable supply after advanced treatment) and 17 mg/l for Class 3 (poor quality water with few fish present) (Fish, 1992). The effect of an input of oxidizable organic pollutant into a stream is shown in Fig. 2.7.

Redox reactions

Redox reactions take place between a reducing agent, which can donate electrons, and an oxidizing agent, which can accept electrons. When these changes occur in aqueous solution, the two reagents act in association and no free electrons participate.

The more electropositive metals readily donate electrons to form stable cations and so are powerful reducing agents. The violent reaction between sodium and water can be written as two half-reactions:

$$\text{reduction } Na(s) \longrightarrow Na^+(aq) + e \tag{2.23}$$

$$\text{oxidation } H_2O + e \longrightarrow {}^-OH + [H] \tag{2.24}$$

In this context water accepts electrons and is acting as an oxidizing agent;

O-Dealkylation

R-O-Me ⟶ R-OH+HCHO

Examples:
(i) methoxychlor

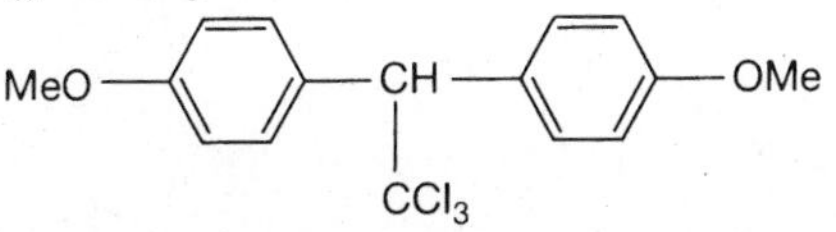

MeO—⬡—CH—⬡—OMe
        |
       CCl$_3$

(ii) phosphorus insecticides

C-Dealkylation

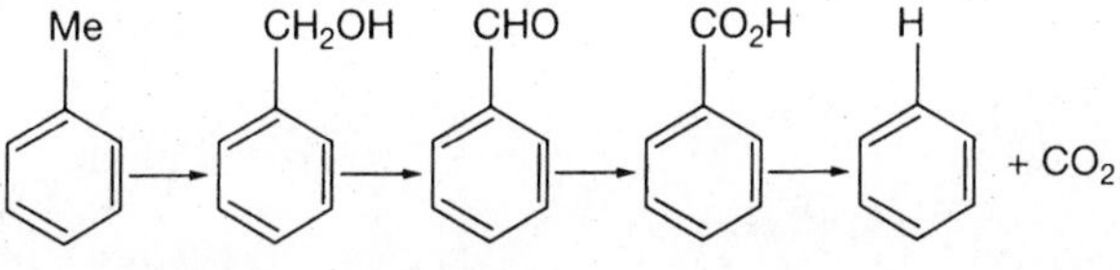

Me   CH$_2$OH   CHO   CO$_2$H   H

⬡   →   ⬡   →   ⬡   →   ⬡   →   ⬡   + CO$_2$

Epoxidation

cf. aldrin ⟶ dieldrin

Aerobic hydroxylation

benzene $\xrightarrow[+2H]{O_2}$ phenol + H$_2$O

Anaerobic hydroxylation

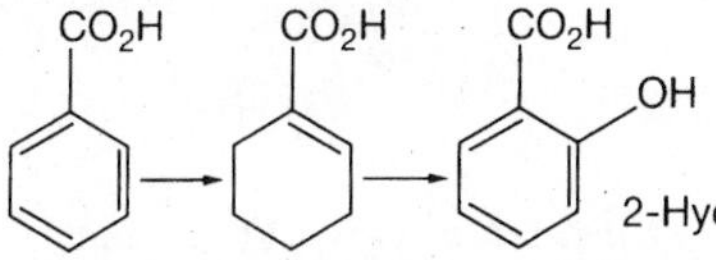

CO$_2$H   CO$_2$H   CO$_2$H

⬡   →   ⬡   →   ⬡—OH   2-Hydroxycyclohexene-1-carboxylate

Fission

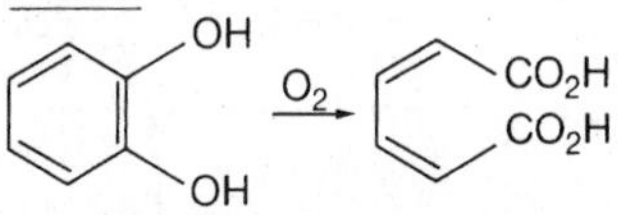

⬡(OH)(OH) $\xrightarrow{O_2}$ (CO$_2$H)(CO$_2$H)

Examples: carbaryl, alkanoate herbicides

Hydrolysis of carbamates, organophosphates and anilide herbicides

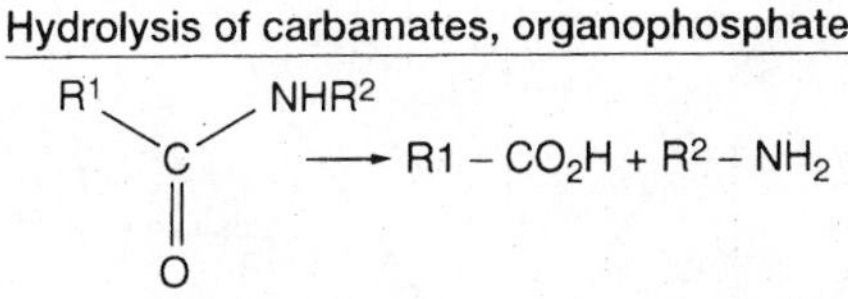

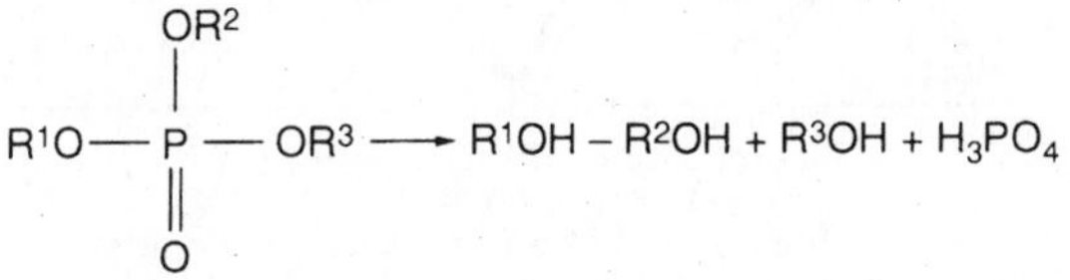

$$\begin{array}{c} R^1 \quad NHR^2 \\ \diagdown C \diagup \\ \parallel \\ O \end{array} \longrightarrow R1 - CO_2H + R2 - NH_2$$

$$\begin{array}{c} OR^2 \\ | \\ R^1O - P - OR^3 \\ \parallel \\ O \end{array} \longrightarrow R^1OH - R^2OH + R^3OH + H_3PO_4$$

Figure 2.6 Biodegradation of organic compounds.

Hydrolytic dehalogenation

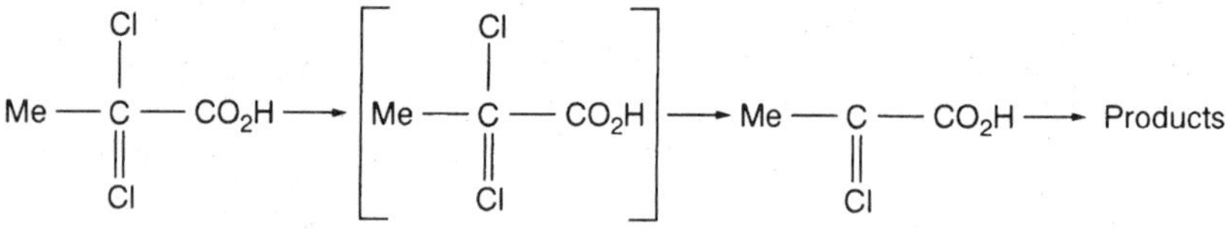

$$Me-\underset{\underset{Cl}{\|}}{\overset{\overset{Cl}{|}}{C}}-CO_2H \longrightarrow \left[Me-\underset{\underset{Cl}{\|}}{\overset{\overset{Cl}{|}}{C}}-CO_2H \right] \longrightarrow Me-\underset{\underset{Cl}{\|}}{\overset{}{C}}-CO_2H \longrightarrow Products$$

Reductive dehalogenation

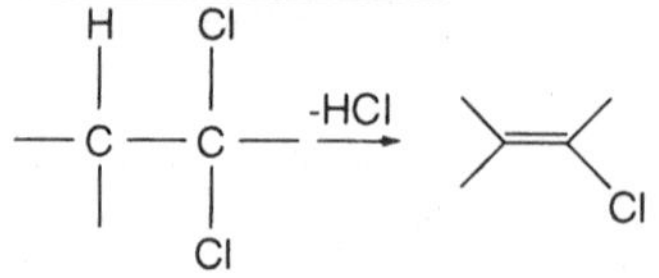

$$-\underset{\underset{}{|}}{\overset{\overset{H}{|}}{C}}-\underset{\underset{Cl}{|}}{\overset{\overset{Cl}{|}}{C}}- \xrightarrow{-HCl} \quad \diagup\!\!\!\diagdown\!\!=\!\!\diagdown_{Cl}$$

Examples: p, p'–DDT(lindane) → DDE

Nitroreduction

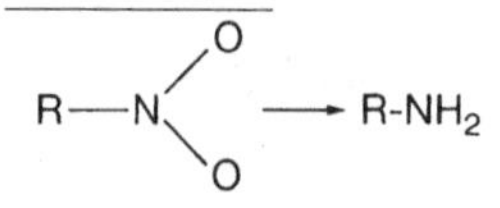

$$R-N\underset{O}{\overset{O}{\diagdown}} \longrightarrow R\text{-}NH_2$$

Example: parathion

Figure 2.6   (*Contd.*)

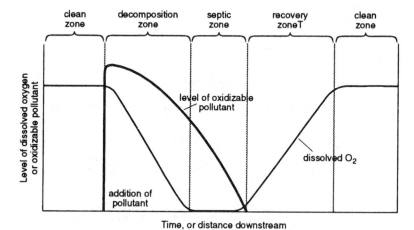

Figure 2.7   The decrease in oxygen concentration resulting from the addition of oxidizable pollutants to a stream (from Manahan, 1991).

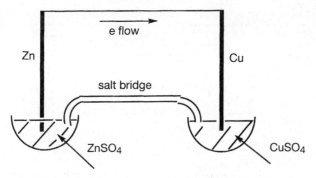

Figure 2.8 An electrochemical cell of Zn/Cu.

by summation of the two half reactions 2.23 and 2.24 we get the redox reaction 2.25:

$$Na(s) + H_2O(l) \longrightarrow Na^+(aq) + {}^-OH(aq) + [H]$$

or

$$2Na(s) + 2H_2O(l) \longrightarrow 2Na^+(aq) + 2\ {}^-OH(aq) + H_2(g) \qquad (2.25)$$

The relative strengths of oxidizing and reducing agents can be related to their behaviour as electrodes in an electrochemical cell. For example, Zn, the more electropositive metal will donate electrons to the less electropositive Cu. In Fig. 2.8 these two metals are shown as electrodes immersed in solutions of their sulphates to maintain electrical neutrality. The bridge is filled with a conducting gel to keep electrical contact between the electrolytes. The half-reactions are:

$$Zn \rightarrow Zn^{2+} + 2e\ (2.04) \quad \text{and} \quad Cu^{2+} + 2e \rightarrow Cu \qquad (2.26)$$

and the overall cell reaction is:

$$Zn + Cu^{2+} \longrightarrow Zn^{2+} + Cu \qquad (2.27)$$

If a potential difference is applied to the cell so as to nullify the electron flow from Zn to Cu, this gives a measure of the difference in their electrode potentials (E^0). The value of E^0 for the cell reaction 2.27 is:

$+ 0.76$ for the *reverse* Zn half reaction

$+ 0.34$ for the copper half reaction

$1.10\,V$ in total

The standard zero value of E^0 has been taken as that for hydrogen gas at $1/atm$. in contact with a platinum electrode in a solution containing H^+ at a concentration of $1\,mol/dm^3$.

Table 2.9 Some standard reduction potentials (E^0 at 25°C)

Half-cell reaction	$E^0(V)$
$Na^+(aq) + e \longrightarrow Na(s)$	-2.71
$Zn^{2+}(aq) + 2e \longrightarrow Zn(s)$	-0.76
$Fe^{2+}(aq) + 2e \longrightarrow Fe(s)$	-0.44
$2H^+(aq) + 2e \longrightarrow H_2(g)$	zero
$Cu^{2+}(aq) + 2e \longrightarrow Cu(s)$	$+0.34$

Table 2.9 is helpful in that it indicates that, if a half reaction for a metal has a more negative E^0 value, it can act as a reducing agent and donate electrons to metals with a less negative value. Thus, Zn can reduce Cu^{2+} but H^2 cannot reduce Zn^{2+}.

It is important to note that although a redox reaction may be thermodynamically possible it may not proceed at a realistic rate. In order to achieve significant reaction it may be necessary to apply a potential above the zero current reduction potential. This is not required for the reduction of Cu^{2+} by Zn with a zero current potential of 1.10 V, but Zn and Fe will not reduce water at pH 7 because of the requirement for *overpotential*.

The stability field of water. This is defined in Fig. 2.9 as the range of values of reduction potential and pH within which water is thermodynamically stable to both oxidation and reduction. The overpotential is the difference between the value of E under practical conditions and the value under reversible conditions.

The effective reduction potential (E) and the standard reduction potential (E^0) in acid solution are given by the Nernst equation:

$$E = E^0 - 0.059 \text{ V} \times \text{pH at 25°C and 1 bar } O_2 \text{ pressure}$$

The upper boundary in Fig. 2.9 separates species which are stable in water from those which will oxidize water; the half reaction is:

$$O_2(g) + 4H^+(aq) + 4e \longrightarrow 2H_2O \quad E^0 = +1.23 \text{ V} \tag{2.28}$$

This is effective at a pH value of 0; therefore, at the lower boundary pH value of 4.0:

$$E = +1.23 - 0.059 \times 4 = +1.00 \text{ V}$$

and at a pH value of 9.0:

$$E = +1.23 - 0.059 \times 9 = 0.70 \text{ V}$$

The lower boundary in Fig. 2.9 separates species stable in water from those which will reduce water; the half reaction is:

$$2H^+(aq) + 2e \longrightarrow H_2(g) \quad E^0 \text{ by convention} = 0 \tag{2.29}$$

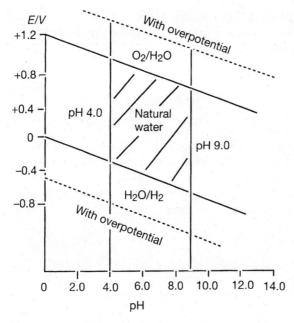

Figure 2.9 The stability field of water.

Therefore at a pH value of 4.0:

$$E = 0 - 0.059 \times 4 = -0.24\,\text{V}$$

and at a pH value of 9.0:

$$E = 0 - 0.059 \times 9 = -0.53\,\text{V}$$

Examination of Fig. 2.9 shows that the couples Fe^{2+}/Fe and Fe^{2+}/Fe^{3+} with standard potentials -0.44 and $+0.77$, respectively, fall within the stable field for water. Hence one would not expect Fe^{2+} to undergo reduction to Fe(s) or oxidation to Fe^{3+}. Moreover, for the reaction:

$$4Fe^{2+}(aq) + O_2(g) + 4H^+(aq) \longrightarrow 4Fe^{3+}(aq) + 2H_2O(l) \qquad (2.30)$$

the standard potential of $+0.46\,\text{V}$ also falls within the stability field of water. Although the positive sign indicates spontaneity, this oxidation is extremely slow at 25°C with $t_{1/2}$ of 1000 days (Singer and Stumm, 1970). However, in natural waters oxidative changes are mediated by bacteria, which greatly accelerates the oxidation of wastes from mines and disposal sites. Pyrite (FeS_2) is a good example, being rapidly oxidized by bacteria such as *Thiobacillus ferrooxidans* (Evangelou and Zhang, 1995):

$$4FeS_2 + 15O_2 + 2H_2O \longrightarrow 2Fe_2(SO_4)_3 + 2H_2SP_4 \qquad (2.31)$$

As the solution pH falls towards 4.0, Fe^{3+} becomes the effective oxidizing agent. In order to write a balanced equation for this oxidation, it is helpful to write out the half reaction:

$$Fe^{3+} + e \longrightarrow Fe^{2+} \tag{2.32}$$

$$FeS_2 + 2e \longrightarrow Fe^{2+} + 2S^{2-} \tag{2.33}$$

$$2S^{2-} + 8H_2O \longrightarrow 2SO_4^{2-} + 16H^+ + 16e \tag{2.34}$$

We are justified in writing Equation 2.34 in this form because it is known that sulphate is an end-product and also that the solution becomes acidic. Notice that in balancing redox reactions, one may involve water molecules as a source of oxygen and protons. Here if eight water molecules are needed to provide oxygen, one must include 16 electrons on the right-hand side to preserve electrical neutrality.

Addition of Equations 2.33 and 2.34 gives 2.35:

$$FeS_2 + 8H_2O \longrightarrow Fe^{2+} + 2SO_4^{2-} + 16H^+ + 14e \tag{2.35}$$

hence 14 Fe^{3+} are required to provide electrons:

$$FeS_2 + 14Fe^{3+} + 8H_2O \longrightarrow 15Fe^{2+} + 2SO_4^{2-} + 16H^+ \tag{2.36}$$

One can formally balance Equation 2.36 by taking the Fe(III) sulphate formed in Equation 2.31 as the source of oxidant:

$$FeS_2 + 7Fe_2(SO_4)_3 + 8H_2O \longrightarrow 15FeSO_4 + 8H_2SO_4 \tag{2.37}$$

Note that through the action of bacteria in producing Fe^{3+} in solution, the slow reaction (2.30) is no longer rate limiting.

Drainage water from polluted sites may have a pH value as low as 2.0, which leads to the solution of other heavy metals, such as Pb, Hg and Cd. When acid drainage water is diluted in the main body of a stream, the pH rises and red-brown ferric hydroxide is precipitated:

$$Fe^{3+} + 3H_2O \longrightarrow Fe(OH)_3(S) + 3H^+ \tag{2.38}$$

Biodegradation of organic compounds in natural waters

The principal changes involved in the biodegradation of organic compounds in natural waters (Samiullah, 1990) are:

1. Reduction of nitrate by microorganisms ($\longrightarrow$ nitrogen and some N_2O) in absence of dissolved O_2.
2. Hydrolysis of pollutants, such as pesticidal esters, amides and organo-phosphate esters, by microorganisms.
3. Dehalogenation of certain chlorinated compounds.
4. Hydroxylation of aromatic ring compounds.

5. Precipitation of metals is determined by the solubility product K_s, where $K_s = a^{M^+} \cdot a^{X^-}$. The activity a = true concentration × the activity coefficient. For salts M^+X^- of low solubility, one can assume that they are completely dissociated and one can simplify this to $K_s = [M^+][X^-]$ where M^+ and X^- are in mol/dm^3.

For copper sulphide $K_s = 3.48 \times 10^{-38}$, so at 25°C the saturated solution concentration of copper is: $\sqrt{3.48} \times 10^{-17}$ mol/dm^3. This will be reduced if there is another source of S^- because their concentration increases the $[X^-]$ term while K_s remains constant. Conversely, the concentration of Cu will be increased if the ionic strength of the solution rises, with acidity say, as this reduces the activity coefficient and so the true concentration of $Cu^{2+} + S^-$ must rise to compensate for this.

6. Complexation of metals (chelating ligands such as NTA in polluted water may mobilize metals adsorbed on sediments).

2.6.1   *The composition and physico-chemical properties of soils*

The concentrations of pollutants in moving air or rivers tend to be diluted fairly rapidly because of mixing and dilution, but in the case of soil many pollutants tend to accumulate. Soils act as a sink for pollutants through adsorption processes which bind inorganic and organic pollutants with varying strengths to the surface of soil colloids. This adsorption inhibits the leaching of pollutants down the soil profile to the water table, reduces their bioavailability to plants and, in the case of organic pollutants, affects their rate of decomposition. In many cases, the reactions occurring in soils are similar to those in sediments but usually only the surface layer of sediments has oxic conditions. The major thickness of sediments has anoxic conditions, which resemble conditions in waterlogged (gleyed) soils.

Soils comprise a mixture of organic, mineral, gaseous and liquid constituents inhabited by a wide range of microorganisms which catalyse many important reactions. Organic matter in soils includes decomposing plant material and humic compounds which have been synthesized by the action of microorganisms on residues of plant material. The mineral constituents of soils can include: particles of weathering rock and discrete rock-forming minerals (such as quartz), clay minerals, hydrous oxides of Fe, Al and Mn, and calcite. The humic substances, clay minerals and hydrous oxides are bonded together in various ways and jointly form the colloidal adsorption complex which plays a very important role in determining the behaviour of pollutants.

The liquid and gaseous phases in soils occupy the system of pores created by the voids between the aggregated solid particles. The aqueous soil solution containing ions and soluble organic compounds forms the liquid phase. The gaseous phase is similar in composition to the atmosphere above the soil surface except that the concentration of CO_2 is often more than eight times higher in the soil air. Some organic compounds are lost from the soil by volatilization and these vapour phases will also be present in the soil air. The relative proportions of water and gases in the soil pores have a very important effect on soil physico-chemical properties (redox and pH), the soil biomass and plant growth. Under conditions of prolonged waterlogging the supply of oxygen is rapidly exhausted. Anaerobic conditions rapidly develop and anaerobic microorganisms replace aerobic organisms and catalyse the chemical reduction of compounds, such as the hydrous oxides of Fe^{3+}, Al^{3+} and Mn^{4+} together with SO_4^{2-} and organic compounds. The decomposition of organic pollutants is strongly affected by the redox conditions and the predominant types of microorganisms.

The concentration of ions in the soil solution is determined by the interacting processes of oxidation, reduction, adsorption, precipitation and desorption. When pollutants reach the soil surface they are either adsorbed with varying strengths on the colloids at the surface of the topsoil, or are washed down through the surface layer into the soil profile in rainwater or snow melt. Soluble pollutants will infiltrate into the topsoil in the system of pores where the adsorption of ions occurs. Insoluble compounds will accumulate on the surface and hydrophobic organic molecules will bind to sites on soil organic matter at the soil surface. These substances become incorporated into the topsoil and deeper profile during mechanical soil movement or down desiccation cracks while being adsorbed on soil particles. Some organic pollutant molecules on the soil surface will undergo photolytic decomposition as a result of exposure to UV wavelengths in daylight.

Several different types of adsorption reaction can occur between the surfaces of organic and mineral colloids and the pollutants. The extent to which the reactions occur is determined by the composition of the soil (especially the amounts and types of clay minerals, hydrous oxides and organic matter), the soil pH, redox status, and the nature of the contaminants. The more strongly pollutants are adsorbed, the less likely they are to be leached down the soil profile or to be available for uptake by plants. Ionic pollutants such as metals, inorganic anions and certain organic molecules, such as the bipyridyl herbicides (e.g. Paraquat), are adsorbed onto soil colloids. Non-ionic organic molecules, which include hydrocarbons, most oganic micropollutants and pesticides, are adsorbed onto humic polymers by both chemical and physical adsorption mechanisms. However, some organic pollutants, such as solvents, tend to be relatively easily leached in regions where there is a marked excess of precipitation relative to evapotranspiration. In many cases, adsorption is a necessary preliminary stage in the decomposition of organic pollutant molecules by bacterial extracellular enzymes.

Soil organic matter plays an important role in cation exchange reactions and in the formation of complexes (mostly chelates) with trace metals. Low-molecular-weight soluble organic molecules can form stable complexes with metals which are mobile in the soil solution and bioavailable because of the protection of the metal from adsorption on soil colloid surfaces by the organic ligand. However, the more highly polymerized, solid-state humus acts as a major adsorbent for metals through the formation of chelates and thus renders them immobile and much less bioavailable.

In addition to the adsorption/desorption processes occurring in soils, the wide range of microorganism species present also have important effects on the behaviour of pollutants. Microorganisms, such as *Thiobacillus* spp. catalyse the oxidation of sulphides. In the case of pollution by tailings from metalliferous mining, particles of ore minerals in the soil, such as PbS, ZnS, and $CuFeS_2$, become oxidized, releasing metal cations such as Pb^{2+}, Cu^{2+}, Zn^{2+} and Cd^{2+} into the soil solution where they undergo adsorption reactions. Sulphide oxidation also causes an increase in soil acidity unless there is a sufficient concentration of carbonates present to buffer it. The decrease in soil pH will diminish the extent of metal adsorption and cause an increase in the concentrations of metals in the soil solution, which can be leached down the soil profile or taken up by plant roots. Bacteria do not tolerate highly acid conditions so plant debris and organic pollutant decomposition will be inhibited at low pH. However, many fungal species are able to tolerate strongly acid conditions although the end-products of organic matter decomposition may differ from those of the bacteria involved in humification.

Some microorganism species can methylate elements such as As, Se and Hg into volatile forms (e.g. CH_3Hg^+), which then diffuse into the atmosphere as part of the gaseous exchanges of soil and atmospheric gases.

2.6.2 Cation and anion adsorption in soils

Ion exchange refers to the exchange between the counter-ions balancing the surface charge on the soil colloids and the ions in the soil solution. Negative charges on soil colloids are responsible for cation exchange. The extent to which soil constituents can act as cation exchangers is expressed as the cation exchange capacity (CEC), measured in $cmol_c/kg$ (previously in milli equiv/100 g). Some examples of the typical CEC values for soil colloidal constituents are (Ross, 1989):

	$cmol_c/kg$
Soil organic matter	150–300
Kaolinite (clay)	2–5
Illite (clay)	15–40
Montmorillonite (clay)	80–10
Vermiculite (clay)	150
Hydrous oxides (Fe, Al, Mn)	4

Soil organic matter has a higher CEC than other soil colloids and plays a very important part in adsorption reactions in most soils even though it is normally present in much smaller amounts (1–10%) than clays (<80%). Sandy soils with low contents of both organic matter and clay tend to have low adsorptive capacities and pose a threat for contaminants infiltrating down to the water table.

The negative charges on the surfaces of soil colloids are of two types: (i) permanent charges resulting from the isomorphous substitution of a clay mineral constituent by an ion with a lower valency, and (ii) the pH dependent charges on oxides of Fe, Al, Mn, Si and organic colloids, which are positive at pH values below their isoelectric points and negative above their isoelectric points. Hydrous Fe and Al oxides have relatively high isoelectric points (>pH 8) and so tend to be positively charged under most conditions, whereas clay and organic colloids are predominantly negatively charged under alkaline conditions. With most colloids, increasing the soil pH, at least up to neutrality, tends to increase their CEC. Humic polymers in the soil organic matter fraction become negatively charged as a result of the dissociation of protons from carboxyl and phenolic groups.

The concept of cation exchange implies that ions will be exchanged between the soil solution and the zone affected by the charged colloid surfaces (double diffuse layer). The relative replacing power of any ion on the cation exchange complex will depend on its valency, its diameter in hydrated form and the type and concentration of other ions present in the soil solution. With the exception of H^+, which behaves like a trivalent ion, the higher the valency, the greater the degree of adsorption. Ions with a large hydrated radius have a lower replacing power than ions with smaller radii. For example, K^+ and Na^+ have the same valency but K^+ will replace Na^+, owing to the greater hydrated size of Na^+. The commonly quoted relative order of replaceability on the cation exchange complex of metal cations is:

$$Li^+ = Na^+ > K^+ = NH_4^+ > Rb^+ > Cs^+ > Mg^{2+} > Ca^{2+} > Sr^{2+}$$
$$= Ba^{2+} > La^{3+} = H^+(Al^{3+}) > Th^{4+}$$

For individual soil constituents, the order of replacement of trace metals is (Alloway, 1995):

Montmorillonite clay: $Ca > Pb > Cu > Mg > Cd > Zn$
Ferrihydrite: $Pb > Cu > Zn > Ni > Cd > Co > Sr > Mg$
Peat: $Pb > Cu > Cd = Zn > Ca$

Anion adsorption occurs when anions are attracted to positive charges on soil colloids. As stated above hydrous oxides of Fe and Al are usually positively charged and so tend to be the main sites for anion exchange in soils. In general, most soils tend to have far smaller anion capacities than cation exchange capacities. Some anions, such as NO_3^- and Cl^-, are not adsorbed to any marked extent but others, such as HPO_4^{2-} and $H_2PO_4^-$, tend to be strongly adsorbed. Some organic pesticides, such as the phenoxyalkanoic acid herbicides, exist as anions at normal soil pH values and are adsorbed to a limited extent by hydrous oxides and by hydrogen bonding to humic polymers.

Specific adsorption is a stronger form of adsorption, involving several heavy metal cations and most anions, which form partly covalent bonds with surface ligands on adsorbents, especially hydrous oxides of Fe, Mn and Al. This adsorption is strongly pH specific and the metals and anions which are most able to form hydroxy complexes are adsorbed to the greatest extent. The order for the increasing strength of specific adsorption of selected heavy metals is:

$$Cd > Ni > Co > Zn >> Cu > Pb > Hg$$

Coprecipitation of metals with secondary minerals, including the hydrous oxides of Fe, Al and Mn, is an important adsorptive mechanism in soils with fluctuating moisture status. Cu, Mn, Mo, Ni, V and Zn are coprecipitated in Fe oxides, and Co, Fe, Ni, Pb and Zn are coprecipitated in Mn oxides. Precipitation of Fe(III) is initially in the form of gelatinous ferrihydrite $[Fe_5(O_4H_3)_3]$, which gradually dehydrates with ageing to more stable forms, such as goethite. Ferrihydrite is more likely to be subsequently dissolved again through a decrease in Eh or pH than goethite. Ferrihydrite coprecipitates other ions and, as a result of its large surface area, acts as a scavenger, sorbing both cations, such as heavy metals, and anions, especially HPO_4^{2+} or $H_2PO_4^+$ and AsO_4^{3-}. Pyrites (FeS_2) forms in severely reducing conditions when sulphate becomes reduced to sulphide, producing H_2S, which then reacts with Fe^{2+} to form FeS and FeS_2. The oxidation of sulphides, such as pyrite, causes marked acidification of soils. Specialized bacteria, such as *Thiobacillus ferrooxidans* and *Metallogenum* spp. (p. 51) are involved in the transformations of Fe and Mn, respectively. Iron and Mn oxides occur as coatings on soil particles, fillings in voids and as concentric nodules. The oxide coatings are usually intimately mixed with the clay and humus colloids and, although mineralogically distinct, form part of the clay-sized fraction. The trace metals normally found coprecipitated with secondary minerals in soils are (Sposito, 1983).

Fe oxides	V, Mn, Ni, Cu, Zn, Mo
Mn oxides	Fe, Co, Ni, Zn, Pb
Ca carbonates	V, Mn, Fe, Co, Cd
Clay minerals	V, Ni, Co, Cr, Zn, Cu, Pb, Ti, Mn, Fe

When reducing conditions cause the dissolution of hydrous Mn and Fe oxides, the concentrations of several other elements in the soil solution are likely to increase. Cobalt, Ni, Fe, V, Cu and Mn are generally more bioavailable from gleyed (periodically waterlogged) soils than from freely drained soils on the same parent material. However, B, Co, Cu, Mo and Zn do not undergo redox reactions themselves but are coprecipitated by the hydrous oxides.

Coprecipitation of trace metals on carbonates (mainly $CaCO_3$) is very important on semi-arid soils and in soils formed on limestone. In the case of Cd, the precipitation of $CdCO_3$ can be accompanied by the chemisorption of Cd, where it replaces Ca in the calcite crystal.

2.6.3   Adsorption and deomposition of organic pollutants

Non-ionic and non-polar organic pollutants are normally adsorbed on soil humic material. Since most soil organic matter is found in the surface horizon, there is a tendency for these pollutants to be concentrated in the topsoil. Migration of organic contaminants down the profile only occurs to any marked extent in highly permeable sandy or gravelly soils with low organic matter contents and where large pores (macropores) and fissures are present.

There are several physico-chemical parameters which are useful in predicting the behaviour of organic chemicals in soils. These include a substance's solubility in water (in mg/l), its soil–water distribution coefficient (k_d), its octanol–water partition coefficient (k_{ow}) and its organic carbon partition coefficient (k_{oc}).

The coefficient k_d is the proportion of the compound bound to solid relative to that remaining in solution at equilibrium:

$$K_d = X/C \qquad (2.39)$$

where X = the amount sorbed per unit weight of soil and C = the equilibrium concentration in solution.

The coefficient k_{ow} provides an indication of the hydrophobicity of a compound:

$$k_{ow} = C_o/C \qquad (2.40)$$

where C_o = the concentration of the substance in octanol, and C = the concentration in water. Low values ($k_{ow} < 10$) indicate a relatively hydrophilic compound (low hydrophobicity) with little likelihood of binding on soil organic matter. The greater the value of k_{ow}, the greater the pollutant substance's affinity for lipids and soil organic matter.

The coefficient (k_{oc}) is the amount of a compound adsorbed per kg of organic carbon.

$$k_{oc} = k_d/f_{oc} \qquad (2.41)$$

where f_{oc} is the fraction of organic carbon in the soil.

From Table 2.10 it can be seen that the herbicides 2,4-D and 2,4,5-T are the most soluble and have low affinities for the soil organic matter; in contrast, the chlorinated hydrocarbon insecticides DDT and Lindane have low solubilities and relatively high affinities for soil organic matter.

Table 2.10 Examples of distribution coefficients and aqueous solubility (S) data for selected organic contaminants

Compound	log k_{ow}	log k_{oc}	S (mg/l)
DDT (insecticide)	5.98	4.32	0.0017
Lindane (insecticide)	4.82	2.96	0.150
2,4-D (herbicide)	1.57	1.30	900
2,4,5-T (herbicide)	0.6	1.72	238
TCDD (dioxin)	6.15	5.67	0.0002

From LaGrega *et al.*, 1994.

However, TCDD shows the greatest hydrophobic organic matter affinity of all the examples shown (LaGrega *et al.*, 1994).

Most organic pollutants are relatively insoluble and do not move down the soil profile, but chlorinated solvents tend to be leached fairly rapidly down the profile of most soil types, including peats. Some pesticides also tend to be relatively easily leached, such as the herbicide atrazine (pp. 320–1). In general, apart from solvents and certain pesticides, most organic pollutants that reach water courses after being in contact with soils, have been transported to the water course while adsorbed on eroded soil particles and not leached through the soil profile.

Adsorption of organic pollutants depends on their surface charge and aqueous solubility, both of which are affected by the soil pH. Adsorption of non-polar organic contaminants onto soil organic matter will not occur in the presence of oils. Microbially synthesized surfactants can help to accelerate the rate of degradation of hydrocarbon oils in contaminated soils. For many organic pollutants, adsorption onto soil colloids and the presence of water are important catalysts of organic micropollutant degradation, which can be of two types (Ross, 1989).

1. Non-biological degradation: includes hydrolysis, oxidation/reduction volatilization and photodecomposition.
2. Microbial decomposition: many pesticide decomposition processes have some biological contribution – there is an initial time lag, while the microorganisms become adapted to the pesticide substrate.

In most cases, non-biological degradation processes, such as photo-decomposition and volatilization, occur at the same time as microbially catalysed reactions. The range of factors affecting the degradation of organic contaminants by microorganisms include: soil pH, temperature, supply of oxygen and nutrients, the structure of the contaminant molecules, their toxicity and that of their intermediate decomposition products, the water solubility of the contaminant and its adsorption to the soil matrix (and, therefore, the organic matter content of the soil). Adsorption tends to decrease and volatilization increase with increasing temperature.

Organic pollutants are decomposed by soil microorganisms but the rate at which this occurs will depend on the nature of the pollutant (its toxicity to microorganisms and that of its decomposition products), the genotype of the microorganisms (whether they have become adapted to decomposing the particular pollutant), the pH and nutrient status of the soil, and its adsorptive properties. Many organic pollutants are more rapidly decomposed after they have been adsorbed onto the soil organic matter. Decomposition of pollutants by microorganisms is brought about by extracellular enzymes and involves an initial lag period while the microorganisms become adapted to the new substrate. Some xenobiotic organochlorine molecules, such as DDT, PCBs and PCDDs, are generally regarded as being highly persistent in soils with residence times of at least 10 years. They have a very slow decomposition rate because the carbon–chlorine bond is not found in nature and so most microorganism species do not possess the enzymes to break this bond. Nevertheless, some species of bacteria and fungi have rapidly evolved the ability to decompose chlorinated organic compounds, and these are being utilized in the bioremediation of contaminated land.

Chlorinated solvents are relatively mobile in soils and some tend to be degraded quite rapidly. Studies of the leaching and degradation behaviour of chlorinated solvents in three Dutch soils showed that 1,4-dichlorobenzene was completely degraded under aerobic conditions and did not reach the groundwater. Chloroform, and to a lesser extent 1,1,1-trichloroethylene, were only poorly degraded under both aerobic and anaerobic conditions but trichloroethylene and tetrachloroethylene were more completely degraded. With the exception of 1,4-dichlorobenzene, all the other compounds broke through into the groundwater at 1 m depth (Loch *et al.*, 1986). Other workers have found that both chlorinated aliphatics and aromatics are leached into groundwaters. Volatilization appears to be the major route for the removal of chlorinated aliphatics from soils, while degradation is not very significant. However, chlorinated benzenes tended to be degraded rather than volatilized (Loch *et al.*, 1986).

Non-biological decomposition of organic pollutants includes photodecomposition when exposed to the UV spectrum of daylight. This usually requires the pollutant to be adsorbed on the surface of soil colloids and obviously only affects the pollutants present on the surface of the topsoil exposed to sunlight. Other non-biological decomposition reactions include oxidation, reduction, hydrolysis and methylation.

Oxidation of organic pollutants occurs by the action of oxygenase enzymes secreted by microorganisms. In alkane hydrocarbons, the initial step in this oxidation is the conversion of a terminal CH_3 group to a CO_2H group (Fig. 3.1, p. 66). Aromatic rings are cleaved by the addition of OH to adjacent carbon atoms. The bacterium *Cunninghamella elegans* is known to attack aromatic ring molecules, such as naphthalene. It should be emphasized that the decomposition products of some organic molecules are more toxic to soil microorganisms, animals and humans than the initial compound. For example, the microbial oxidation products of the PAH molecule – benzo-[a]-pyrene – are carcinogenic because they bind to cellular DNA (Section 6.2.5).

As stated above, many pollutants are sorbed by soil colloids and so in many cases little transport occurs within the soil profile, except down desiccation cracks or worm channels while adsorbed to soil particles. Soil particles can also be transported long distances (thousands of kilometres) and a brown snow event in Arctic Canada caused by soil particles transported from China and eastern Russia, typified by spherical soot particles from the industrial use there of coal. This incident brought about a significant degree of pollution by pesticides (Welch *et al.*, 1991); some of the pollutants identified in the brown snow were:

hexachlorocyclohexane isomers	934 pg/l
DDT isomers	238 pg/l
pentachloroanisole	1442 pg/l
trifluralin (a pesticide)	764 pg/l

Soil particles washed into water courses become part of the sediment load of the stream and may eventually be deposited on the stream or river bed, or in lakes and estuaries where they may undergo reducing reactions which lead to the solubilization of ions and molecules sorbed by hydrous oxides.

However, several solvents such as chloroform, 1,1,1-trichloroethylene and tetrachloroethylene can leach through soil profiles and reach aquifers (Loch *et al.*, 1986). The soil's hydraulic conductivity will be an important factor in determining the rate at which this will occur.

A substantial number of pesticide compounds and their derivatives are present in most agricultural soils because different compounds are used for different pests and crops and new products are continually being introduced as a result of the development of pest resistance to old formulations. There is, therefore, the possibility of interactions between pesticides and differing behaviour in the soil and the groundwater. The EC maximum acceptable concentration of pesticides in drinking water is $0.1 \mu g/l$ but this is often exceeded and may be unenforceable since, during the growing season, rain-out levels can rise to $0.2–0.3 \mu g/l$; a practical limit rarely exceeded in drinking water is $1.0 \mu g/l$. The pesticides most frequently causing problems with drinking water quality are the herbicides atrazine, simazine, meco-prop and isoproturon, and carbamate and chloropropionate insecti-cides. Water pollution by pesticides is exacerbated by preferential flow down through fissures (desiccation cracks and subsoiling fissures). The concentration of soil-acting pesticide in the soil solution will be thou-sands of times greater than the maximum permissible concentration in drinking water (Foster *et al.*, 1991). Typical rates of application of pesti-cides in the UK are around $0.2–0.5 kg/ha$ on agricultural land, but much higher rates are used for non-agricultural applications of herbicides, e.g. defoliation in forestry firebreaks, and weed control on railways, airfields and highways.

2.7
Overview of the
fate of soil pollutants

Figure 2.10 shows the main routes by which organic and inorganic pollutants, respectively, enter soil and their behaviour within the soil. Both types of pollutant reach the soil surface as a result of deposition from the atmosphere or placement, and they are both also deposited on vegetation foliage. The major differences between organic and inorganic pollutants, in terms of these pathways diagrams, are: (i) the organic compounds can be decomposed by photolysis while either in the atmosphere or on the surface of soil or foliage, (ii) when in the soil, organic pollutants tend to be mainly sorbed on the humus material, whereas inorganic pollutants are sorbed to various sites on soil mineral surfaces also, (iii) organic pollutants can be decomposed in the soil as a result of microbial activity, and (iv) although

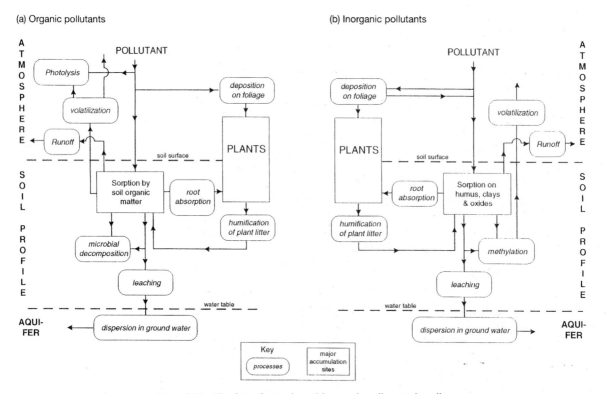

Figure 2.10   The fate of organic and inorganic pollutants in soil.

both types of pollutant can be volatilized from the soil to the atmosphere, in the case of inorganic pollutants, this usually occurs only after the element has been converted to an organometal form, such as a methylated species, e.g. CH_3Hg^+.

With regard to the other parts of the diagrams, both types of pollutant can be washed away sorbed to particles in surface water run-off, both have the potential to be absorbed into the foliage and the root system of plants, and both can be leached down the profile and ultimately enter and become dispersed in the groundwater.

References

Alloway, B. J. (ed.) (1995) In *Heavy Metals in Soils* (2nd edn), Ch. 2. Blackie, Glasgow.

Briggs, G. A. (1969) *Plume Rise*. US Atomic Energy Commission, Oak Ridge, TN.

Crathorne, B. and Dobbs, A. J. (1990). In *Pollution: Causes, Effects and Control* (2nd edn), Ch. 1. (ed. R. M. Harrison). Royal Society of Chemistry, Cambridge.

De Mora, S. J. (1992) In *Understanding Our Environment* (2nd edn), Ch. 4. (ed. R. M. Morrison). The Oceans. Royal Society of Chemistry, Cambridge.

Department of the Environment (DOE) (1988) Inputs of dangerous substances to water: proposals for a unified system of control. The Government's consultative proposals for a unified system of tighter controls over the most dangerous substances entering aquatic environments. *The Red List*, July.

Dix, H. M. (1981) *Environmental Pollution*. John Wiley, Chichester.

Elsom, D. (1987) *Atmospheric Pollution*. Blackwell, Oxford.

Envangelou, V. P. and Zhang, Y. L. (1995) A review: pyrite oxidation mechanisms and acid mine drainage prevention. *Crit. Rev. Environ. Sci. Technol.*, **25**, 141–199.

European Economic Community (1976) *Directive on pollution caused by certain dangerous substances discharged into the aquatic environment of the community*, 76/464/EEC. EEC, Brussels.

Fish, H. (1992) In *Understanding Our Environment* (2nd edn), Ch. 3. (ed. R. M. Harrison). Royal Society of Chemistry, Cambridge.

Foster, S. S., Chilton, P. J. and Stuart, M. E. (1991) *J. Inst. Water Env. Man*, 186.

Freeze, R. A. and Cherry, J. A. (1979) *Groundwater*. Prentice Hall, Englewood Cliffs, N.J.

Hemond, H. F. and Fechner, E. J. (1994) *Chemical Fate and Transport in the Environment*. Academic Press, San Diego.

Hewitt, C. N. and Harrison, R. M. (1986) In *Understanding Our Environment* (1st edn), Ch. 1. (ed. R. E. Hester). Royal Society of Chemistry, London.

Holdgate, M. W. (1979) *A Perspective of Environmental Pollution*. Cambridge University Press, Cambridge.

House of Commons Environment Committee (1990) 1st Report, *Contaminated Land*. HMSO, London.

Identification and Listing of Hazardous Wastes (1986) *Code of Federal Regulations*, **40** (July 1) Part 261, US Government Printing Office, Washington DC, pp. 354–408.

LaGrega, M. D., Buckingham, P. L. and Evans J. C. (1994) *Hazardous Waste Management*. McGraw-Hill, New York.

Loch, J. P. G., Kool, H. J., Lagas, P. and Verheul, J. H. A. M. (1986) In *Contaminated Soil*, (eds. J. W. Assink and W. J. van den Brink), p. 63. Kluwer, Dordrecht.

Manahan, S. E. (1984) *Environmental Chemistry* (4th edn). Brooks/Cole, Monterey, CA.

Manahan, S. E. (1991) *Environmental Chemistry* (5th edn). Lewis Publishers, Chelsea, MC.

Masters, G. M. (1991) *Introduction of Environmental Engineering and Science*. Prentice Hall, Englewood Cliffs, NJ.

Mather, J. (1993) *Land Contam. Reclamat.*, **4**, 187–196.

Miller, D.W. (1985) Chemical contamination of groundwater. In *Groundwater Quality*. Ch. 4. (eds C. H. Ward., W. Giger and P. L. McCarty). John Wiley, New York.

Ross, S. (1989) *Soil Processes*. Routledge, London.

Samiullah, Y. (1990) *Prediction of the Environmental Fate of Chemicals*. Elsevier, Amsterdam.

Singer, P. C. and Stumm, W. (1970) *Science*, **167**, 1121–1124.

Sposito, G. (1983) In *Applied Environmental Geochemistry* (ed. I. Thornton). Academic Press, London.

Tolba, M. K., El-Kholy Osama, A., El-Hinnawi, E., Holdgate, M. W., McMichael, D. F. and Munn, R. E. (eds) (1992) *The World Environment. 1972–1992*. UNEP, Chapman & Hall, London.

Ward, C. H., Giger, W. and McCarty, P. L. (eds) (1985) *Groundwater Quality*. John Wiley, New York.

Welch, H. E., Muir, D. C. G. and Lemoine, B. M. (1991) *Environ. Sci. Technol.*, 280.

Young, P. C. (1990) Quantitative Systems Methods in Evaluation of Environmental Pollution Problems. In *Pollution: Causes, Effects and Control* (2nd edn), (ed. R. M. Harrison). Royal Society of Chemistry, Cambridge.

Further reading

Alloway, B. J. (1992) In *Understanding Our Environment* (2nd edn), Ch. 5. (ed. R. M. Harrison). Royal Society of Chemistry, Cambridge.

ENDS (1992) *Dangerous Substances in Water: a practical guide*.

Franklin, D., Hawke, N. and Lowe, M. (1995) *Pollution in the United Kingdom*. Sweet & Maxwell, London.

Murley, L. (1995) *Pollution Handbook*. National Society for Clean Air, Brighton.

Shriver, D. F., Atkins, P. W. and Langford, C. H. (1994) *Inorganic Chemistry* (2nd edn). Oxford University Press, Oxford.

Toxicity and risk assessment of environmental pollutants 3

Human and other animals are exposed to chemicals via water, air, soils, dusts and their diets. These chemicals enter the body by ingestion (mainly in the diet and in water, but also on hands, in soil on vegetables and from dust swallowed in mucus), inhalation (air and dusts) and by dermal contact (soil, air and water). In severe cases, with highly corrosive chemicals, these exposure routes can result in localized damage to the cells of the mouth, trachea and digestive system, nostrils and respiratory system, the skin and the sensitive eyes. However, in most cases, the toxic effects only occur after the pollutant has entered the bloodstream following absorption through either the gut, the lungs or the skin. Once in the bloodstream the chemicals are circulated around the body and undergo metabolism, usually in the liver, or are stored in various organs. Some of the products of this metabolism may be excreted via the kidneys in urine, the digestive tract in faeces, the lungs in exhaled air or sweat from the skin (Rodricks, 1992).

In humans and higher animals, metabolic conversion of compounds not essential for normal biological functions takes place mainly in the liver, but some metabolism can occur in the lungs, intestines, kidneys and the skin. These conversions are usually catalysed by enzymes, but possession of the right enzymes depends on the similarity of the pollutant to commonly encountered substances and to evolutionary adaptations. New xenobiotic compounds may remain unaltered or only very slowly metabolized because of the lack of previous exposure to the chemicals. An example of a metabolic process which reduces the toxicity of a pollutant is the conversion of toluene, a neurotoxin which is absorbed through the lungs, into benzoic acid, which is much less toxic and more easily excreted than toluene (Rodricks, 1992) (Section 6.2.4 and Fig. 3.1). Examples of the production of more toxic metabolites include Dieldrin from Aldrin (p. 286) and chloroethylene oxide from vinyl chloride (p. 293).

Toxicology is the study of the effects of poisonous substances on living organisms, including the way in which they gain entry into the organisms. Above a certain concentration, the toxicant has detrimental effects on some biological function. The concentration at which a significant detrimental

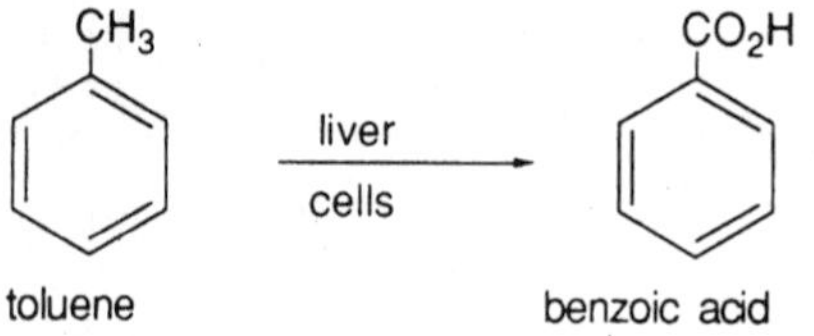

Figure 3.1 The transformation of toluene to less toxic benzoic acid in liver cells (from Rodricks, 1992).

effect occurs is determined by the *dose response*. The critical (or threshold) dose at which toxicity occurs differs between species, sexes and individuals within a species as a result of genetic and other factors, such as the composition of the diet and some illnesses (Manahan, 1991; BMA, 1991) (Section 5.3 gives dose–response curves of metals).

The *dose*, or degree of exposure of an organism to a toxicant, can be expressed as:

- the amount of toxicant present in the organism (units of mass of toxicant per unit weight of body mass of the organism)
- the amount of the toxicant entering the organism (in the diet, drinking water, or inhaled air in animals and absorbed through the roots or through the leaf cuticle in the case of plants)
- the concentration in the environment of the organism (duration of exposure is important).

The effect of the toxicant dose is called the *response*, and this can vary from no discernible effect to death. Toxicity is commonly categorized on the basis of the duration of exposure, that is acute, chronic and subchronic. Acute exposure involves a single dose whereas chronic exposure refers to exposure over a long time period (often almost a lifetime – 2 years in the case of test rodents). Subchronic exposure is dosing over a shorter time period – fraction of a lifetime, such as one eighth of an experimental rodent's lifetime (Rodricks, 1992).

Acute toxicity is caused by fast poisons, which include both synthetic and naturally occurring compounds, such as the botulinum toxins produced by the soil bacterium *Clostridium botulinum*, the venom of certain snakes (e.g. rattlesnake and cobra) or species of spider (e.g. black widow spider), plant-derived toxins, such as strychnine and nicotine, and some synthetic chemicals, including organo-phosphorus compounds (Section 6.5), phosphine (PH_3), phosgene ($COCl_2$) and sodium fluoroacetate. These substances are classed as supertoxins because they cause lethal effects in humans at doses of less than 5 mg/kg body weight (as shown in Table 3.1). In general, the toxicity of any chemical depends on its **a**bsorption, **d**istribution, **m**etabolism and **e**xcretion (ADME) (Rodricks, 1992).

Table 3.1 A classification of toxins on the basis of lethal doses for humans (Rodricks, 1992)

Toxicity rating	Probable lethal dose for humans (mg/kg/body wt)
Practically non-toxic	> 15 000
Slightly toxic	5000–15 000
Moderately toxic	500–5000
Very toxic	50–500
Extremely toxic	5–50
Supertoxic	< 5

Table 3.2 Lethal doses (LD_{50}) of TCDD (dioxin) for different animal species (from BMA *Hazardous Waste and Human Health* (1991), by permission of Oxford University Press; Section 6.4)

Animal	LD_{50} (μg/kg/body wt)
Guinea pig	1.0
Rat (male)	22
Rat (female)	45
Monkey	> 77
Rabbit	115
Mouse	114
Dog	> 300
Bullfrog	> 500
Hamster	5000

Acute toxic effects are quantified by controlled experiments to determine the dose causing the immediate death of 50% of the organisms exposed (LD_{50} value). Estimates for humans are extrapolated from values for small mammals (and are subject to many possible errors, such as marked gentoypic variations in susceptibility to a toxicant, as shown for TCDD in Table 3.2). The subchronic effects of chemicals are determined by investigating the biochemical and other changes which take place over a period of months. Investigations on chronic effects will examine effects on the lifespan of the organism, cancer induction, changes in geriatric conditions and effects on the offspring caused by exposure of the parent to toxic chemicals. However, it is important to note that the acute and subacute effects of toxins determined in laboratory experiments cannot always be relied on to predict responses to the same chemicals in the environment. This is because of interactions between pollutants (antagonistic and synergistic effects) and reactions of the toxicants with the components of the environment (such as adsorption, photodecomposition, acidification and dissolution). In the case of non-carcinogens, it is possible that there may be safe or threshold dose levels of toxins (NOAEL = no observed adverse effect level). However, carcinogens are not considered to have safe or threshold concentrations because a single genetic change may lead to an uncontrolled reaction. Nevertheless, carcinogens differ considerably in potency (e.g. aflatoxin B_1 is a million times more potent than trichloroethylene).

3.2
Effect of pollutants on animals and plants

Space does not permit a detailed discussion of the effects of pollutants on animals and plants and so the following brief lists of effects are provided. The reader is recommended to consult specialized toxicological texts for more details of the subject.

3.2.1   Effect of pollutants on humans and other mammals

The types of response to toxicants which occur in humans and other mammals include (Manahan, 1991):

- alterations in the vital signs of temperature, pulse rate, respiratory rate and blood pressure
- abnormal skin colour
- unnatural odours
- effects on the eye, which include:
    miosis (excessive contraction of pupil)
    mydriasis (excessive pupil dilation)
    conjunctivitis (inflammation of the membrane covering the front of the eyeball)
    nystagmus (involuntary movement of the eyeballs)
- gastrointestinal effects: pain, vomiting, paralytic ileus (stoppage of normal peristalsis)
- central nervous system effects: convulsions, paralysis, hallucinations, ataxia and coma.

Subclinical (recondite) effects of toxicants in humans and other mammals include (Manahan, 1991):

- damage to the immune system
- chromosomal abnormalities
- modification of the functions of liver enzymes
- slowing of the conduction of nervous impulses.

The major types of biochemical effect of pollutants on animals are (Manahan, 1991):

- impairment of enzyme function by the binding of the toxicant to enzymes, coenzymes, metal activators, or enzyme substrates
- alteration of cell membrane or carriers in cell membranes

- interference with lipid metabolism, resulting in excess lipid accumulation
- interference with respiration
- interference with carbohydrate metabolism
- stopping or interfering with protein biosynthesis through toxic effects on DNA
- interference with regulatory processes mediated by hormones or enzymes.

3.2.2 Teratogenesis, mutagenesis, carcinogenesis and immune system defects

Teratogenesis is the creation of birth defects arising from damage to embryonic or foetal cells or from mutations in egg or sperm cells. The biochemical mechanisms of teratogenesis include: enzyme inhibition by xenobiotics, deprivation of essential nutrients and alteration of the placental membrane.

Mutagenesis is the creation of mutations by chemicals or ionizing radiation, which bring about alterations to DNA to produce inheritable traits. Although mutations can occur naturally in the absence of xenobiotic substances, most mutations are harmful. Mechanisms of mutagenicity are similar to those of carcinogenesis and teratogenesis.

Chemical carcinogenesis occurs when xenobiotic substances cause uncontrolled cell replication (i.e. cancer). From the public's viewpoint, carcinogenesis is the most commonly associated toxic effect of hazardous substances. Tumours may be benign (when they are contained within their own boundaries) or malignant (when they undergo metastasis, i.e. they break apart and invade other parts of the body) (Masters, 1991).

There are two major steps by which xenobiotic substances can cause cancers: the *initiation stage* and the *promotional stage*. Many chemical carcinogens have the ability to form covalent bonds with DNA, thus altering the DNA so that cell replication becomes uncontrolled. Many chemical carcinogens are either alkylating agents, which attach alkyl (e.g. CH_3) groups, or arylating agents, which attach aryl moities (such as phenols) to DNA through the nitrogen and oxygen atoms in the nitrogenous bases (pyrimidine and purine) of DNA.

Chemicals that cause cancer directly are called *primary carcinogens*; however, most xenobiotics involved in carcinogenesis are *pre-* or *procarcinogens* which undergo phase I reactions (lipophilic xenobiotic species undergo enzyme catalysed reactions involving attachment of polar groups such as OH to become more water soluble) or phase II reactions (in which the polar functional groups in phase I provide sites for conjugation reactions) to produce *ultimate* carcinogens. Vinyl chloride ($CH_2=CHCl$) is a primary carcinogen which can cause a rare form of liver cancer in humans, especially those employed in PVC manufacture (p. 277).

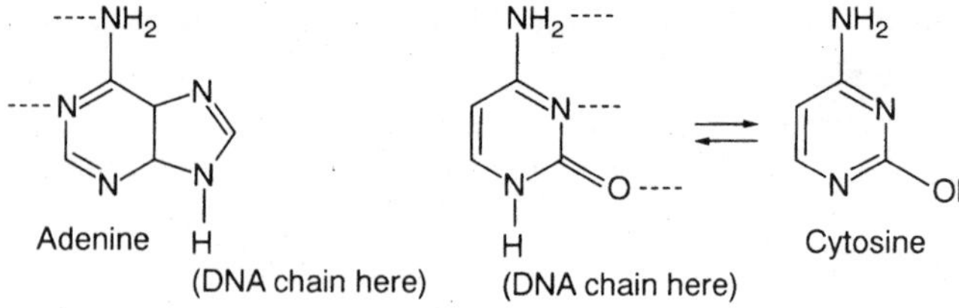

Figure 3.2 Pyrimidine (adenine) and purine (cytosine) bases.

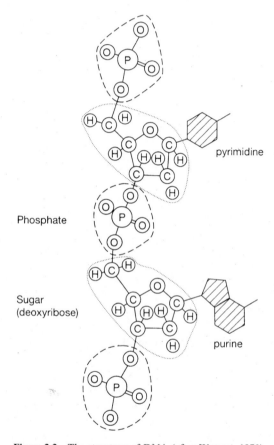

Figure 3.3 The structure of DNA (after Watson, 1970).

The Ames test is the most frequently used procedure to test for mutagenic properties of chemicals. The Ames test is based on the use of liver enzymes to convert procarcinogens to ultimate carcinogens. Histidine-requiring *Salmonella* bacteria are inoculated onto a medium that does not contain histidine. Procarcinogen chemicals will increase the chances of mutations

Table 3.3 The ten most important causes of cancer risk in the USA (USEPA 1986, in Masters, 1991)

Rank	Cause of cancer
1 (=)	Occupational exposure to chemicals (*c.* 20 000 substances)
1 (=)	Indoor radon (< 20 000 lung cancers/year from exposure in home in the USA)
3	Pesticide residues in foods (6000 cancers/year)
4 (=)	Indoor pollutants (non-radon) – mostly caused by tobacco smoke (< 6500 cancers/year)
4 (=)	Consumer exposure to chemicals (cleaning fluids, pesticides, particleboard and asbestos-containing products) (*c.* 10 000 chemicals)
6	Hazardous/toxic air pollutants
7	Depletion of stratospheric ozone – UVB radiation (caused by pollutants, e.g. CFCs)
8	Hazardous waste sites (inactive) *c.* 25 000 sites in US
9	Drinking water – radon and trihalomethanes (from chlorination)
10	Application of pesticides (e.g. agricultural workers: high individual risk)

occurring which will give rise to forms of the bacteria which can synthesize histidine (Manahan, 1991).

The immune system acts to protect the body from xenobiotic chemicals, infectious agents (viruses and bacteria) and neoplastic cells which give rise to cancerous tissue. Adverse effects on the body's immune system are being increasingly recognized as a consequence of exposure to hazardous substances, UV radiation, etc.

Another reaction is allergy or hypersensitivity, when the immune system over-reacts to foreign agents or their metabolites in a self-destructive manner, e.g. Be, Cr, Ni, formaldehyde, pesticides, resins and plasticizers (Manahan, 1991).

Of the ten most important causes of cancer in the US, listed in Table 3.3, six of the causes are related in some way to environmental pollution. In addition, the occupational and consumer exposure to chemicals (equal first and equal fourth causes, respectively) will generally involve substances which are also important environmental pollutants. The distinction between these two causes is that the chemicals have not necessarily undergone emission and transport in the environment, although there is a strong likelihood that their manufacture, and/or disposal, may have led to some environmental pollution.

However, Ames and Gold (1990) in an article entitled *Misconceptions on pollution and the causes of cancer* stress that environmental and dietary exposure to most pollutants, especially synthetic pesticides which are often suspected of being linked to the onset of cancer, is less important in causing cancer than many other factors.

The International Association for Research on Cancer (IARC) classifies chemicals on the basis of their potential risk for causing cancer in humans. This classification has four groups, which are defined as follows:

Group 1: the agent is carcinogenic to humans (e.g. As, Cr(VI), vinyl chloride, benzene)

Group 2A: the agent is probably carcinogenic to humans (e.g. Be, Cd)

Group 2B: the agent is possibly carcinogenic to humans (e.g. dichloromethane, 1,2-dichloroethane, tetrachloroethene)

Group 3: the agent is not classifiable as to its carcinogenicity to humans (where there is inadequate evidence of carcinogenicity in either humans or experimental animals) (e.g. Cr(III), inorganic fluorides, Se, 1,1,1-trichloroethene)

Group 4: the agent is probably not carcinogenic to humans.

Oestrogenic chemicals and declining male fertility

Untimely exposure to oestrogens (**1**, **2**) can reduce male reproductivity and may induce breast cancer. They are members of a group known as

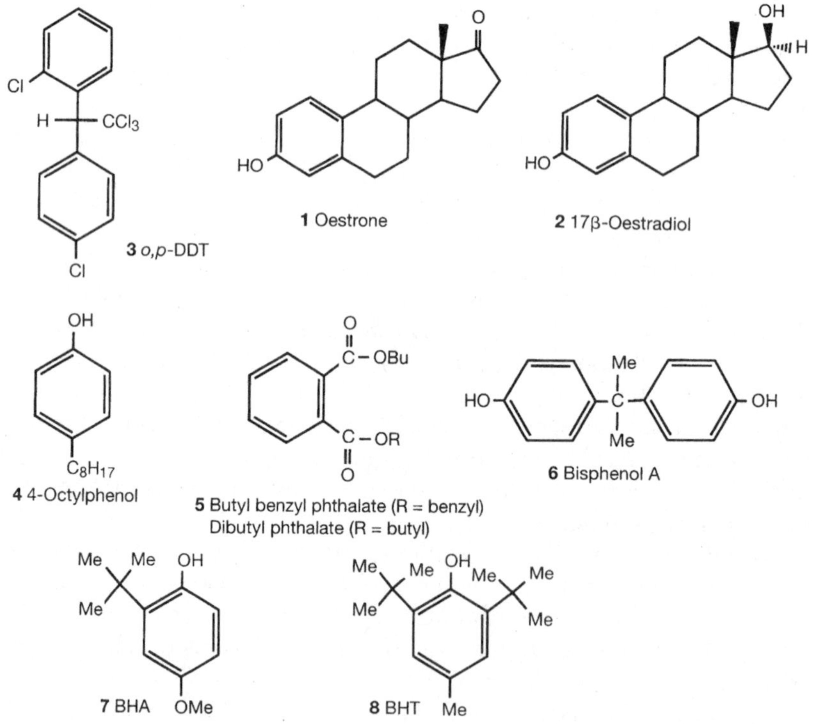

Figure 3.4   Oestrogenically active compounds.

'endocrine disruptors', which are also linked with other cancers and Alzheimer's disease (ENDS, 1995a) (Fig. 3.4).

There is growing concern in technologically advanced countries about a decline in sperm counts in human males (Sharpe and Shakkebaeck, 1993). This was given impetus in the UK by popularization on television (BBC, 1993) but it has been pointed out (Safe, 1995) that some of the conclusions were statistically insecure. An extensive recent study (Auger *et al.*, 1995) in Paris showed that there has been a decline of 2.6%/year in sperm counts during 1973–92. It is considered that reduced sperm counts, sexual deformities and testicular cancer are linked to environmental or lifestyle factors acting in early childhood or in the foetus. The decline in the 1990s is the result of chemicals disrupting the hormonal system which were present in the environment 20–40 years before. The effects of present-day conditions will not be apparent until 20–30 years time (ENDS, 1995a)

Safe (1995) exonerates the principal organochlorine pollutant DDT, its degradation product DDE and the PCBs, although some activity in those of lower chlorine number is attributed to their hydroxylation products (Section 6.4.12) *o,p*-DDT (3) is oestrogenically active while Endosulfan, Toxaphene and Dieldrin (p. 286) induce human breast cancer cells. Some pollutants, including TCDD and PAH oxidation products, combat these effects; it was shown at Seveso that breast cancer decreased in areas with high TCDD levels.

Studies on fish in rivers receiving sewage effluent have shown that oestrogenic effects were apparent up to 4–5 km downstream from discharge points. Alkylphenol ethoxylates (APEs) used in wool scouring were responsible and they have been found in polluted water at levels up to 330 µg/l. About 18 000 t of the related nonylphenol ethoxylates (NPEs) are used annually in the UK for a variety of purposes, including agrochemicals, textiles and metal finishing. Some industries have agreed to phase out APEs and NPEs over the next few years (ENDS, 1995b).

The relative activity of a group of oestrogenic chemicals found in sewage effluent has been evaluated (Jobling *et al.*, 1995) in terms of their ability to bind to the oestrogen receptor in the liver of rainbow trout. Table 3.4 shows the response relative to the maximum induced by the most active natural oestrogen, 17β-oestradiol (2). It can be seen that octylphenol is the most active of the artificially made compounds, but a molar concentration 10^3 times that of the natural oestrone is required to achieve the maximum response. The phthalate esters are of concern because they are the most abundant artificially made chemicals in the environment and occur in natural waters at levels ranging between nanograms and milligrams per litre. They are widely used to impart flexibility to plastics; thus, butyl benzyl phthalate (BBP) is found in vinyl floor tiles and adhesives whilst dibutyl phthalate (DBP) occurs in food packaging and PVC. The monomeric esters tend to permeate foodstuff within the packaging, and levels in the USA are

Table 3.4 Relation between oestrogenic response and concentration (Jobling *et al.*, 1995)

Compound	Percentage of maximum response	Molar concentration
17β-Oestradiol (**2**)	10	10^{-12}
	80	10^{-10}
	100	10^{-8}
4-Octylphenol (**4**)	10	10^{-7}
	70	10^{-6}
	100	10^{-5}
Butyl benzyl phthalate	20	10^{-6}
(BBP (**5**) R = benzyl)	50	10^{-5}
Dibutyl phthalate	30	10^{-5}
(DBP (**5**) R = butyl)		

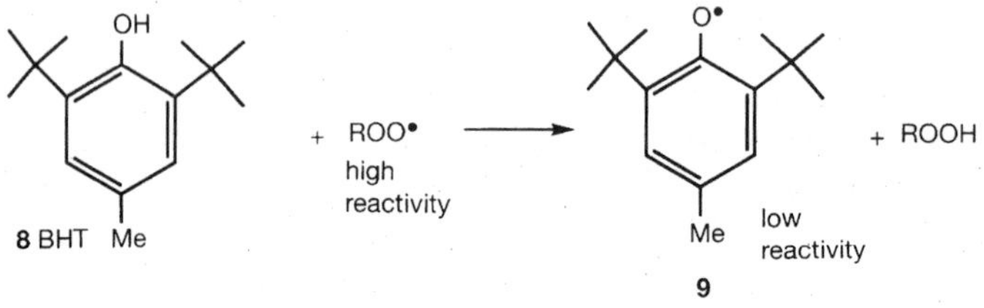

Figure 3.5 Inactivation of oxidizing free radicals.

50–500 μg/kg. In the UK, the intake in foodstuffs is typically 230 μg/day and in chocolate bars wrapped in polypropylene reached 14 mg/kg (14 ppm). Consumers are also exposed to epoxy resins used to line water pipes and tin cans.

Other compounds which exhibit oestrogenic activity include bisphenol A (**6**) and also hindered phenols such as BHA (**7**) and BHT (**8**), which are added deliberately to foodstuffs to protect them from oxidation by free radicals. Figure 3.5 shows how a peroxy radical capable of oxidizing food is inactivated by exchange (Perkins, 1994) to give the more stable, hindered and much less active aryloxy radical (**9**).

The US Health Effects Research Laboratory tested 30 000 compounds in a predictive model and found that 1% could show some endocrine disruptive activity (ENDS, 1995a). It must be borne in mind that the relative activity determined by *in vitro* experiments, such as those mentioned above, is likely to vary from that *in vivo* because of alternative metabolic pathways in animals. It is also significant that the phenols BHA and BHT are 10^6 times less active than the 17β-oestradiol.

3.2.3   Ecotoxicology

Ecotoxicology is not just concerned with the effects of toxicants on one species but on a wide range of interacting species present in an ecosystem. Behavioural effects, therefore, will be an important consideration in addition to biochemical and physiological responses. Pollutants can affect living organisms in two ways: (i) by being directly toxic or (ii) by causing an adverse change in the organism's habitat which then adversely affects the organism (Smith, 1986). Directly, toxic pollutants first gain entry into an organism and follow an internal pathway which either results in excretion or the disruption of a biochemical process and the onset of toxicity. Indirectly, toxic pollutants alter the physical or chemical conditions of the environment to such an extent that the survival of the organisms is threatened. For example, the depletion of the dissolved oxygen content of a stream, which occurs with the discharge of untreated sewage or silage effluent, can kill many organisms by asphyxiation.

Direct sublethal effects of pollutants also vary between organisms in an ecosystem and some examples of these in animals and plants include (Connell and Miller, 1984):

- Physiological effects on: metabolism, photosynthesis and respiration, osmoregulation, feeding and nutrition, heartbeat rate, blood circulation, body temperature, water balance.
- Behavioural responses: (individuals) sensory capacity, rhythmic activities, motor activity, motivation and learning (groups and between individuals) migrations, intraspecific attraction, aggression, predation, vulnerability, mating.
- Effects on reproduction: viability of eggs and sperm, breeding and mating behaviour, fertilization and fertility, survival of offspring.
- Genetic effects: chromosome damage, mutagenic and teratogenic effects.
- Effects on growth: body and organ weights, developmental stages.
- Histopathological effects: abnormal growths, membrane abnormalities (respiratory and sensory), reproductive organs.

The organisms within an ecosystem will differ markedly in their tolerance to both direct toxicity and the indirect effects of pollutants. Differences in susceptibility to direct toxicants are clearly illustrated in the LD_{50} data for the lethal dose of dioxin (TCDD) shown in Table 3.2. These data show that not only do species differ in susceptibility but that even males and females of the same species vary, the LD_{50} for the female rat being twice as high as that for males.

3.3
Assessment of toxicity risks

Risk assessment of environmental pollutants is concerned firstly with identifying the hazards which the substances or energy may pose to the health of people, animals and plants, and any damage which they may cause to structures and commodities and, secondly, with estimating the probability of these types of harm occurring. It is important to know the pathways and dose–response relationships for all hazardous substances and possible target organisms (including humans, farm livestock, crop and other plants as well as surrounding ecosystems). Much of this information is now available for most of the commonly encountered pollutants although it is constantly being added to and improved. Nevertheless, it is important to recognize that there is a great deal of uncertainty inherent in the characterization of risk.

Risk assessment is a difficult subject because the general public's perception of the importance of risks is strongly influenced by the media. A study in the USA published by Allman (1985) compared the general public's ranking of hazards with those of risk assessment experts. The public ranked nuclear power as the greatest hazard while the experts ranked it number 30. However, accidents from motor vehicles were ranked 2 and 1, respectively, with the experts placing smoking in second position and the public ranking it fourth after handguns (a hazard relatively unique to the USA) (Allman, 1985). The data given below relate to an objective assessment of risks to the general public in the UK. The statistical risk of death from the following causes is (Connor, 1995):

Any type of cancer	1 in 4
Smoking 10 cigarettes/day	1 in 200
All natural causes	1 in 850
Violence/poisoning	1 in 3300
Accident in the home	1 in 7700
Road accident	1 in 8000
Civilian nuclear power accident:	
existing nuclear power plants	1 in 10 000 000
new nuclear plants	1 in 100 000 000

The widely recognized disparity of the general public's assessment of risk compared with an objective analysis by experts indicates that the 'acceptability of risk' is a very important aspect. Although smoking is clearly the most serious 'avoidable risk' which many people expose themselves to, it appears to be more acceptable than civilian nuclear power, which poses a very low risk. The responsibility for these contrasts between real and perceived risk is largely the result of the media, which sensationalizes isolated incidents. In the context of environmental pollution, natural toxins including aflatoxins produced by the fungus *Aspergillus flavus* in mouldy peanuts and the botulism toxin synthesized by the bacterium *Clostridium botulinus* are more powerful than almost any industrial chemical or its decomposition product (Connor, 1995).

3.3.1 Pollutants in contaminated land

Taking contaminated land as an example, Table 3.5 summarizes the commonly encountered types of risk associated with the pollutants.

Empirical transfer coefficients have been determined and models have been developed for quantifying dose–response relationships. In practice, many states or countries have established lists of critical concentrations (or 'trigger concentrations') for the risk assessment of site and environmental survey data. The basis for these different sets of values varies according to the target groups they are intended to protect from the effects of the pollutants.

Examples of tables of critical (trigger) concentrations for organic pollutants used in different countries are given below (the equivalent data for heavy metals are given in Section 5.3.6).

1. The Netherlands previously had a scheme which gave three indicative values for a wide range of pollutants: A, the 'normal' reference value; B, the value at which it is necessary to conduct further investigations into the form and bioavailability of the pollutant; and C, the intervention value above which the soil definitely needs cleaning-up (Moen *et al.*, 1986). This system was superseded by an effect-oriented scheme of 'Environmental Quality Standards for Soil and Water' (Netherlands Directorate General for Environmental Protection, 1991). These standards are based on ecological function and comprise 'target values' (TV) for soils and waters which represent the final environmental quality goals for the Netherlands. In the case of surface and groundwaters, both target and limit values are given. The limit value is equivalent to the maximum permissible risk level and is

Table 3.5 Examples of the risks from chemicals in contaminated land (adapted from Alloway (1992) based on Beckett and Sims (1986) and ICRCL (1987))

Human and animal toxicity and carcinogenicity
- the direct ingestion of contaminated soil (mainly by children and grazing livestock) e.g. CN, As, Pb, PAHs
- inhalation of dusts, toxic gases and vapours from the contaminated soil, e.g. benzene, solvents, Hg, CO, HCN, H_2S, PH_3 and asbestos
- uptake by plants of contaminants hazardous to animals and people through the food chain, e.g. Cd, As, Pb, Tl, PAHs
- contamination of drinking water supplies, e.g. phenols, CN^-, SO_4^{2-}, soluble metals, pesticides, e.g. atrazine in groundwater and permeation of water pipes by solvents
- skin contact, e.g. tars, phenols, asbestos, radionuclides, PAHs, PCBs and PCDDs

Phytotoxicity
- SO_4^{2-}, B, Cu, Ni, Zn, herbicide residues

Fire and explosion
- CH_4 and high calorific wastes from landfills
- coal dust
- petroleum, solvents

Deterioration of building materials and services
- SO_4^{2-}, SO_3^{2-}, Cl^-, coal tar, phenols, mineral oils, solvents

intended to indicate the environmental quality to be achieved in a given period. It is intended that limit values would be progressively reduced until they reach the target value. However, the authors of the standards recognized that pollutants generally have long residence times in soils and so they have not given limit values for soils, only target values. The intervention values (C values) of the 1986 scheme still remain because if the concentrations of a pollutants reached this level there would definitely be a need for a thorough investigation and remedial action. The target values are given for a 'standard soil' with 10% organic matter and 25% clay, but formulae are provided to allow the values to be calculated for soils with a wide range of clay and organic matter contents. For example, the Pb target value is determined by the formula: $50 + L + H$ (where $L = \%$ clay and $H = \%$ organic matter).

Examples of reference values from the Netherlands Environmental Quality Standards for Soil and Water are given in Table 3.6 (data for metals in both soils and waters in this Netherlands list are given in Chapter 5).

Table 3.6 Guide values (in μg/g) and quality standards used in the Netherlands for assessing soil contamination by organic and inorganic substances. See Sections 6.2 for PAHs, 6.3 for benzene, 6.5 for PCBs, PCDDs and PCDF and Chapter 4 for PCP

Category	A	B	C	TV
Inorganic pollutants				
CN⁻ (total free)	1	10	100	1
CN⁻ (total complex)	5	50	500	5
Br	20	50	300	
S	2	20	200	
Polycyclic aromatics				
PAHs (total)	1	20	200	
Naphthalene	0.1	5	50	15
Anthracene	0.1	10	100	50
Benzo-[a]-pyrene	0.05	1	10	25
Chlorinated hydrocarbons				
CH total	0.05	1	10	
PCBs	0.05	1	10	
Chlorophenols (total)	0.01	1	10	
Pentachlorophenol	–	–	–	2
Pesticides				
Pesticides (total)	0.1	2	20	
Aromatic compounds				
Aromatics (total)	0.1	7	70	
Benzene	0.01	0.5	5	
Toluene	0.05	3	30	
Phenols	0.02	1	10	
Other organic compounds				
Cyclohexane	0.1	5	60	
Pyridine	0.1	2	20	
Gasoline	20	100	800	
Mineral oil	100	1000	5000	

A = reference value, B = test requirements, C = intervention value, from 1986 scheme; TV = target value in 1991 Environmental Quality Standards for Soils and Waters.

2. In the UK, the Department of the Environment set up an Interdepartmental Committee for the Redevelopment of Contaminated Land (ICRCL) to draw up a list of trigger concentrations for contaminants (1987). These trigger concentrations are more pragmatic than the Netherlands Environmental Quality Standards and are based mainly on the risk to human health. Unlike the Dutch standards, the ICRCL values vary for different proposed uses of the contaminated land. The lowest values are given for garden soils where vegetables are likely to be grown, with higher values for parks and open spaces, and the highest values for land to be developed for industrial uses where the transfer of pollutants from the soil to plants is not likely to be significant in terms of its impact on human health. Examples of the ICRCL values for both general sources of pollution and more specifically for land affected by coal carbonization (gas and coke works) are given in Table 3.7. These are due to be replaced in 1996 by values based on probalistic risk assessment.

3. In Canada, the National Contaminated Sites Remediation Program published interim Environmental Quality Criteria in 1991 for use in the evaluation of contaminated sites (Table 3.8). The values given are intended to enable sites to be classified as high, medium or low risk according to their

Table 3.7 UK Department of the Environment (ICRCL) trigger concentrations for contaminants associated with former coal carbonization sites

Contaminant	Proposed use	Trigger concentrations (μg/g)	
		Threshold	Action
PAHs	Gardens, allotments	50	500
	Landscaped areas	1000	10 000
Coal tar	Gardens, allotments	200	–
	Landscaped areas, open space, buildings, hard cover	500	–
Phenols	Gardens, allotments	5	200
	Landscaped areas	5	1000
Free cyanide	Gardens, allotments	25	500
	Buildings, hard cover	100	500
Complex	Gardens, allotments	250	1000
cyanides	Landscaped areas	250	5000
	Buildings, hard cover	250	NL
Thiocyanate	All uses	50	NL
Sulphate	Gardens, allotments	2000	10 000
	Landscaped areas, buildings	2000	50 000
	Hard cover	2000	NL
Sulphide	All uses	250	1000
Sulphur	All uses	500	20 000
Acidity	Gardens etc.	pH < 5	pH < 3

NL = no limit set because contaminant does not pose a particular hazard when land used for this purpose.

Table 3.8 Selected data from the Interim Canadian Environmental Quality Criteria for Contaminated Sites (values in µg/g) (Canadian Council of Ministers of the Environment, 1991; reproduced with permission of the Minister of Supply and Services Canada, 1993)

	Land uses			
	Background	Agricultural	Residential	Industrial
(a) **Metals**				
As	5	20	30	50
Ba	200	750	500	2000
Be	4	4	4	8
Cd	0.5	3	5	20
Cr^{6+}	2.5	8	8	–
Co	10	40	50	300
Cu	30	150	100	500
CN (free)	0.25	0.5	10	100
CN (total)	2.5	5	50	500
Pb	25	375	500	1000
Hg	0.1	0.8	2	10
Mo	2	5	10	40
Ni	20	150	100	500
Se	1	2	3	10
Ag	2	20	20	40
Sn	5	5	50	300
Zn	60	600	500	1500
(b) **Organic pollutants**				
Monocyclic hydrocarbons				
Benzene	0.05	0.05	0.5	5
Chlorobenzene	0.1	0.1	1	10
Toluene	0.1	0.1	3	30
Phenols				
Phenols (each)	0.1	0.1	1	10
Chlorophenols (each)	0.05	0.05	0.5	5
PAHs				
Benzo-[a]-pyrene	0.1	0.1	1	10
Naphthalene	0.1	0.1	1	10
Chlorinated hydrocarbons				
Chlorinated aliphatics (each)	0.1	0.1	5	50
Chlorobenzenes (each)	0.05	0.05	2	10
Hexachlorobenzene	0.1	0.05	2	10
PCBs	0.1	0.5	5	50
PCDDs and PCDFs	0.00001	0.0001	0.001	–

impact (current or potential) on human health and ecosystems. It is a screening system and is not intended to be a quantitative risk assessment for individual sites.

The 1991 Environmental Quality Standards for the Netherlands (Table 3.6) give target values for soils which will be very difficult and expensive to achieve, especially in the case of some of the ubiquitous contaminants, such as Pb. The threshold value for Pb under the ICRCL scheme used in the UK

(Table 3.7) is 500 µg/g. This implies that concentrations below this figure should not cause problems. In contrast, the Dutch target value for Pb is 85 µg/g (for a standard soil) and the Canadian background value (benchmark level) is even lower at 25 µg/g, but the trigger concentration for the remediation of agricultural land is 375 µg/g. The Dutch scheme is based on the ecological effects of contaminants and, therefore, soils meeting the target values would be suitable for any use, such as food production or nature conservation. However, the UK and Canadian values take into account different uses of land. The UK value of 500 µg/g Pb is the most achievable, because many urban soils in the UK would come within this range and the more excessively polluted sites would be considered unacceptable.

Of particular relevance is the intended future use of the contaminated land. In some countries, such as the Netherlands, the intention is to ameliorate the site to a specification which will allow the land to be used for any purpose (multifunctionality). The more pragmatic approach in other countries, such as the UK, is to relate the site quality specification to the intended use. For example, a contaminated site required for development as a warehouse complex would not need to have such low concentrations of toxic compounds as a site to be used for housing where the residents may grow vegetables in their gardens. However, any explosive hazard, such as methane release, would need to be removed for both new uses of the site. Even though the use of a site may not require a high degree of clean-up, the remaining contaminants may migrate within the soil to the ground-water, undergo chemical changes or remain a potential problem for future uses of the site.

3.3.2   Pollutants in drinking water

The concentrations of pollutants in drinking water are very important since adults drink on average 1.5 l/day and are, therefore, likely to be more rapidly affected by pollutants from this route than from the diet. A selection of some of the guideline values currently in use are given in Tables 3.9 and 3.10.

The organic pollutants in Table 3.10 are acknowledged to be those closely related to human health. It is interesting to note that for the two isomers of dichloroethane, the 1,1-dichloroethane isomer is considered to be 33 times more hazardous than the 1,2-isomer. The lowest guideline values, which imply a high toxicity hazard, are those for the PAH benzo-[a]-pyrene and hexachlorobenzene.

After the National Rivers Authority (NRA) was set up under the Water Act of 1989, prosecutions for infringements rose dramatically from an average of only 39/year during 1984–89 to 370 in 1990 and then to 959 in the following year.

Table 3.9 Guideline and maximum acceptable concentrations for metal and other inorganic pollutants in water for human consumption (µg/l) (from: WHO, 1993; Murley, 1995; Manahan, 1991; Canada Council of Ministers of the Environment, 1993) (Maximum permissible concentrations except for WHO and EC guideline values)

	WHO guide	EC guide	EC max	Canada	UK	USA
As	10	–	–	25	50	50
B	300	1000	–	5000	2000	1000
Ba	700	100	–	1000	1000	–
Cd	3	–	–	5	5	10
Cu	2000	3000[a]	–	< 1000	3000	1000
Cr	50	–	–	50	50	50
Fe	–	50	200	< 300	200	50
Hg	1	–	–	1	1	–
Mn	500	20	50	< 500	50	50
Pb	10	–	–	10	50	5
Zn	–	5000[a]	–	< 5000	5000	5000
Ca	–	100000	–	–	250000	–
Mg	–	30000	50000	–	50000	–
CN	70	–	–	200	50	–
F	1500	–	700–1500[b]	1500	1500	800–1700[b]
NO$_3$$^-$	50000	25000	50000	–	50000	–
NO$_2$$^-$	3000	–	100	–	100	–
SO$_4$$^-$	–	25000	250000	–	250000	–

[a] Cu and Zn in water after it has been standing for 12 hours at point of consumption (first draw from tap). A guide value of 100 µg/l for both Cu and Zn in water leaving the treatment works.
[b] Maximum concentrations of F depend on ambient temperature (lower values at higher temperatures).

Table 3.10 World Health Organization guideline values for selected organic micropollutants in drinking waters (WHO, 1993)

Organic pollutant	Guideline value (µg/l)
Pesticides	
Alachlor	20
Aldicarb	10
Aldrin/Dieldrin	0.03
Atrazine	2
Bentazone	30
Carbofuran	5
Chlordane	0.2
Chlorotoluron	30
DDT	2
1,2-Dibromo-3-chloropropane	1
2,4-D	30
1,2-Dichloropropane	20
1,3-Dichloropropane	n.a.d.
1,3-Dichloropropene	20
Ethylene dibromide	n.a.d.
Heptachlor and heptachlor epoxide	0.03
Hexachlorobenzene	1
Isoproturon	9
Lindane	2
MCPA	2

(Contd.)

Table 3.10 *(Contd.)*

Organic pollutant	Guideline value (μg/l)
Methoxychlor	20
Metolachlor	10
Molinate	6
Pendimethalin	20
Pentachlorophenol	9
Permethrin	20
Propanil	20
Pyridate	100
Simazine	2
Trifluralin	20
Chlorophenoxy herbicides other than 2,4-D and MCPA	
2,4-DB	90
Dichlorprop	100
Fenoprop	9
MCPB	*n.a.d.*
Mecoprop	10
2,4,5-T	9
General organic pollutants	
Chlorinated alkanes	
Carbon tetrachloride	2
Dichloromethane	20
1,1-Dichloroethane	*n.a.d.*
1,2-Dichloroethane	30
1,1,1-Trichloroethane	2000
Chlorinated ethenes	
Vinyl chloride	5
1,1-Dichloroethene	30
1,2-Dichloroethene	50
Trichloroethene	70
Tetrachloroethene	40
Aromatic hydrocarbons	
Benzene	10
Toluene	700
Xylenes	500
Ethylbenzene	300
Styrene	20
Benzo-[a]-pyrene	0.7
Chlorinated benzenes	
Monochlorobenzene	300
1,2-Dichlorobenzene	1000
1,3-Dichlorobenzene	*n.a.d.*
1,4-Dichlorobenzene	300
Trichlorobenzenes (total)	20
Miscellaneous	
Di(2-ethylhexyl)adipate	80
Di(2-ethylhexyl)phthalate	8
Acrylamide	0.5
Epichlorohydrin	0.4
Hexachlorobutadiene	0.6
Edetic acid (EDTA)	200
Nitrilotriacetic acid	200
Dialkyltins	*n.a.d.*
Tributyltin oxide	2

n.a.d. = no adequate data available.

Table 3.11 Critical concentrations for common gaseous pollutants (% v/v) (Smith, 1991)

Contaminant	Toxicity	Flammability	Trigger for potential hazard
Carbon dioxide	0.5	–	0.125
Carbon monoxide	0.005	12–75	0.00125
Sulphur dioxide	0.0005	–	0.000125
Hydrogen sulphide	0.001	4.3–45.5	0.0025
Methane	14	5–15	0.25
Petrol	0.1	1.4–7.6	0.025

3.3.3 Toxic or explosive gases and vapours

Although a wide range of toxic gases and fumes can be released in accidents in chemical plants or stores, there are several gases and vapours which are commonly encountered in many pollution situations, including contaminated land, waste disposal and fuel combustion. The safe concentrations for some of these gases and vapours are given in Table 3.11 (Smith, 1991). It is interesting to note that the most hazardous of these, when judged by the concentration causing toxicity, is SO_2.

With regard to combustibility, calorific values greater than 10 MJ/kg are normally regarded as a potential combustion hazard but materials with values between 2 and 10 MJ/kg could also pose a problem in some circumstances. Materials with calorific values below 2 MJ/kg are unlikely to burn (Fleming, 1991).

References

Allman, W. F. (1985) We have nothing to fear. *Science*, **85**, 41.
Alloway, B. J. (1992) In *Understanding Our Environment* (2nd edn), Ch. 5. (ed. R. M. Harrison). Royal Society of Chemistry, Cambridge.
Ames, B. N. and Gold, L. S. (1990) *Agnew. Chem. Int. Ed. Engl.*, **29**, 1197.
Auger, J., Kunstmann, J. M., Czyglic, F. and Jouannet, P. (1995) Decline in semen quality among fertile men in Paris during the past 20 years. *New Engl. J. Med.*, **332**, 281–285.
BBC (1993) Assault on the Male. *Horizon*, 31 October.
Beckett, M. J. and Sims, D. L. (1986) In *Contaminated Soil* (ed. J. W. Assink and W. J. van den Brink). Martinus Nijhoff, Dordrecht.
British Medical Association (1991) *Hazardous Waste and Human Health*. Oxford University Press, Oxford.
Canada Council of Ministers of the Environment (1991) *Interim Canadian Environmental Quality Criteria for Contaminated Sites*. Report CCME EPC-CS34, Winnipeg, Manitoba.
Connell, D. W. and Miller, G. J. (1984) *Chemistry and Ecotoxicology of Pollution*. John Wiley, New York.
Connor, S. (1995) *Independent on Sunday*, 3 September.
EC (1980) *Quality of water for human consumption*. 80/778/EEC. Commission of European Communities, Brussels.
ENDS (1995a) *Environmental Data Services, Report No. 243*, April 1995, pp. 15–20.
ENDS (1995b) *Environmental Data Services, Report No. 246*, July 1995, pp. 5–7.
Fleming, G. (ed.) (1991) Chapter 1 in *Recycling Derelict Land*. Thomas Telford, London.
ICRCL (Interdepartmental Committee on the Redevelopment of Contaminated Land) (1987) *Guidance on the Assessment and Redevelopment of Contaminated Land*. Department of the Environment, London.

Jobling, S., Reynolds, T., White, R., Parker, M. G. and Sumpter, J. P. (1995) A variety of environmentally persistent chemicals, including some phthalate plasticisers, are weakly estrogenic, *Environ. Health Perspect*, **103**, 582–587.

Manahan, S. E. (1991) *Environmental Chemistry* (5th edn). Lewis Publishers, Chelsea, MI.

Moen, J. E. T., Cornet, J. P. and Evers, C. W. A. (1986) In *Contaminated Soil* (ed. J. W. Assink and W. J. van den Brink), Martinus Nijhoff, Dordrecht.

Murley, L. (ed.) (1995) *Pollution Handbook*. National Society for Clean Air and Environmental Pollution, Brighton.

Netherlands Directorate General for Environmental Protection (1991) *Environmental Quality Standards For Soil and Water*. Ministry of Housing, Physical Planning and Environment, Leidschendam, Netherlands.

Perkins, M. J. (1994) *Radical Chemistry*, p. 122. Ellis Horwood, Chichester.

Rodricks, J. V. (1992) *Calculated Risks*. Cambridge University Press, Cambridge.

Safe, S. H. (1995) Environmental and dietary estrogens and human health: is there a problem? *Environ. Health Perspect*, **103**, 346–351.

Sharpe, R. M. and Shakkebaeck, N. E. (1993) Are estrogens involved in falling sperm counts and disorders of the male reproductive tract? *Lancet*, **341**, 1392–1395.

Smith, M. (1991) In *Recycling Derelict Land*, Ch. 5. (ed. G. Fleming) Thomas Telford, London.

Smith, S. (1986) *Understanding Our Environment* (1st edn), Ch. 5. (ed. R. E. Hester). Royal Society of Chemistry, London.

Watson, J.D. (1970) *The Double Helix*. Weidenfeld & Nicholson, London.

WHO (World Health Organization) 1993 *Guidelines for Drinking Water Quality* (2nd edn), Vol. 1. WHO, Geneva.

Further reading

Cockarham, L. G. and Shane, B. S. (1994) *Basic Environmental Toxicology*. CRC Press, Boca Raton, FL.

Francis, P. M. (1994) *Toxic Substances in the Environment*. Wiley Interscience, New York,

Masters, G. M. (1991) *Introduction to Environmental Engineering and Science*. Prentice Hall, Englewood Cliffs, NJ.

Sittig, M. (1994) *Worldwide Limits For Toxic and Hazardous Chemicals in Air, Water and Soil*.

Smith, S. (1992) *Understanding Our Environment* (2nd edn), Ch. 8. (ed. R. M. Harrison). Royal Society of Chemistry, Cambridge.

4 Analysis and monitoring of pollutants

4.1
Introduction The range of environmental pressures has been reviewed in the introduction to this book. This chapter deals with chromatographic techniques for the detection and identification of organic pollutants and with atomic absorption emission spectroscopy for the determination of polluting metals.

Organic pollution was first manifested following the growth in the use of pesticides in the years immediately after the Second World War and through the 1950s. The first organochlorine pesticides were DDT, Lindane and Dieldrin (Section 6.4) and over the period 1950–70 the numbers of individual herbicides rose from about 20 to 140; over the same period, insecticide numbers rose from about 40 to 170 (Conway and Petty, 1991). Overuse and misuse of these compounds led to the death of wildlife, especially species at the head of food chains, including raptorial birds, foxes and badgers (Section 6.4.7). Public concern required that such compounds were detected and controlled.

Thin-layer chromatography (TLC) developed in the 1960s finds some applications, but gas–liquid chromatography (GLC) (Section 4.4) was predominant through the 1950s and subsequently. Pollutants could initially be traced using thermal conductivity detection at 1 ng/l, but continuous improvements in the sensitivity and range of detectors raised the detection level to better than 1 pg/l, that is by six orders of magnitude.

The mass spectrometer (see below) was introduced as a detector in 1957 and in current practice this enables the analyst to approach the parts per trillion level of detection and to identify polluting substances in the one operation. It must, however, be emphasized that the highest level of detection can only be achieved in a laboratory protected from airborne interference. The use of mobile laboratories capable of sampling and analysing airborne pollutants in real time has grown in the last decade, prompted in part by pressure arising from legislation such as the UK Environmental Protection Act of 1990.

The technique of high-performance liquid chromatography (HPLC) was initiated by Horvath in 1967, but its full potential was not realized until suitable pumps and packing materials became available in the 1970s. With its wide application to involatile materials and range of eluting solvents it has a position comparable to GLC. This is especially true now that direct linkage to a mass spectrometer is possible using thermospray equipment (Voyksner, 1994).

Inorganic pollution arises from mining and smelting of metals, coal burning and chemical production coupled to widespread applications in engineering and electronics. These activities lead to demands for the quantification of pollutant metals in rainfall, rivers and groundwater, also from mines, waste disposal sites and former gasworks. Experiments with mercury in 1939 formed the basis of the modern technique of atomic absorption (AA, see below).

Current methods of atomic absorption are based on work by Walsh in 1955, who showed that metal ions could be reduced to metal atoms in flames; this led to widespread use towards the end of the next decade. The resolution depends on the absorption by the metal atoms of light from a monochromatic source and the necessary sensitivity was achieved through the use of photomultiplier detectors. Recent improvements include the use of graphite furnaces for analysis of small solid samples and the replacement of flames by plasma sources, an important development which can detect refractory elements such as uranium. A further refinement is the use of the mass spectrometer to resolve metal ions produced in plasma. The detection limit for this last combination varies with the metal but lies in the range of 1 ppt to 1 ppb.

In reading the following description of analysis methods, it must be appreciated that without a properly executed sampling regimen the extreme sensitivity of any analytical procedure will be wasted and may even be misleading.

4.2 Chromatography

Chromatography is the dominant analytical technique for the identification and quantification of organic pollutants. Classical chromatography was first used by Tswett in 1906 for the analysis of plant extracts, which were charged on to the top of a column of chalk and eluted with a downward flow of petrol. Chromatograms of this sort depend on the different affinities of the components of a mixture for the stationary phase; this could be chalk, alumina, kieselguhr or other surface active solid. Partition into the downflowing solvent then carries the separated compounds to the end of the column in sequence. There are a number of practical details which are essential for efficiency:

1. The analyte should be concentrated at the column head, since ideally it would be located as a point source.
2. Sufficient solid phase must be provided so as to ensure that it is not overloaded with material.

3. Care must be taken in choosing the solvent, which must have sufficient strength or affinity with the charge to ensure that this becomes desorbed from the stationary phase. If too polar a solvent is chosen, it will limit interaction with the solid phase, which will no longer differentiate between one component and another; that is why they will not be resolved.

4. The flow rate must be on a reasonable time scale, but not so fast as to detract from the efficient 'mass transfer' of components between the stationary solid phase and the flowing solvent (Section 4.4).

Modern chromatography has many specialized applications including paper for the separation of amino acids and peptides and ion-exchange for the analysis of mixtures of metal ions. Gel chromatography is based on the principle of the molecular sieve, which retains smaller molecules and excludes larger molecules which pass rapidly through the system. For our purpose, it is necessary to concentrate on those methods of most use for the analysis of pollution problems.

**4.3
Thin-layer
chromatography**

TLC was first used in 1938 by Izmailov and Shraiber. It is similar in application to paper chromatography and has replaced it to a considerable extent. The adsorbent or stationary phase is spread evenly on a plate of glass or a sheet of metal foil. For analytical purposes, the layer will be about 0.1 mm thick and will only accept a small point charge if the adsorbent is not to be saturated. For chromatography of larger amounts, layers 1–2 mm thick can be used and they can be charged as a basal streak at the rate of about 1 mg/cm length.

The plate is usually developed by standing it in a reservoir containing a small quantity of solvent or mixture of solvents (Fig. 4.1)

The tank must be covered to ensure that the space around the plate is saturated with vapour and under these conditions the solvent will rise by capillary action to a height of about 15 cm. This will be complete in about 20 minutes and the speed and simplicity of the technique accounts for its widespread application. One important use is in monitoring the eluent from larger-scale column chromatograms and another is in evaluating solvents for use in HPLC.

The adsorbent most used is silica gel, but alumina, cellulose and ion-exchange resins can also be applied. The range of solvents is extensive and accounts in considerable part for the power of the method.

Detection.   Unlike paper this presents no problem. Colourless substances of low reactivity can be visualized on silica and alumina by spraying the developed plate with a 10% solution of concentrated sulphuric acid in ethanol and leaving in an oven until oxidized traces appear. Other strong oxidizing agents such as permanganate can be used. Most often prepared

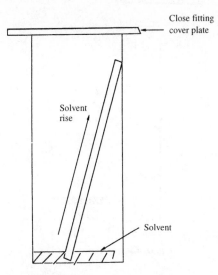

Close fitting
cover plate

Solvent
rise

Solvent

Figure 4.1 Typical tank for TLC.

plates are used which include fluorescent markers and they reveal the eluted
spots on placing under a UV lamp. The spots appear as darker zones against
a green (254 nm) or purple (366 nm) background.

4.3.1 Separation of pesticides by TLC

Figure 4.2 shows the end result of eluting a mixture of five organochlorine
pesticides on silica gel (Stahl, 1969). The solvent chosen was cyclohexane
80% : chloroform 20%. A suitable alternative would have been the use of *n*-
hexane with alumina as the stationary phase. The spots show up clearly with
a fluorescent marker and may be identified by running a wider plate on
which the individual reference compounds are spotted out separately along
the base. Alternatively, the spots can be scraped off and extracted with a hot
solvent and the extract identified by mass spectrometry. The technique
works well for this group of complex individual compounds; only for
Dieldrin and methoxychlor (Section 6.4.5) is separation incomplete.

4.3.2 Separation of metal cations by TLC

Figure 4.3 is a block diagram which represents the movement of a group of
heavy metal ions dissolved as their nitrates using *n*-butanol containing 1.5 M
hydrochloric acid (15%) as solvent and acetylacetone (0.5%) for complexation. The record shows the mixture running on the right and individual ions

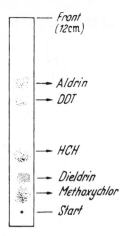

Figure 4.2   TLC of pesticides.

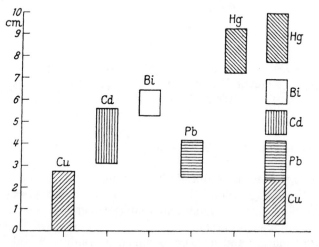

Figure 4.3   TLC of heavy metals.

which were spotted on for identification by reference. After development, the spots are detected by drying, exposure to NH_3 and then to H_2S. The metals are revealed as their black or brown sulphides, save for cadmium, which is yellow. Metal anions, for example CrO_4^{-2} and AsO_3^{-3}, can also be run on TLC with eluents such as acetone : water or alcohol : water and after extraction from the plate their identity may be confirmed by AA spectroscopy (Section 4.6).

4.4
Gas–liquid
chromatography   For the environmental scientist GLC is of first importance (Braithwaite, 1990a). It was originally developed in 1952 by Martin and James, and as the

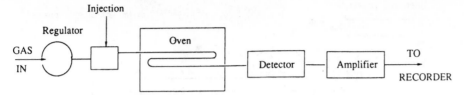

Figure 4.4   Diagram of a GLC column and essential hardware.

name implies it depends on partition between a flowing gaseous phase and a stationary liquid phase. Although more sophisticated than TLC, the same provisions must be met if the column is to work efficiently.

Figure 4.4 shows a glass column up to 4 m long with an internal diameter (i.d.) of 2–4 mm. It is usually in the form of a spiral so that it will fit conveniently within the oven. The entry port is adapted to connect a flow of inert gas, nitrogen or, better, helium, which has passed through a flow control, and a sieve to remove traces of moisture. The entry port also includes a re-sealing silicone rubber septum cap, so that the sample may be injected through it into the material at the beginning of the column; this ensures that the analyte is initially concentrated as the narrow band essential for efficient separation.

The molecular mass of the analyte is restricted to about 500 Da and if it is of low volatility it must be run at a temperature high enough to ensure that it has a significant vapour pressure. The containing oven includes a fan to give as even a temperature distribution as possible. For practical purposes, the maximum operating temperature is about 260°C and the injection port must be maintained 50°C above this to prevent condensation of the sample.

Clearly attention must be paid to the nature of the analyte. Volatile hydrocarbons are readily examined, but polar substances like alcohols, amines and phenols may require derivatization. Carboxylic acids must be esterified before the analysis. In this context silylation is important; here a polar substance is converted to a silyl ether derivative by reaction in a volatile solvent with trimethylchlorosilane (Equation 4.1) or hexamethyldisilazane (Equation 4.2):

$$Ph-OH \; + \; Cl-SiMe_3 \; \rightarrow \; Ph-O-SiMe_3 \qquad + \; HCl$$

$$\text{phenol} \qquad \text{trimethylsilyl} \quad \text{phenyl trimethylsilyl} \qquad\qquad (4.1)$$

$$\qquad\qquad \text{chloride} \qquad\quad \text{ether}$$

$$2 \, Ph-NH_2 + Me_3Si-N{=}N-SiMe_3 \rightarrow 2 \, PhNH-SiMe_3 + N_2 \qquad (4.2)$$

$$\text{aniline} \qquad\qquad\qquad\qquad\qquad\quad \text{trimethylsilyl aniline}$$

These reactions take only a few minutes for completion at room temperature; the solvent is then evaporated in a stream of nitrogen and

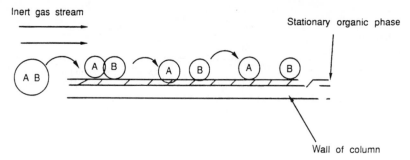

Figure 4.5 Diagrammatic impression of the progress of separation.

replaced with that chosen for injection onto the column. Silylation is also an essential pretreatment for the column before it is packed to ensure that 'hot' spots are blocked and do not spoil the separation.

Packed columns require a solid particulate support that is inert and has a large surface area/unit volume. Suitable materials may be firebrick derived, such as Chromosorb P or diatomaceous substances derived from filter aids. They need to be acid-washed and silylated before the stationary phase is applied to them as a dilute solution in a volatile solvent. After removal of the solvent, the surface saturated material is packed into the column; Chromosorb W has pores of $8\,\mu m$ diameter, a surface of $1\,m^2/g$ and it can be packed at a density of $0.24\,g/cm^3$.

Figure 4.5 shows how, as the gas flow sweeps the charge along the column, a succession of equilibria is established between analyte in the stationary phase and that in the gaseous phase. Given that the components have different affinities for the stationary phase, those with the greater affinity will lag behind and eventually separate so that a volume of pure carrier gas lies between them. The process bears some relation to fractional distillation, whose efficiency was measured in theoretical plates, that is the number of points in the column where equilibrium was established between ascending vapour and the condensed liquid.

The enhanced efficiency of GLC arises because the stationary phase can be chosen so as to bias the partition in favour of particular compounds. Here use is made of the activity term, which will be close to unity when the substrate is similar to the stationary phase but will diverge if the substance differs in type. This is true for differences in polarity, and Table 4.1 gives examples. With this low-polarity hydrocarbon as the stationary phase, the alkanes are eluted in the order of their boiling points, as shown down the table. However, considerable divergence is seen for the polar carbonyl compound and still more for the hydrogen-bonded propanol.

An example of how this may be put to practical use is in the separation of benzene (b.p. 80.1°C) from cyclohexane (b.p. 81.4°C). This is a near impossibility by fractionation but is straightforward by GLC with the high-

Table 4.1   Activity coefficients in n-hexadecane ($C_{16}H_{32}$) at 30°C

Compound	Coefficient	b.p. (°C)
1. n-Pentane	0.88	36
2. n-Hexane	0.89	69
3. Cyclohexane	0.77	81
4. Propionaldehyde (C_2H_5CHO)	4.0	49
5. Acetone (CH_3COCH_3)	6.3	56
6. n-Propanol	31.5	98

boiling ester dinonyl phthalate (**1**) since the common aromatic structure favours benzene and hence its activity towards this phase diverges from that of cyclohexane.

Relative retention.   In the above example, the retention volume for benzene (V_a) is the volume of carrier gas which passes before benzene is eluted from the head of the column. Similarly the retention volume of cyclohexane is V_b. In practice the ratio $V_a/V_b = 1.6$, very different from the b.p. ratio, clearly shows how benzene is retained by the phase with a common structural element. The nonyl phthalate would evidently be useful for the separation of carboxylic esters.

Polar substances such as alcohols and amines are best chromatographed on polar phases such as polyethylene glycol (**2**) and ethylene glycol succinate (**3**).

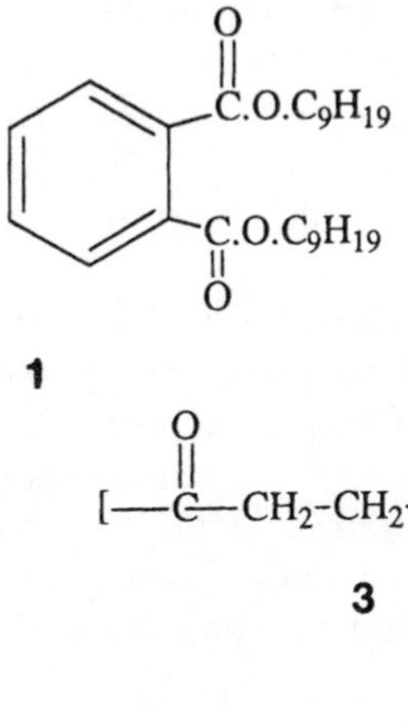

1

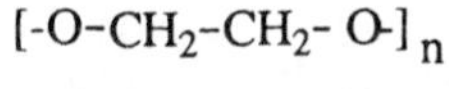

$$[-O-CH_2-CH_2-O-]_n$$

2

$$[-\overset{\overset{\displaystyle O}{\|}}{C}-CH_2-CH_2-\overset{\overset{\displaystyle O}{\|}}{C}-O-CH_2-CH_2-O-]_n$$

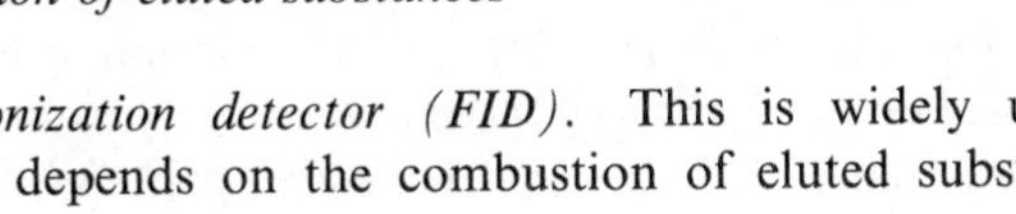

3

4.4.1   Detection of eluted substances

The flame ionization detector (FID).   This is widely used, extremely sensitive and depends on the combustion of eluted substances within a

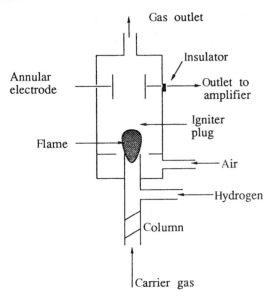

Figure 4.6   Diagram of a flame ionization detector.

small flame of hydrogen burning in an air stream at the head of the column (Fig. 4.6).

When eluted carbon compounds burn, free radicals are released which combine with oxygen, releasing electrons to produce an ion current. This is proportional to the amount of eluant and may be amplified and used to drive a pen recorder. To be significant, the visual response has to be at least twice that of recorder background noise. The FID is extremely sensitive and responds to a concentration of one picogram $(10^{-12}\,\text{g})/\text{ml}$ of carrier gas; the limit of detection is about a microgram. This will vary with the analyte: hydrocarbons respond more strongly than alcohols and other compounds with electronegative substituents.

The response varies with individual compounds and if a quantitative result is needed a calibration plot must be made setting out detector response over a range of concentrations of each component. Alternatively the chromatogram can be re-run after 'spiking' the sample with a known sample and comparing the peak areas obtained in the two instances. A superior method is the introduction of a known quantity of substance similar to the analyte to act as an internal standard and this is illustrated in the example of Fig. 4.30 (p. 122).

The electron capture detector (ECD).   The sensitivity of the FID falls away for compounds like the PCBs (Section 6.4.10) containing electronegative atoms because they do not burn well (note that the polybromobiphenyls are used as fire retardants). The FID response is also poor for polluting gases

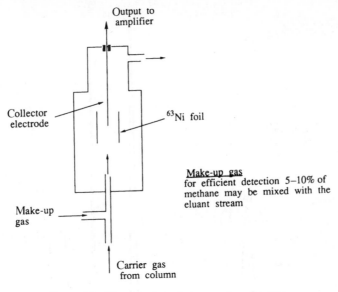

Figure 4.7   Diagram of an electron capture detector.

such as CO, CO_2 and SO_2. Very fortunately, it is just this type of electro-negative character which enhances the response of the ECD (Fig. 4.7).

The detector is set up with a steady electron current flowing from a radioactive source, usually a piece of ^{63}Ni foil. When molecules with an affinity for electrons enter the detector space, some are captured and the current falls and this imbalance can be amplified and recorded.

The sensitivity of the ECD can be two orders of magnitude better than the FID and in use care must be taken not to inject excessive amounts of the analyte because this will overload the detector. If a solvent such as chloroform or carbon tetrachloride is used to extract the sample, it must be completely removed in a stream of inert gas (nitrogen) and replaced by a hydrocarbon before making the injection. The response of the ECD to chloroform and carbon tetrachloride is about 10^4 times that of a mono-chloroalkane.

4.4.2   Principal parameters

Figure 4.8 shows a typical pen recorded gas chromatogram of three components. It is clear that a satisfactory separation has been achieved, but note that the peak with the longest retention time (t_r) or with the largest elution volume (flow rate $\times$ t_r) is appreciably broadened. This is the direct result of the progress of partition between phases; each transfer, or each plate, broadens the sample distribution and as a result the peaks achieve the

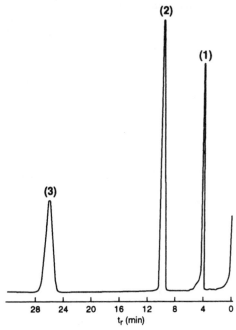

Figure 4.8 Typical pen-recorder gas chromatogram (Gopalan, 1977).

shape of a Gaussian curve. If the width at the base (w) results from n transfers (plates) then:

$$n = 16(t_r/w)^2$$

or for width at half-height

$$n = 5.54(t_r/w)^2$$

For a column of length L cm: $L/n =$ HETP, the height equivalent to a theoretical plate. For an efficient column this will be small and of the order of 0.05 cm or 2000 per metre.

When the analyte contains a range of substances of varying polarity and molecular weight some will inevitably be broadened by long retention. In these circumstances, the record can be improved by temperature programming. Here the chromatogram is run at a temperature consistent with the efficient elution of the more volatile components and thereafter the oven is programmed for a temperature rise at some 2–10°C/min until it reaches an upper temperature range suitable for the separation of the less volatile components. The retention time of the latter and the spread is thereby reduced, as can be seen in Fig. 4.9. This shows the separation of three alkylbenzenes and 17 n-alkanes; their thermal stability permits running to a high maximum temperature.

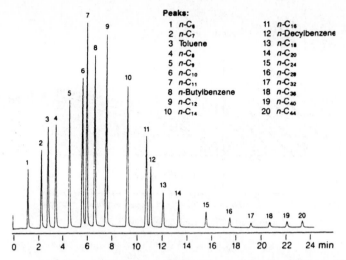

Peaks:

1	$n\text{-}C_6$	11	$n\text{-}C_{16}$
2	$n\text{-}C_7$	12	$n\text{-Decylbenzene}$
3	Toluene	13	$n\text{-}C_{18}$
4	$n\text{-}C_8$	14	$n\text{-}C_{20}$
5	$n\text{-}C_9$	15	$n\text{-}C_{24}$
6	$n\text{-}C_{10}$	16	$n\text{-}C_{28}$
7	$n\text{-}C_{11}$	17	$n\text{-}C_{32}$
8	$n\text{-Butylbenzene}$	18	$n\text{-}C_{36}$
9	$n\text{-}C_{12}$	19	$n\text{-}C_{40}$
10	$n\text{-}C_{14}$	20	$n\text{-}C_{44}$

Figure 4.9 Temperature programmed separation of a hydrocarbon mixture. 50.8 mm × 0.3 mm Me silicone on Chromosorb column, programmed from − 25°C at 15°C/min to 350°C. Carrier gas He at 30 ml/min with FID detection. Total neat sample 0.1 μl (approximately 80 μg) (SUPELCO, 1990).

4.4.3 Optimum operating conditions

The factors for optimum operating conditions are defined within the Van Deemter equation:

$$\text{HETP}(H) = A + B/\bar{u} + C \cdot \bar{u} \qquad (4.3)$$

where $\bar{u} =$ the mean velocity of the mobile phase; $A =$ the component of eddy diffusion; $B =$ the component of longitudinal diffusion; and $C =$ the component of mass transfer.

For efficiency these variables must be chosen so as to minimize H and this introduces an element of compromise.

Eddy diffusion (A).   This is a constant factor which arises from the range of path lengths which the solute molecules can take around the particles of the stationary phase, including some movement into the pores. The best remedy is the use of particles of regular shape and as small as possible, consistent with pressure being available to maintain the flow through them.

Longitudinal diffusion (B).   Varies with the flow rate; when this is low there is every opportunity for the solute to spread through the gaseous phase from points of high concentration to those of lower concentration. This natural redistribution runs counter to the progress of partition and it can be seen from Equation 4.3 that it varies inversely with flow rate.

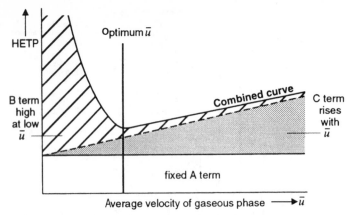

Figure 4.10   The variation in H with flow rate.

Mass transfer effect (C).   This is concerned with the resistance to transfer of material from the liquid to the flowing gaseous phase. This can only reach a true equilibrium if the phases are stationary, an unrealizable condition although low flows are favourable. With rising flow rate, the two phases get further out of step and the $C \cdot \bar{u}$ terms get progressively larger; hence it is clear that to minimize the combined effect of $B/\bar{u}$ and $C \cdot \bar{u}$ an intermediate rate must be established. Figure 4.10 shows the summation of the three factors and the minimum in the curve corresponds to the best compromise. In practice, the flow is set slightly above this because even a small fallback leads to a steep rise in H.

4.4.4   Capillary columns for GLC

These were first developed by Golay in 1958 and they now represent the most effective columns available. They are of steel or flexible fused silica with an i.d. of 0.2–0.5 mm and length in the range of 10–50 m. They may be wall-coated open tubular (WCOT), which are made by forcing a solution of the stationary phase slowly through while gradually raising the temperature, or support-coated open tubular (SCOT), which include a fine layer of support material and can accept rather more charge than the WCOT, although this is still limited to about 10 µg.

The HETP of capillary columns is about 0.03 mm or 30×10^3/m, which is about 10 times that obtainable with packed columns. Capillary columns have other advantages including:

1. neglible broadening by eddy diffusion
2. shorter elution times than packed columns
3. narrow peaks, which ensure a low detection limit

4. low back pressure because of the open end and hence very long columns can be used.

4.4.5   Analysis of urban air pollution

Even allowing for the sensitivity of capillary GLC, the direct measurement of all pollutant levels in air is not possible and so pre-concentration is carried out by pumping a measured volume of air through an adsorptive trap. This will be a column packed with a material such as Tenax (**4**), a polymer of 2,6-diphenyl-*p*-phenylene oxide:

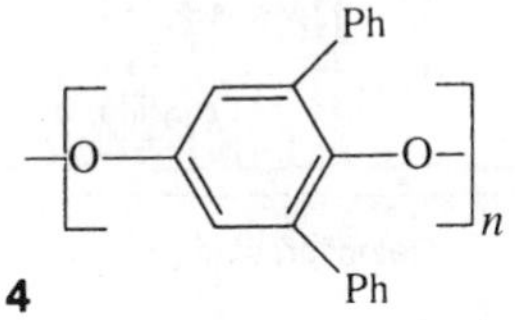

4

The procedure for desorption is critical and quite complex equipment has been engineered for the purpose. In outline, the Tenax column is placed in a preheated oven and the analyte swept out in an inert gas stream into a cooled trap. This trap is then placed in the input line to the GLC column and heated for a few seconds only. This time restriction is necessary because the charge must be as concentrated as possible and prolonged tailing must be avoided, although this means that injection will be incomplete. Figure 4.11 shows the recorded pollutants in air at Bilthoven in May 1990. Notice the intense peaks (1, 2, 3, 6) arising from the most volatile compounds and the significant peaks from the toxic components benzene (27) and hexane (24). Prolonged sampling of the air may not be possible because the most intense peaks would overload the capillary column. If a quantitative result is required, the injection procedure will have to be calibrated by loading known amounts of each component in turn in the Tenax column and evaluating the proportion discharged through the GLC column.

4.4.6   Detection by mass spectrometry

Mass spectrometry is not only a very sensitive recorder of peaks as they are eluted from the column but it can also identify the component (Suelter and Watson, 1991).

In its basic construction, the mass spectrometer focuses ions produced by bombardment of the parent molecules with a high-energy beam of electrons (*c.* 70 eV.) A fruitful collision:

$$M + e \rightarrow M_+^{\cdot} + 2e \tag{4.4}$$

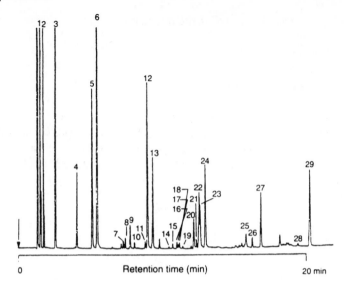

Peaks:

1.	ethane	16.	2-methyl-2-butene
2.	ethene	17.	1-pentene
3.	propane	18.	2-methyl-1-butene
4.	propene	19.	*cis*-2-pentene
5.	isobutane	20.	methylcyclopentane
6.	*n*-butane	21.	cyclohexane
7.	*trans*-2-butene	22.	2-methylpentane
8.	1-butene	23.	3-methylpentane
9.	isobutene	24.	hexane
10.	*cis*-2-butene	25.	2/3-methylhexane
11.	cyclopentane	26.	*n*-heptane
12.	2-methylbutane	27.	benzene
13.	*n*-pentane	28.	*n*-octane
14.	3-methyl-1-butene	29.	toluene
15.	*trans*-2-pentene		

Figure 4.11 Hydrocarbons in smog at Bilthoven. 50 m × 0.5 mm fused silica capillary column, temperature programmed at 40°C initially and then at 10°C/min to 200°C. Air sample 0.41 and FID detector (CHROMPAK, 1992).

dislodges an electron leaving a molecular ion (M_+^+), which can be focused on a slit because the radius (r) of the path it follows within a magnetic field is a function (Equation 4.5) of its mass, the field strength (B) and the accelerating voltage (V):

$$M/e = B^2 r^2 / 2V \qquad (4.5)$$

So by holding either variable B or V constant and varying the other the mass spectrum can be scanned; in practice it is usual to scan magnetically. Figure 4.12 is a diagrammatic representation. The ion source is maintained at a high vacuum to prevent unproductive collisions with air. Following a productive collision between a molecule of eluent and the electron stream

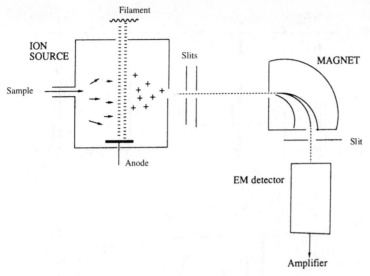

Figure 4.12 Single-focusing mass spectrometer.

from the filament, the molecular ion moves under the accelerator potential. After focusing it passes the collector slit and the ion current is enhanced by an electron multiplier when it may be recorded.

The highly energized ions live very short lives and undergo fragmentation within the mass spectrometer, although in the great majority of examples some of the molecular ion survives. The daughter peaks which result from fragmentation are typical of the individual mother ion and can be used to identify it. Figure 4.13 shows the principal daughter peaks formed by four organohalogen compounds, the ion current of each is related to the strongest as 100%. Since chlorine has two isotopes ^{35}Cl and ^{37}Cl, in a relative abundance of 3:1 (Table 4.2), ethyl chloride has a molecular ion of mass 64 (with ^{35}Cl) and one of one-third the intensity of mass 66 (with ^{37}Cl). In this spectrum, a peak is formed by loss of a methyl radical (mass 15) to leave a daughter ion of mass 49:

$$CH_3-CH_2-Cl_{\dot{+}} \rightarrow {}^{\bullet}CH_3 + {}^{+}CH_2-Cl$$
$$\text{lost} \qquad \text{collected}$$

(4.6)

Only the charged species will be collected.

In the spectrum of vinyl chloride, the twin molecular ion peaks and a major peak [M-35(37)] at 27 are seen. For trichloroethylene, the record is more complex, as the molecular ion has four components including a peak of low intensity (136) in which all three of the less common ^{37}Cl atoms have survived. In theory, a compound with n chlorine atoms will give rise to $(n + 1)$ peaks, but for polychlorocompounds like DDT and the PCBs, the

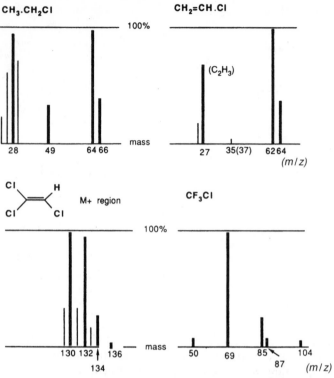

Figure 4.13   The mass spectra of some organohalogen compounds.

Table 4.2   The accurate masses of some common elememts

^{12}Carbon	12.00000	^{13}Carbon	13.00335
^{1}Hydrogen	1.00782	^{16}Oxygen	15.99491
^{35}Chlorine	34.96885	^{37}Chlorine	36.96590
^{19}Fluorine	18.99840	^{14}Nitrogen	14.00307

relative ion currents of the less probable fragments may be too weak to detect.

In the spectrum of CFC 12 the molecular ion is much less stable than the most abundant CF_3^+ ion and the peak from CF_3. ^{37}Cl was not detected. Fluorine has only one nuclide of atomic mass 19 and this is why it is included in compounds used as references for accurate mass measurements (p. 103).

The link between the GLC column and the spectrometer has been made in various ingenious ways to overcome the presentation of a sample in a gas stream to a detector operated under high vacuum. It is a common practice

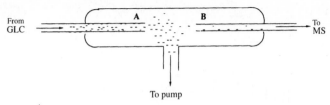

Figure 4.14   Glass jet separator.

to divide the eluent gas stream and pass one fraction into another detector. However, only a small proportion of the remainder could be directly injected into the mass spectrometer and this would seriously impair its sensitivity. A favoured way of overcoming the incompatibility in the gas pressures is the insertion of a glass jet separator (Fig. 4.14). The eluate from the GC, helium or nitrogen diffuses from the narrow orifice (A, *c.* 0.1 mm) and is pumped from the envelope much faster than the large molecules of the analyte. Hence a highly enriched gas stream continues into the mass spectrometer via the skimmer capillary (B, *c.* 0.2 mm).

Determination of accurate mass

If an electrostatic field is first applied to ions after they emerge from the source (Fig. 4.12), it acts as an energy focus. Subsequent magnetic focusing then gives much higher resolution and fragment ion masses can be recorded at better that one-thousandth of a unit. This measurement requires a scale and this is obtained by injecting a perfluoroalkane to provide peaks of standard mass. The whole spectrum is scanned comparatively, the controlling computer then subtracts the standard spectrum and that of the analyte is printed with peaks assigned to four significant decimal places.

Given an accurate mass of say 92.03928, the possible molecular formula can be read from tables, and this value corresponds to one of the isomeric chlorobutanes, C_4H_9Cl. Even allowing for experimental error, this mass is clearly differentiated *at high resolution* from the nearest alternative, which is glycerol, $C_3H_5O_3$ ($=92.04735$). Note that typical fragments are associated with each of these molecules, loss of ^{35}Cl and ^{37}Cl being unmistakable.

Another example of the value of an accurate mass would be that typical of a long-chain alcohol such as nonanol, $C_9H_{19}OH = 144.15142$. This differs significantly from the accurate mass of the naphthols, $C_{10}H_8O = 144.05752$. Here the nonanol spectrum includes a daughter ion [M-18] from loss of a water molecule, while naphthol differs in that it has a major fragment ion [M-28] from loss of CO.

The library of mass spectra includes a compendium of up to 10 major peaks of known compounds. With modern computerized data processing,

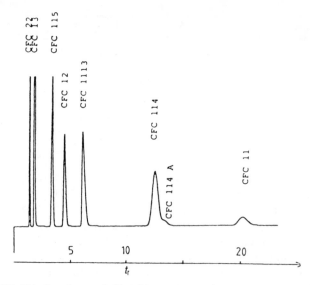

Figure 4.15 GC–MS of a mixture of chlorofluorocarbons. The t_r values in the spectrum were assigned from a parallel run using a WCOT column 25 m × 0.32 mm, temperature programmed from 30–200°C. The individual peaks were identified by a computerized comparison with a library of MS spectra (Rastogi, 1990).

those obtained from the analyte can be compared with these data and the best fit read out directly. An example of this in practice was the analysis of CFCs in aerosol cans (Fig. 4.15). The aerosol products were examined by spraying into the collection valve of a solvent-resistant Tedler bag. Gas samples of 0.1 ml were taken from the bag for analysis and 54 out of 448 were found to contain over 1% of CFC. A proportion of these were claimed to be 'safe to the environment'.

Chemical ionization and fast atom bombardment

When investigating the mixtures commonly found in environmental samples, the production of high-energy ions by the electron beam can be disadvantageous, the spectra have a multiplicity of peaks and are difficult to interpret. This problem can be surmounted by producing the molecular ion by chemical means.

In chemical ionization methane is ionized by electron impact to produce a primary ion (**5**) which transfers hydrogen on collision to give the secondary ion (**6**):

$$CH_4 + e \longrightarrow CH_4^{\cdot+} \ (\mathbf{5})$$

$$CH_4^{\cdot+} + CH_4 \longrightarrow CH_3 + CH_5^{+} \ (\mathbf{6}) \qquad (4.7)$$

CH_5^+ is then directed to the sample, which becomes ionized with hydrogen transfer:

$$Sample\ M + CH_5^+ \longrightarrow Sample\ (M + H)^+ + CH_4 \qquad (4.8)$$

These sample ions carrying one additional mass unit behave similarly to those produced by electron impact but undergo limited fragmentation. The spectra include more intense molecular ions and few, if any, daughter peaks.

Fast atom bombardment (FAB) is carried out with high-energy atoms of an inert gas. For example, these are obtained by first producing ions $A^{\bullet +}$ and accelerating the argon ions through an electric field before exchanging the charge:

$$A^{\bullet +}(fast) + A(slow) \longrightarrow A(fast) + A^{\bullet +}(slow) \qquad (4.9)$$

Most of the kinetic energy here is in the fast atoms, which are directed to the sample dispersed in glycerol, or a polymeric glycol, on a plate at ambient temperature. This also gives rise to $(M + H)^+$ ions, in addition to those formed from the glycerol.

Reverse geometry mass spectrometry (MS/MS)

MS/MS experiments are conducted using a reverse-geometry instrument (Fig. 4.16) in which magnetic focusing precedes electrostatic or energy focusing. This reverse geometry instrument will perform in the same way as conventional instruments and has additional advantages, notably the magnetic field (B, Equation 4.5) and accelerating voltage (V) can be set so as to pass a specific ion (often the intense molecular ion from chemical ionization).

The separated ion (M_2) now enters a collision chamber placed between the magnetic and electrostatic ionizers. This chamber is filled with argon at a

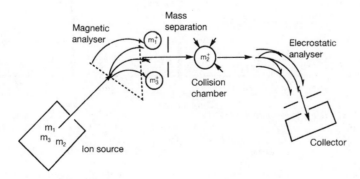

Figure 4.16   Reverse geometry mass spectrometer.

pressure in the range of $10^3 - 10^2$ Torr and glancing collisions with the ion beam lead to loss of typical fragments. As the ion is no longer undergoing acceleration, its kinetic energy is shared between the fragment (x) and the daughter ion formed $(M_2 - x)$.

As a numerical illustration, consider n-octane (C_8H_{18}, $M = 114$) whose normal spectrum includes the ion of mass 85, formed by loss of $CH_3 - CH_2^{\bullet}$. In the MS/MS experiment, this daughter ion will have $85/114 = 74\%$ of the energy of the parent ion. With further fragmentation, other daughters will be formed with successively reduced kinetic energies. One is, therefore, now viewing an energy spectrum, rather than a mass spectrum, and the daughter ions will all be collected, provided that the electric sector voltage is scanned downward from the value (E_0) required to focus the parent ion. In this way one obtains a spectrum of an individual ion free from chance coincidence of ions originating from other components of the mixture.

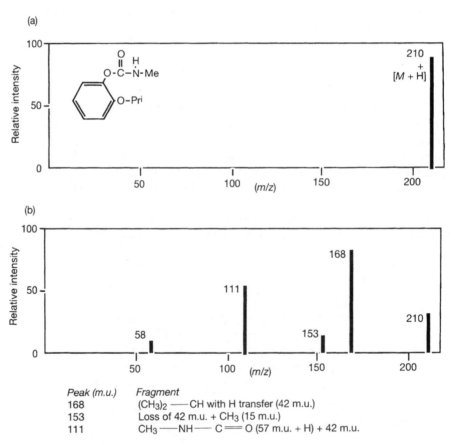

Peak (m.u.)	Fragment
168	$(CH_3)_2$ ——CH with H transfer (42 m.u.)
153	Loss of 42 m.u. + CH_3 (15 m.u.)
111	CH_3 ——NH—— C══O (57 m.u. + H) + 42 m.u.

Figure 4.17 Spectrum of Propoxur. (a) Normal chemical ionization spectrum. (b) MS–MS spectrum.

Figure 4.17 shows the lower MS/MS spectrum obtained from the pesticide Propoxur, which is used as a greenhouse fumigant against whitefly and aphids and as a domestic insecticide against flies and mosquitoes. In the normal MS spectrum (Fig. 4.17a) only the parent ion ($M = 209 + 1$) would be detectable, but in the MS/MS spectrum (Fig. 4.17b) the energetically separated daughter peaks can be seen. The relatively weak 58 m.u. peak arises from the alternative cleavage of the molecular ion (Equation 4.6) to generate the $CH_3-NH-C=O^+$ ion.

Peak (m.u)	Fragment
168	$(CH_3)_2-CH$ with H transfer (42 m.u.)
153	Loss of 42 m.u. + CH_3 (15 m.u.)
111	$CH_3-NH-C = O$ (57 m.u. + H) + 42 m.u.

We see that in this application the mass spectrometer has been used both to separate and to analyse the sample. It is also applicable to electron ionization MS and more recently, its use has been extended to the analysis of eluents from HPLC columns (Section 4.5.2). Electrically heated thermospray interfaces are capable of evaporating flows of up to 2 ml/min and maintaining an MS pressure of 10^{-5} torr (Voyksner, 1994).

Direct air sampling

Mobile equipment is available (Lane, 1982) which samples air directly at a rate of 1 l/s and so gives the results in real time. For this application, the ions are produced by chemical ionization rather than by electron impact. Here ions are produced by electron bombardment of air components and their charge is transferred to the substrate (M) by charge transfer, typically from hydrated protons:

$$H_3O^+ + M \longrightarrow H.M^+ + H_2O \qquad (4.10)$$

This technique is well suited to the production of simple spectra in which molecular ions tend to survive because the activation is less severe. An alternative to magnetic guidance is used whereby the ions $H.M^+$ pass into a quadrupole device, with two pairs of electrodes to which one direct and one alternating potential is applied. These may be varied so that only the ion of required mass can maintain a stable path to the slit, others are lost either by collision with an electrode or by ejection from the analyser. Figure 4.18 shows isopleths, or pollution contours, collected with the equipment on a van which was driven around the neighbourhood of an industrial source. It shows an intense source of the selected ion of benzthiazole at a chimney (over 106 ppb) grading downward towards a general dispersion at 2 ppb. Clearly the application of this type of mobile analysis is a powerful way of resolving legal disputes about pollution sources.

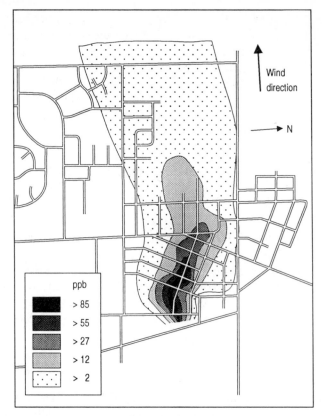

Figure 4.18   Pollution levels in an industrial plume (reprinted with permission from Lane, D. A., *Environ. Sci. Technol.*, 16, 45A, © 1982 American Chemical Society).

The Joint Research Centre of the EC employs mobile monitors of this type to determine air pollution in cities (Section 6.2.3). They have also come into prominence in the UK because industrial air monitoring of factories and process emissions has become a vital part of air quality management programmes following full enforcement of the 1990 Pollution Act in 1996.

Figure 4.19 shows a sample probe and the seven main analysers of such a system (Symonds Travers Morgan). The probe will sample a gas stream in the temperature range 30–1200°C. When the gas streams are below 300°C, or when the stack wall is thicker than 5 cm, the stem of the probe may be heated to prevent corrosion by condensate. The location in the chimney is crucial because of swirling of the gases and of stratification caused by bends. The sampling lines must also be maintained at 180–190°C, well above the dewpoints of sulphuric and nitric acids. Any condensation formed would act as a solvent for soluble gas components, which would not then reach the analysers. The transport lines are made from PTFE to minimize absorption and reaction effects.

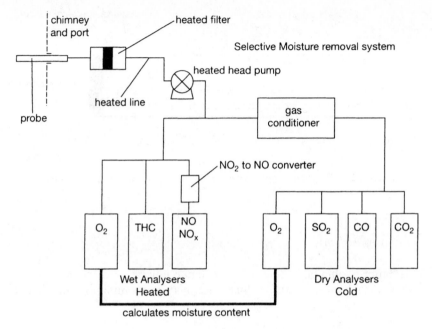

Figure 4.19 A continuous emission monitoring (*CEM*) system.

All particulates must be removed before the sample enters the analysers. Those shown (Fig. 4.19) are in two banks, the left-hand side runs hot to prevent condensation, while water is removed from the right-hand side by a chiller unit. If data need to be reported on a dry basis, the volume of this condensed sample can then be deducted from the sample volume. In the UK, different processes require the samples to be expressed under a variety of standard conditions. For incinerators, data are given at 273 K, 101.3 kPa, 11% O_2 on a dry gas basis. Non-combustion sources are usually given at 272 K, 101.3 kPa on a wet gas basis.

4.4.7   Analysis of common pollutants

The oxygen analyser

The oxygen analyser (Fig. 4.20) depends on the mobility of oxygen ions in zirconia (ZrO_2) in the temperature range 600–1200°C. Consequently, if the sensor is exposed to the sample gas on one face and to a reference stream of dry air on the other, a potential difference is established which is a function of the difference in the oxygen concentrations. Platinum electrodes communicate this to a recording instrument.

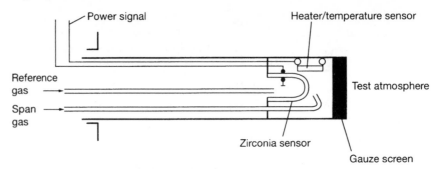

Figure 4.20   The zirconia potentiometric oxygen sensor (Hartmann and Braun).

Total hydrocarbons

This figure is obtained using a heated flame ionization detector (Fig. 4.6) with a 40% hydrogen/60% helium fuel mixture to minimize any oxygen synergism effects.

Oxides of nitrogen

NO is determined by its chemiluminescent reaction with ozone:

$$NO + O_3 \longrightarrow NO_2^* + O_2$$
$$\downarrow$$
$$NO_2 + \text{a photon} \tag{4.11}$$

About 10% of the NO_2 produced in this reaction is electronically excited and on returning to the ground state emits light in the wavelength range 590–2600 nm. A photomultiplier tube converts this to a signal which is proportional to the flow rate of the NO through the reaction chamber.

A figure for NO_2 is obtained using the converter. The gas is passed through a stainless steel tube at 600°C, when the dioxide is reduced to NO. The total NO in the gas stream is then analysed as above, and the NO_2 component obtained by difference.

The ozone analyser

Although not shown in Fig. 4.19, ozone analysis will be required in studies of air pollution. It also depends on signals to a photomultiplier from a chemiluminescent reaction, namely that between ethylene and ozone:

$$n - CH_2{=}CH_2 + O_3 \longrightarrow 2HCHO^* \text{ and other products} \tag{4.12}$$

The reagents are mixed with air and steadily supplied to the reaction chamber with ethylene in a large excess, the light emitted is then proportional only to the concentration of the minor component, the ozone.

Sulphur dioxide, carbon dioxide and carbon monoxide

All three gases, SO_2, CO_2 and CO, are determined using non-dispersive infrared (NDIR) (Fig. 4.21). In contrast to dispersive IR, when a broad band of light energy is used to plot a spectrum (p. 169), a wavelength which is strongly absorbed by the analyte gas is preselected using a filter or narrow band source. The radiation is split into two beams by the chopper and passed intermittently but simultaneously through both cells. The reference cell is filled with a *dry* non-absorbing gas while the *dry* analyte is passed through the other. The two compartments of the detector cell are filled with the gas under examination; when any IR-absorbing gas enters the sample cell, the detector goes out of balance and the linking diaphragm moves. This diaphragm is one plate of a condenser and so its capacitance changes, giving rise to an electrical impulse proportional to the concentration of the absorbing gas. Carbon dioxide has a suitable IR peak at 3000 cm^{-1}, SO_2 at 1330 cm^{-1} and CO at 2150 cm^{-1}.

Monitoring of a clinical waste incinerator

The CEM system described above has been used to evaluate a hospital incinerator (R. S. Brown of Symonds Travers Morgan, personal commu-

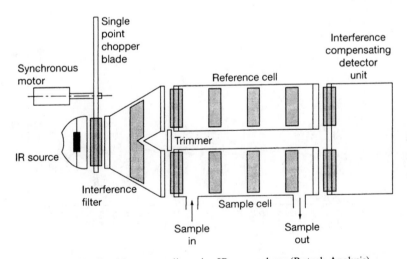

Figure 4.21 Dual-beam non-dispersive IR gas analyser (Rotork Analysis).

nication 1993). The data in Fig. 4.22 cover the same time period 10.00–16.60 hours and are only separated to avoid coincident plots. They represent a clinical waste incinerator operating under proper control. Figure 4.22a displays emissions of CO and THC, while Fig. 4.22b displays emissions of NO_x and SO_2. They show oxygen supply at about 15% and CO_2 at 5% (by

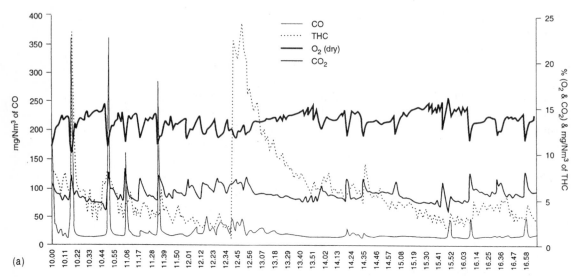

Figure 4.22a   Emissions data on 19.11.93 for a batch clinical waste incinerator (data expressed at 273 K, 101.3 kPa, 11% O_2 and dry gas).

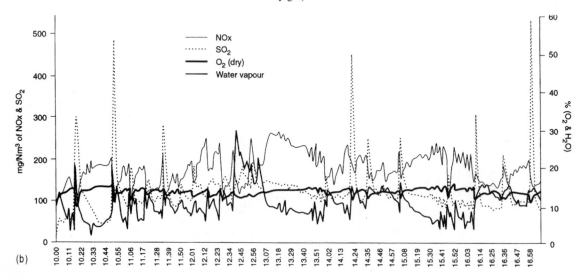

Figure 4.22b   Emissions data on 19.11.93 for a batch clinical waste incinerator (data expressed at 273 K, 101.3 kPa, 11% O_2 and dry gas). (By courtesy of Symonds Travers Morgan.)

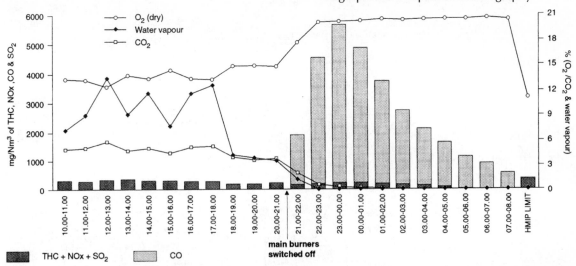

Figure 4.23 Hourly average emission data for a clinical waste incinerator (data expressed at 273 K, 101.3 kPa, 11% O_2 and dry gas, except for CO_2/O_2 or H_2O which are uncorrected for oxygen) on 19.11.93 and 20.11.93. (By courtesy of Symonds Travers Morgan.)

volume on right-hand scale); there are brief surges in emissions of THC and CO to 350 mg/Nm^3 (left-hand scale of Fig. 4.22a). Reference to Fig. 4.22b shows that SO_2 surges briefly to 300–500 mg/Nm^3 with NO_x emission largely in the range of 200–300 mg/Nm^3 (left-hand scale).

Figure 4.23 shows the pattern over an extended timescale, including 21.00 hours when the main burners were prematurely extinguished. This reveals inefficient performance since, with adequate oxygen supply, CO_2 fell to a very low level but CO rose to over 5000 mg/Nm^3 and remained well above the HMIP limit for 8 hours.

This example illustrates the relevance of continuous emission monitoring to integrated pollution control (Section 2.2). It has also been applied to the real-time analysis of air pollution on city streets (Table 6.10).

4.5
High pressure liquid chromatography

Classical liquid chromatography suffers from the disadvantage that it is slow in use and fractions are eluted in substantial volumes of solvent. This is the result of imprecise loading of the analyte and of the major eddy diffusion term arising from the large irregular particles of the stationary phase. It was realized early on that considerable improvement would result from efficient loading of columns of small uniform particles. However, such columns require solvent to be pumped at a stable pressure up to 6000 psi and pumps did not become available until the late 1960s. In the following years, HPLC grew in popularity and capability. A block diagram of a typical system is shown in Figure 4.24.

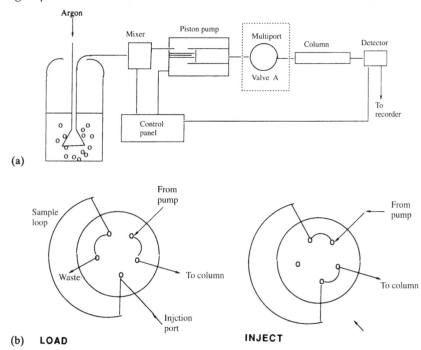

Figure 4.24   Typical components of an HPLC system (Braithwaite, 1985).

4.5.1   The components

Columns are made of steel to withstand the high pressure and are 10–50 cm long with i.d. of 4–6 mm, although short capillary columns with an i.d. of 1–2 mm have come increasingly into use for the analysis of small quantities. *Packing material* is almost always of spherical silica of 5–10 μm diameter and pores about 10 nm. Good results can be obtained using low-polarity solvents operated in the classical manner, but increasing use is made of the reverse phase mode. Here a low-polarity stationary phase is attached to the column material rather in the manner of the stationary phase of GLC. To prevent loss of the stationary phase, it is bonded to the silica by reaction of polar centres with a trialkylsilylchloride used in the manner described for masking in GLC (Fig. 4.25).

Solvents must be degassed. Figure 4.24 shows the bubbling of a pure inert gas through the solvent in the reservoir when traces of other dissolved gases diffuse into the stream and are swept out to air. Provision can be made for two or more solvents to be delivered to a mixer ahead of the column. The control can regulate the proportions and, if required, programme them by progressively enriching that of lower polarity. The plumbing must be rigorously airtight. A selection of solvents in common use is shown in

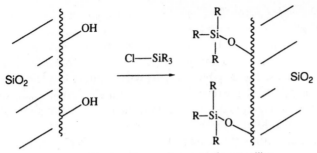

Figure 4.25 Attachment of reverse phase to silica.

Table 4.3 Some properties of common solvents for HPLC

Solvent	Viscosity (cp)	UV cutoff (nm)
Pentane	0.23	205
Cyclohexane	1.00	205
Toluene	0.59	285
Diethyl ether	0.24	220
Chloroform[a]	0.57	245
Ethyl acetate	0.45	260
Acetonitrile	0.36	210
Methanol	0.57	210

[a] Normally contains 0.5% ethanol.

Table 4.3 arranged with the weakest at the top. The viscosity is given since this bears upon back pressure and the UV cutoff because this limits UV detection.

The *pump* is a crucial and highly engineered component and several types are available (Ahuja, 1992). It is essential that the high pressure is delivered pulse-free in order to avoid unacceptable levels of baseline noise in the detector. For most purposes a pressure of about 1500 psi is adequate, but this will be exceeded for a high viscosity solvent.

Sample injection should be through a port (Fig. 4.24b) fitted with one of a range of loops into which the charge may be injected from a syringe. This is designed so that the *full* loop may be switched into the main stream from the pump and the contents delivered in a limited volume to the column. A good system will have a minimum of free volume between the port and the column head.

4.5.2   Detectors

There have been as many as 20 types of detector but most have specialized applications and comment is reserved here for those of widest application.

Refractive index (RI) detectors depend on dual light beams focused on two balanced photocells. One beam passes through a cell full of pure solvent while the other monitors the flow from the column; once eluant enters the latter the RI changes and the photocells go out of balance. This type of detector can be used widely but its sensitivity is limited to about 300 ng/ml.

UV detectors are of great importance although their use is restricted by solvent cutoff and confined to molecules which absorb UV above this point. In practice this limitation is not so serious because many polluting substances fall within this category. It is also often possible to form a derivative of the substrate which is UV absorptive. For example, aliphatic alcohols react readily with aromatic acid chlorides.

$$R\text{-OH} + \underset{Cl}{\overset{O}{\|}}\!\!\diagdown\!\!\langle\bigcirc\rangle\!\!-NO_2 \longrightarrow \underset{RO}{\overset{O}{\|}}\!\!\diagdown\!\!\langle\bigcirc\rangle\!\!-NO_2 + [HCl] \tag{4.13}$$

The sensitivity towards a solute molecule can be gauged from the molecular extinction coefficient (ε) at a maximum. The detector can be turned to respond to this narrow wavelength band and when absorption occurs in the eluent path the balance with a pure solvent passing is lost, in the manner of the RI detector.

The relation between incident (I_0) and emergent light is given by:

$$\log I_0/I = \varepsilon\, l \cdot c \tag{4.14}$$

where l = length of light path (cm) and c = molar conc./1. From this we see that ε is equivalent to the reciprocal of the concentration required to reduce the incident light to one-tenth. For a nitrobenzene standard $\varepsilon = 9000$ at 251 nm and if we assume that a 50 ng sample enters the detector with a 1 cm path in a volume of 0.2 ml, the change in through light is given by:

$$\log I_0/I = 9 \times 10^3 \times 50 \times 10^{-9}/M \times 0.2 \times 10^{-3}$$

with $M = 123$; $\log I_0/I = 0.018$ whence $I_0/I = 1.04$, which would cause a detectable 4% change.

The structures of a group of polycyclic aromatic hydrocarbons are shown in Fig. 6.6 (p. 268). They are original products of coal tar distillation and are present in diesel exhaust and tobacco smoke. They are the subject of concern because the group includes cancer-inducing substances. The activity of benzo-[a]-pyrene, the most toxic, is compared with that of the weakly active chrysene in Table 4.4. These compounds absorb UV light more intensely than nitrobenzene because of their extended conjugated systems and the UV detector is very sensitive to them (Fig. 4.26). These compounds and some others can be detected with still greater sensitivity by inducing their fluorescence with a laser light source.

Electrochemical detectors are extremely sensitive and respond to as little as 1 ng/ml. They depend for their action on the electro-oxidation or reduction of the eluent and the potential change which then occurs in the

Table 4.4   some polycyclic aromatic hydrocarbons

Benzo-[a]-pyrene	Strongly carcinogenic and mutagenic at 5 µg/plate Occurs at 2 µg/100 cigarettes Fuel oil at 10 mg/t London air 20 ng/m^3
Chrysene	Weak carcinogen Occurs at 6 µg/100 cigarettes Indoor air 4 ng/m^3

cell. Unless they are derivatized, ethers, alcohols and carboxylic acids are not sensed.

The *criteria for efficiency* are similar to those discussed for GLC, but it should be noted that eddy diffusion is limited by particle size and longitudinal diffusion is less important because it occurs so much more slowly in the flowing liquid phase than in the gas.

Speed is of the essence and consequent inefficiencies in mass transfer are more than compensated for by the wide range of solvents available, especially as they can be mixed and varied continuously throughout the elution. These advantages mean that separations can be achieved on a short column (*c.* 25 cm) that would require many metres of a GLC column. Comparisons are, however, odious where GLC and HPLC are concerned since both are essential for the analysis of the wide range of environmental samples, but some pros and cons are given in Table 4.5.

4.5.3   *Analysis of polluted air*

Phenols are trapped by bubbling air through a weakly alkaline solution containing a diazonium salt (**7**); the solution of diazotized phenol (Equation 4.15) was made up to a standard volume and a sample injected into the HPLC. Figure 4.27 shows the separation of those phenols which occur in industrial emissions, car exhausts and tobacco smoke.

$$\text{Ph-OH} + \text{O}_2\text{N}\text{---}\langle\text{---}\rangle\text{---N}\equiv\text{N}^+ \text{ BF}_4^- \longrightarrow \text{O}_2\text{N}\text{---}\langle\text{---}\rangle\text{---N}=\text{N---}\langle\text{---}\rangle\text{---NO}_2 \quad (4.15)$$

7

Amines in ambient air can be detected at concentrations as low as 0.1 µg/m^3 although volumes of the order of 100 l must be drawn through an adsorption column impregnated with phosphoric acid. Once the sample has been collected, the free amines are released by addition of alkali and then derivatized to ensure UV detection. One suitable method is conversion into the *m*-toluamides (**9**) by reaction with *m*-toluoyl chloride (**8**), as shown in

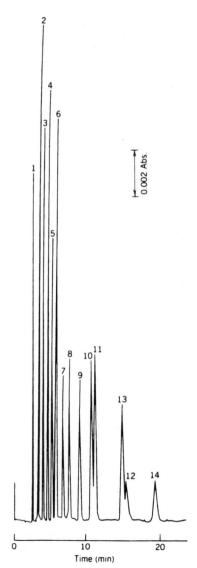

Figure 4.26   Separation of PAH on a microcolumn of silica ODS. Column 10 cm × 0.25 mm, flow rate 3 μl/min, detection at 254 nm. Sample: 1, benzene; 2, naphthalene; 3, biphenyl; 4, fluorene; 5, phenanthrene; 6, anthracene; 7, fluoranthrene; 8, pyrene; 9, *p*-terphenyl; 10, chrysene; 11, 9-phenylanthracene; 12, perylene; 13, 1,3,5-triphenylbenzene; 14, benzo-[a]-pyrene. Wavelength of UV detection: 254 nm. Injection volume: 0.02 μl (Takeuchi and Ishii, 1981; Ahuja, 1992).

Equation 4.16. Figure 4.28 shows typical responses of reference samples of those amines found in polluted air. The sensitivity may be gauged from peak 6, which corresponds to 48 pmol *n*-propylamine and peak 9 corresponding to *n*-butylamine (40 pmol). Isocratic elution, that is with a solvent of

Table 4.5 Comparative behaviour of GLC and HPLC

Feature	GLC	HPLC
Speed	Adequate	Superior
Involatiles	Fails	Acceptable
Metal ions	Inapplicable	Acceptable
Gases	Acceptable	Inapplicable
Temperature programming	Acceptable	Unnecessary
Flowing phase	Limited, inert	Extensive, active
Detection	Excellent	Depends on substance

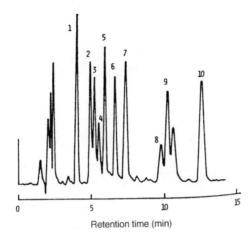

Figure 4.27 Separation of phenolic pollutants. 20 cm column in reverse phase, 85% MeOH: 15% water at 1.1 ml/min, 1500 psi, detection at 365 nm. Phenols 50–400 ng. 1, Phenol; 2, *m*-cresol; 3, *o*-cresol; 4, α-naphthol; 5, 3,5-xylenol; 6, 2,3- 2,5- and 2,6-xylenol; 7, *p*-cresol; 8, β-naphthol; 9, 3,5-xylenol; 10, 2,4-xylenol. (Reprinted with permission from Kuwata, K. *et al.*, *Anal. Chem.*, © 1980 American Chemical Society.)

constant composition, is quicker than gradient elution but here fails to separate ethylamine from dimethylamine (peaks 2 and 3).

$$R\text{-}NH_2 + \quad \underset{\textbf{8}}{\text{[8]}} \quad \xrightarrow{-\text{HCl}} \quad \underset{\textbf{9}}{\text{[9]}} \qquad (4.16)$$

This technique is capable of detecting pollutants from tobacco smoke. A typical value for methylamine is 1.6 µg/m^3.

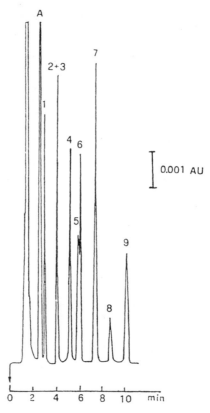

Figure 4.28 Separation of aliphatic amine air pollutants. A, NH$_3$ (excess); 1, methylamine (55 pmol); 2 and 3, ethylamine (61 pmol) + dimethylamine (40 pmol); 4, allylamine (53 pmol); 5, isopropylamine (47 pmol); 6, *n*-propylamine (48 pmol); 7, ethylenediamine (58 pmol); 8, diethylamine (38 pmol); 9, *n*-butylamine (40 pmol) (Possanzini and Di Palo, 1990).

4.5.4   Analysis of polluted water

HPLC columns can be used for trace enrichment of these samples. A measured volume of water is drawn through the column when the organic contaminants are taken up preferentially by the ODS layer. Once sufficient volume has been passed through the column, a graded change of solvent is made leading to elution of the analyte. In practice, it is best to concentrate the sample in a short pre-column, which is then connected to an analytical column by replacing one loop of the injector (Fig. 4.24). To concentrate the charge the pre-column is reversed so that the charge goes directly into the analytical column. Some diffusion of the sample is inevitable in this procedure and, as can be seen (Fig. 4.29), broad peaks are obtained. Note that the peak at 1.37 min marks the point in time when the analyte solvent

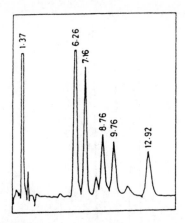

Figure 4.29 Separation of a pesticide mixture after trace enrichment. 5 cm pre-column of 10 μm ODS silica eluted with MeOH/water (3:1) through a 25 cm ODS analytical column at 2 ml/min with UV detection at 220 cm (Braithwaite and Smith, 1990b).

Table 4.6   Individual pesticides from trace enrichment

Compound	Peak (t_r)	UV max. (nm)	Content (μg)
Methoxychlor	One, 6.26	–	13
Dieldrin	Two, 7.16	216	23
Heptachlor	Three, 8.76	213	22
DDT	Four, 9.76	206	14
Aldrin	Five, 12.92	213	17

enters the detector cell. It can be seen from their unsaturated structures (Figs 6.13, 6.15; pp. 284, 286) that all five of the component pesticides absorb UV light in the region of 220 nm, but the maxima will not coincide at this wavelength (Table 4.6) and the molecular extinctions will differ; hence for quantification, each response must be compared with that of a reference standard.

4.5.5   Trace enrichment followed by GLC analysis

An example of this application is the detection of chlorophenols accidentally released from a sawmill into lake water. A sample (100 ml) taken downstream was injected through a plug of ODS silica, which was then reversed and eluted with acetone. Use of this powerful solvent ensures that most of the analyte is eluted in a volume of only 4 ml, and since UV is not used for detection there will be no interference.

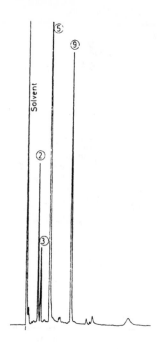

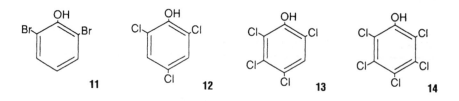

Compound	Peak	Concentration (μg / ml)
11 2,6-Dibromophenol	3	Internal standard
12 2,4,6-Trichlorophenol	2	0.50
13 2,3,4,6-Tetrachlorophenol	5	1.8
14 Pentachlorophenol	9	0.3

Figure 4.30 GLC of acetyl chlorophenols from polluted lake water. A 25 m × 0.2 mm SE-30 quartz capillary column was used at 200°C with a He flow rate of 0.6 ml/min. In order to avoid overloading, the injection was split so that only 1/20 entered the column, the rest going to waste (Renberg and Linstrom, 1981).

To ensure good resolution, the polar OH group in the phenols (e.g. **10**) was acetylated with acetic anhydride. Excess of this reagent was removed by hydrolysis and the o-acetates extracted into 1.0 ml of hexane. This procedure gives good resolution (Fig. 4.30) since the analyte is contained in a small volume.

$$\text{Ph—OH} + (Ac)_2O \longrightarrow \text{Ph—OAc} + AcOH \qquad (4.17)$$

10

Quantitative results were obtained by the addition of the same amount of 2,6-dibromophenol to all samples, when it acts as an internal standard (i.s.). Then if a substance (s) gives rise to a peak of area (A_m) and the standard to a peak of area (A_{im}) then the concentration (C_m) of the substance is given by:

$$C_m = A_m/A_{im} \times C_{im} \times R_{im}/R_m \qquad (4.18)$$

the ratio of the response factors R_{im}/R_m of the internal standard and each component of the analyte must be determined independently by plotting the relative peak areas for a series of mixtures of known composition (also p. 94).

4.6 Pollution by metals– atomic absorption spectroscopy

Attention will be concentrated on atomic absorption (AA) spectroscopy (Lajunen, 1992), which is the principal method of determining the levels of heavy metals in environmental samples.

4.6.1 Historical introduction

The technique had its beginnings in 1814 with Fraunhofer's observation of line spectra in the sun. In 1859, Kirchhoff and Bunsen established the key principle that 'Matter absorbs light at the same wavelength at which it emits light'.

Atomic absorption was first used by Woodson in 1939 for the analysis of mercury in air. In 1955, Walsh showed how metal ions could be reduced to the parent atoms within flames and that their concentration could then be found by the absorption of monochromatic incident light, although it was almost 10 years before a commercial instrument became available. Since then the use of AA spectrometers has expanded rapidly and with the introduction in 1970 of the plasma source there has been an increase in the related method of atomic emission spectroscopy (AES).

4.6.2   Basic theory of atomic absorption and emission

When a solution containing metal ions is introduced as fine droplets into the gas stream entering the burner (Fig. 4.33), the metal ions are reduced to neutral atoms by electron capture.

Figure 4.31 is a simplified diagram of the energy levels in the lithium atom, which has a pair of electrons in the 1s orbital of lowest energy and a single bonding electron in the higher 2s orbital. When the atom is excited by an input of energy, as experienced in the flame, the electron in the 2s orbital may enter an orbital of higher energy. The various levels are restricted by quantum theory and subject to Planck's equation (4.19):

$$\text{Energy} = h/\text{line wavelength} \tag{4.19}$$

where h is Planck's constant. The energy difference between the 2s and the 2p orbitals is almost 2 electron volts (eV) and this precise quantum will be absorbed from the flame to promote an electron to the upper level. We see from Fig. 4.31 that this corresponds to a line in the spectrum of wavelength 670.8 nm. The lines for three other transitions are shown corresponding to wavelengths of 323 ($2s \rightarrow 3p$), 812 ($2p \rightarrow 3s$) and 610 nm ($2p \rightarrow 3d$) so that the lower energy corresponds to the longer wavelength.

The frequencies on the right of Fig. 4.31 are in reciprocal centimetres and

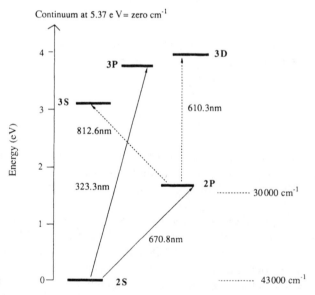

Figure 4.31   Energy levels for the Li atom. 1 unit on the voltage scale = 8000 cm^{-1}. A low-energy transition corresponds to a long wavelength. For the $2s \rightarrow 2p$ transition, wave number = 1/wavelength (cm^{-1}) = $10^7/670.8 = 14\,900$ cm^{-1}.

Table 4.7 Variation of excited to ground state atoms with temperature

	2000 K	3000 K	4000 K	5000 K
Sodium (589 nm)	1 in 10^5	6 in 10^4	4 in 10^3	2 in 10^2
Calcium (423 nm)	1 in 10^7	4 in 10^5	6 in 10^4	3 in 10^3

bear an inverse relation to the electron volt scale on the left. For the $2s \rightarrow 2p$ transition:

$$\text{frequency (cm}^{-1}) = 1/\text{wavelength (cm)} = 10^7/670.8 = 1.49 \times 10^4 \, \text{cm}^{-1}$$

The strongest line in the lithium spectrum is that arising from promotion from the lowest $2s$ level to the next highest $2p$ level. The intense red colour of the flame arises from emission at 670 nm when the excited electron falls back into the lower level. It follows from Kirchhoff and Bunsen's principle that we may evaluate the concentration of atoms in flames by the intensity of either absorption or emission at a chosen wavelength.

In practice, only a small proportion of atoms are excited at any one time, even for the most probable transition, and Table 4.7 shows how this rises with temperature for the easily excited golden yellow line of sodium (589 nm) and for the blue calcium line (423 nm). It follows that flame temperatures of at least 2000 K are needed if sufficient sensitivity is to be achieved and this is possible when acetylene (ethyne) burns in air. Acetylene/ oxygen flames exceed 3000 K and acetylene/N_2O reaches 3400 K. At high temperatures, an electron may be expelled altogether and this is a source of inefficiency as the ions are not recorded. An input of 5.37 eV is required to ionize lithium but potassium with an ionization potential of 4.34 eV gives ions more readily.

4.6.3   The Lambert–Beer law

This states that at a given wavelength the absorbance is proportional to the number of absorbing species in the light path. With absorbance defined as $\log I_0/I$, where I_0 = the energy emitted and I = the energy entering the detector, we have:

$$\log I_0/I = a \cdot l \cdot c \tag{4.20}$$

where a = a constant, l = length of light path and c = the required concentration. This relation is given in another form in Equation 4.14. Therefore, by plotting absorbance against a series of samples of known concentration a calibration curve can be produced (Fig. 4.32) from which any atomic concentration can be read, given a known detector response.

At the normally low excitation levels the law is usually obeyed and a linear plot results. However, at high concentrations, some light may pass

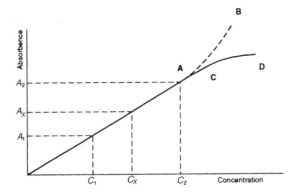

Figure 4.32   Typical calibration curve for AA spectroscopy.

through unabsorbed and the plot then turns along CD towards the concentration axis. If ionization is occurring, the plot turns along AB as ionization is reduced at higher concentrations. Provided it has been verified that the curve is linear at lower concentrations, an unknown concentration (C_x) can be obtained from the its absorbance (A_x) by comparison with measurements for two reference standards $(C_1,\ A_1)$ and $(C_2,\ A_2)$.

4.6.4   Instrumental details

The basic AA spectrometer (Unicam) is shown in Fig. 4.33. The *lamp* operates in the range of 100–400 V and is filled with either argon or neon. When high-energy electrons leave the cathode they generate positive ions of the inert gas (Ar^+ or Ne^+) which are accelerated to strike the *hollow* cathode and sputter atoms from it. These atoms are excited in their turn as was shown for lithium (Fig. 4.31) and as these fall back into the ground state the emission spectrum is produced and escapes through the hollow cathode to produce the incident beam. A range of these lamps (HCL) is available and the one whose cathode element corresponds with the one to be determined is selected.

The *monochromator* for AA spectroscopy need not normally be capable of high resolution because the emission of the lamp matches the spectrum of the analyte. It may be adjusted to select any suitable spectral line and should be able to select emitted lines separated by 0.2 nm (but see Section 4.6.5).

The *detector* is a photomultiplier tube which responds to the incident emission by releasing electrons from its cathode. This initial current is enhanced by interaction with a series of anodes and finally used to activate a recorder.

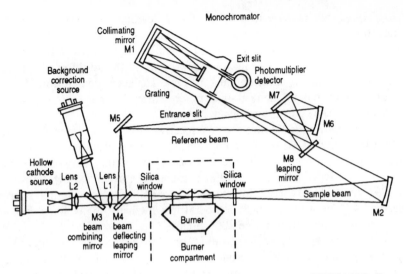

Figure 4.33   AA spectroscopy: diagram of the basic instrument (UNICAM Ltd).

The *recorder* may be a simple ammeter but modern instruments can store and reveal readings on a video display.

Double beam operation is used (Fig. 4.33) whereby the output from the lamp is divided by a rotating mirror-chopper combination M_3–M_4 into a sample beam M_4, M_2, which passes through the flame, and a reference beam, M_5. M_6 is the recombined beam after passage. The system's electronics takes the ratio of the two, I_{ref}/I_{sample}, so balancing out any transitory changes in the source.

The system avoids the possible error that can occur with a single beam instrument where the source output may change between measurement of the standard and measurement of the sample. Double beam operation has the disadvantage of reducing the incident light on the sample to half and it cannot compensate for variations within the flame. For emission measurement, the lamps are turned off as the flame is the energy source.

4.6.5   Interferences

Apart from variation in the source, other factors may introduce errors of measurement and these are summarized in this section.

Stable compounds may be formed by oxidation in the flame to refractory oxides which are incompletely atomized. Metals which behave in this way include Al, Ca, Mg and Ti; phosphates and silicates of alkaline earth metals may also lead to underestimation. The remedy is to employ the hotter N_2O/ acetylene flame.

Ionization will increase in a hotter flame, leading to a reduction in the concentration of unionized absorbing atoms of the analyte. It may be countered by the addition of a metal which is more readily ionized; for example the ionization of lithium is suppressed by the addition of 0.1% of potassium chloride (*cf.* Section 4.6.2, p. 125). The accurate determination of lithium is required in medical research because this metal is a powerful depressant of the central nervous system.

Spectral interference is rare for AA spectroscopy, given the well resolved output from the HCL, so that coincidence in the wavelength of absorbance of two species is unlikely. Should this occur, as for that between the lines for Cu at 213.859 and Zn at 213.856 nm, then Cu can be determined at 217.894 nm. This type of interference is significant for plasma sources (Section 4.7) because the high temperatures intensify many lines that are weak in the temperature range 2000–3000 K.

Non-specific absorption arises from absorption and scatter of incident light by species other than the analyte and can be allowed for by making a background correction. The Smith–Hiefte method is one way of doing this; a measurement is taken at the normal lamp operating current and then another well above this current. At the higher level, the emission lines are broadened so that most of the incident light is not absorbed by the analyte. Only the non-specific absorption is recorded and it may then be subtracted from the normal measurement. Alternative background correction methods employ either the deuterium arc or the Zeeman technique.

Physical interference may arise if the standard and sample are taken up at different rates into the flame from the nebulizer. This can be avoided by preparing standard and sample in a common solvent.

Matrix interferences are related to compound formation between the analyte and other species in the sample. In the following example, this interference is overcome by equating the matrix concentration in both standard and sample solutions.

4.6.6   The determination of sodium in concrete

Sodium levels in concrete can be determined by AA spectroscopy (Adams, personal communication, 1992). One cause of failure in concrete results from the uptake of NaCl used for de-icing and its conversion into the strongly alkaline oxide Na_2O. One example of this problem is to be found in raised sections of the M6 motorway around Birmingham in the UK.

The analysis is complicated by matrix interference because the critical oxides are at a low concentration (Na_2O normally 0.1%, K_2O normally 0.5% with CaO about 30%). The large excess of calcium interferes with the determination of both alkali metals and its effect is counteracted by the addition of aluminium as nitrate. As a result, calcium becomes combined as the aluminate, $CaAl_2O_4$; this is stable at the temperature of the air/acetylene flame and so the calcium is not atomized and measurement of Na and K may proceed. This is an example of an advantageous chemical interaction.

The experimental procedure entails the extraction of powdered concrete or a sample of cement (c. 0.5 g) with boiling dilute nitric acid. The cool extract is diluted to 500 ml in a volumetric flask and a pipetted sample (25 ml) mixed with aluminium nitrate (10 ml $\equiv$ 0.2 g Al) and made up to 100 ml. These dilutions are chosen so as to give concentrations of analyte on the linear section of the standard plot, assuming the low levels given above.

Aluminium salts always contain some sodium and allowance for this must be made by preparing a blank solution of the aluminium nitrate, using the same dilution procedure, and obtaining an absorbance reading (B_{Al}) for its sodium content.

Because of the matrix interference, a standard blank must also be prepared having the same concentrations of Al and Ca as the sample. It will give a further absorbance figure for the standard blank B_{stand}.

For a sample (0.5009 g) of ordinary Portland cement, the following absorbances were obtained (*cf*. Fig. 4.32):

	Absorbance
Portland cement	0.420
Aluminium blank (B_{Al})	0.026
By difference the sample absorbance	0.394
Na_2O standard 0.5 µg/ml absorbance	0.272
Standard blank (B_{stand})	0.090
By difference standard absorbance	0.182

Therefore, by proportion, Na in cement $0.5 \times 0.394 \div 0.182$ µg/ml and the Na content in the original extract of 500 ml from 0.5009×10^6 µg is

$$\frac{0.5 \times 0.394 \times 500 \times 100}{0.182 \times 0.5009 \times 10^6}\% = \mathbf{0.109\%}$$

A figure which falls within acceptable limits.

4.6.7   Sample preparation

In many analyses, the method used for cement will be satisfactory and, if need be, concentrated nitric acid can be used to break down organic material and to oxidize metals which resist other acids. For highly resistant samples such as plant seeds, a mixture of nitric, perchloric and sulphuric acids (3:1:1 by vol.) is used, the sulphuric acid is added to ensure that, on concentration, the lower boiling perchloric acid evaporates first because of the risk of explosion should it accumulate.

Solutions can also be obtained by fusion of material with fluxes such as sodium carbonate or sodium peroxide. This method suffers from the disadvantage of introducing a large concentration of ions into the matrix.

4.6.8　Precision and accuracy of measurement

A distinction must be made between these two factors.

Precision refers to the range within which results may be reproduced, which for AA spectroscopy should be 0.5%, provided that the measurements are made at concentrations well above the detection limit.

Accuracy refers to the difference between the true and measured result. It will be affected by random errors arising during sampling, preparation, dilution and injection of the sample.

Measurements could be made with high precision but would still be inaccurate if a systematic error was being made, for example failure to recognize spectral interference.

Analysis of small samples.　Flame AA spectroscopy aspirates the sample at about 4 ml/min and this is acceptable when a large volume is available, as in the example above. However, consider the determination of cadmium, a rare metal which can cause kidney damage if taken in excess of WHO maximum tolerable levels of 400 µg/week. Although this metal occurs at average levels of only 0.015 µg/g in UK soils, it is concentrated by plants, especially vegetables such as lettuce and spinach, reaching the order of 1 µg/g in leaves of dry weight some 50 mg; such a sample, therefore, contains $1 \times 50/1000$ µg or 1/20 µg.

The detection limit for Cd by flame AA spectroscopy (Fig. 4.36, below) under ideal conditions is < 1 µg/l and the leaf sample contains 1/20 of this, enough to provide 50 ml at the detectable concentration. To ensure precision, at least ten times this concentration is necessary, that is a sample volume of only 5 ml, which is inadequate. This type of problem has been overcome by the use of a graphite furnace to atomize the sample.

4.6.9　Graphite furnace atomization

Figure 4.34 is a simplified diagram of a graphite furnace. It was shown by L'vov that atomization could be improved by aspirating the sample in a stream of inert gas on to the graphite platform (A), within the heated graphite cylinder (B). The platform is supported in poor thermal contact with the outer cylinder; this ensures that the platform temperature lags behind, so that when the analyte is atomized it enters a stable gas phase at a steady temperature.

A temperature programme is followed, which is repeated for each sample. A typical sequence would be:

1. Slow drying up to 400 K for 30 s (matrix modifiers may be added here if needed).
2. Thermal pretreatment to bring the temperature to about 700 K. This

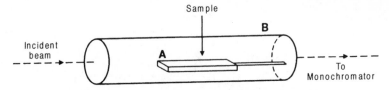

Figure 4.34 Simplified diagram of a graphite furnace.

allows for the dispersal of volatile decomposition products, which would interfere if present during the absorption measurement.

3. The gas flow is stopped so that the atomized analyte is not swept out of the light path. Cd is volatile and its spectrum will appear in only 0.3 s; the absorption will be measured by summation over the next 2–3 s.

4. The furnace temperature is raised 100°C above the highest atomization temperature to clean the system before cooling.

A valuable application of the furnace is in the atomization of small solid samples.

4.7
A plasma source

The argon plasma that serves as an atomic excitation source is driven by a 40.68 MHz free-running RF generator, which sends an alternating current through water-cooled coils around the quartz torch (Fig. 4.35) that supports the plasma. After initiation by sparks from a Tesla coil, the oscillating magnetic field around the coils accelerates electrons in the plasma and collisions with argon atoms cause further ionization. When recombination occurs, the emitted radiation raises temperatures and the geometry directs the plasma into a doughnut-shaped volume at 7000–9000 K.

The system is supported by outer, intermediate and inner quartz tubes, which introduce argon into the torch. The inner tube carries the sample aerosol through a zone at *c.* 4000–5000 K, which ensures that solvent is stripped and atomization occurs prior to entering the viewing zone at 6000 K; at low-power operation this lies at 10–20 mm above the coil. The tangential outer flow serves to maintain and stabilize the plasma while thermally insulating the outer envelope. Gas flow through the intermediate tube is not essential to the operation of the torch, but is useful when the sample includes a high concentration of organic compounds. The conditions in the viewing zone are not suitable for AA spectroscopy and so emission is determined (ICP-OES). This has become a technique of first importance and has the advantages of:

1. speed, precision and simplicity of operation
2. intensity of emission is linear and is not attenuated
3. freedom from matrix effects

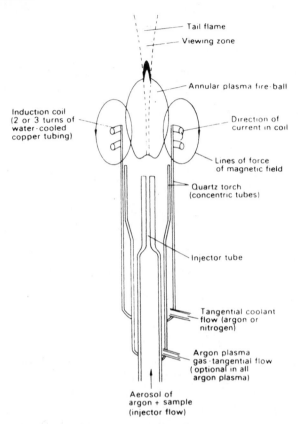

Figure 4.35   Diagram of the inductively coupled plasma torch (Ebdon, 1982).

4. refractory elements such as uranium and boron (Section 4.6.5) are atomized.

Figure 4.36 compares detection limits by various spectroscopic techniques.

Introduction of the sample

The sample is introduced following formation of the aerosol by the more common gas nebulizer or else by ultrasonic nebulization. For general purposes, a Gem Tip cross-flow nebulizer is used, which resists HF and is not clogged, provided the solution contains less than 5% salts. For higher salt concentrations, a cone spray nebulizer is preferred, while for solution concentrations below 1% a Meinhard type gives good sensitivity. Such low

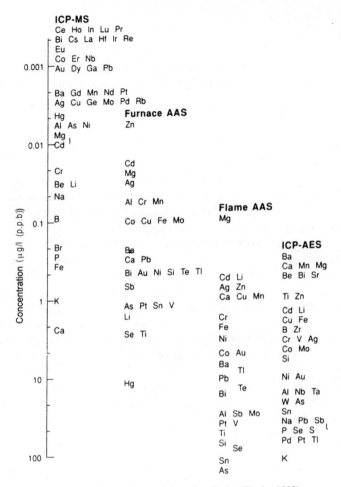

Figure 4.36 Comparison of detection limits (Slavin, 1992).

concentrations are necessary if the more expensive ultrasonic nebulizer is used, but as a result the sensitivity is increased by a factor of 10 to 50 for most elements compared with that of pneumatic nebulization. At the higher sensitivity, the use of a graphite furnace may be avoided.

4.7.1   Method development in ICP-OES (ICP-AES)

The high excitation temperature results in spectra with many emission lines and interference may well result from other elements, especially those which are line-rich. These include Fe, Co, Ni, V, Cr, Ti, W, Mo and rare earth elements; evaluation becomes most difficult when these are major components or when interference arises from elements such as Ca and Mg with intense emissions.

In the absence of interference, one may select the most intense line from the tabulated literature, otherwise a compromise must be accepted and measurement becomes dependent on a weaker line in an emission window.

Optimization

For good performance, it is necessary to optimize the power input to the ICP, the gas flow rates and the observation height above the coil. These parameters will vary with both the type of nebulizer and the sample solvent.

The parameters will also vary somewhat from element to element and for different lines in the same spectrum. This presents no problem if individual elements are measured in sequence, but when analytes are determined simultaneously a compromise will again have to be accepted. Furthermore, ICP-OES is mainly used to determine many elements simultaneously, and in this event the recommended parameters for the Perkin Elmer Optima instrument are as given in Table 4.8. When calibrating instruments for simultaneous measurements, commercial standards are employed which contain the analytes at concentrations of about 500–1000 times their detection limits. It is usual to employ only one such standard for each analyte. The samples are mostly in 2–5% (by vol.) nitric or hydrochloric acid. Sulphuric and phosphoric acids are best avoided because many metals form insoluble sulphates and phosphates and clear, filtered, solutions need to be fed to the nebulizer.

Table 4.8   Typical ICP operating parameters

	Nebulization/sample solvent		
	Pneumatic/ aqueous	Ultrasonic/ aqueous	Pneumatic/ organic
RF power (kW)	1.1	1.2	1.5
Plasma argon flow (1/min)	15	15	15
Auxiliary argon flow	0.75	0.85	1.0
Nebulizer argon flow	0.5	0.5	0.7
Sample solution uptake (ml/min)	1.0	2.5	0.7
Observation height above coil (mm)	15	15	15

4.7.2   Determination of metals in environmental samples

The concentration (C_{un}) of each unknown analyte can be determined as a proportion of the standard (C_{st}) when the two emission intensities are known. Each will be reduced by the intensity (I_{rb}) found for a reagent blank; this is likely to be small compared with the intensities found for the unknown (I_{un}) and standard (C_{st}). Then:

$$C_{un} = \frac{I_{un} - I_{rb}}{I_{st} - I_{rb}} \times C_{st} \tag{4.21}$$

The total emission intensities measured for the unknown $I_T(un)$ and the standard $I_T(st)$ will both include background emissions, $I_B(un)$ and $I_B(st)$, which arise from the other elements in solution, including argon. These are not related to the analyte concentration and they must be determined and subtracted to give the true emission intensities for entry in Equation 4.21; hence:

$$I_{un} = I_T(un) - I_B(un) \tag{4.22}$$

$$I_{st} = I_T(st) - I_B(st) \tag{4.23}$$

Making the background correction

Figure 4.37 shows the total emission for molybdenum at a wavelength of 284.823 nm with satellites arising from co-components.

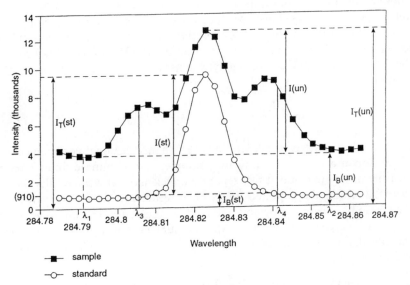

Figure 4.37   Emission spectra at Mo 284.823 nm. Cocomponents: Ca, Co, Cr, Fe, Na, Ni, V & W. See text for definition of symbols.

Table 4.9 Determination of concentration (μg/1) of metals in a rainwater sample by ICP-OES

Metal	Ist line (nm)	Conc.	2nd line (nm)	Conc.
Al	396.152	55	309.271	49
Ba	455.403	10.1	233.527	10.1
Be	313.042	0.025	–	–
Ca	396.847	3.25×10^3	317.933	3.32×10^3
Cd	214.438	13.3	226.502	13.5
Co	238.892	0.56	228.616	0.32
Cr	205.552	0.21	267.716	0.26
Cu	324.754	16.7	327.396	17.0
Fe	259.940	15.3	238.204	15.3
K	766.491	2.46×10^3	–	–
Na	589.592	2.17×10^3	588.995	2.17×10^3
Mg	279.553	543	285.213	541
Mn	257.610	19.6	259.373	19.4
Ni	231.604	1.3	341.476	1.1
Pb	220.353	164	216.999	164
Sr	407.771	8.98	421.552	8.80
Zn	213.856	1.09×10^3	202.548	1.08×10^3

The background emission of the standard $I_B(st)$ is observed independently and can be read directly from the steady baseline as 910 units. The background emission of the unknown $I_B(un)$ must be evaluated from measurement of intensity at wavelengths which are unaffected by co-components. Thus, a figure based on readings at λ_3 and λ_4, while suitable for the standard, is unsuitable for the unknown. For a reliable figure, readings must be taken from outside the co-component band at λ_1 and λ_2: the average of these two intensities gives a figure for $I_B(un)$ of 3850 units. A more accurate result can be obtained for the background (I_B) at the analyte emission wavelength (λ) from the relation:

$$I_B = \frac{I(\lambda_2) - I(\lambda_1)}{\lambda_2 - \lambda_1} \times (\lambda - \lambda_1) + I(\lambda_1) \tag{4.24}$$

Table 4.9 lists a set of concentrations (P. Zhang, personal communication 1995) for metals in a rainwater sample. To eliminate interference, measurements are made at two separate wavelengths and the results should then fall within the expected limits of experimental error. The power of the method is illustrated by both the number and concentration range of the detected analytes. In practice, the instrument software can complete the calculations of analyte concentrations, but an example for copper in the above sample follows.

For the copper concentration based on emission at 327.396 nm:

For the standard: $I_T = 44\,350$ $I_B = 225$; hence $I_{st} = 44\,125$

For the analyte: $I_T = 2716$ $I_B = 526$; hence $I_{un} = 2190$

The concentration of the standard $(C_{st}) = 400$ µg/l

The reagent blank intensity $I_{rb} = 333.9$

Therefore,

$$C_{un} = \frac{2190 - 333.9}{44125 - 333.9} \times 400 = 17.0 \text{ µg/l}$$

ICP-OES is necessary for the accurate monitoring of trace elements in soil. For example, boron is an essential plant nutrient and cereal crops have a soil requirement of 5–20 µg/g. The range in the UK is 7–70 µg/g but toxic symptoms appear at levels only slightly above this. In semi-arid areas, irrigation water can carry toxic concentrations of boron which arise from contact with the water-soluble minerals kernite ($Na_2O \cdot 2B_2O_3 \cdot 4H_2O$) and colemanite ($2CaO \cdot 3B_2O_3 \cdot 5H_2O$). In these regions, the sunflower crop is at risk; below 35 µg/g the plant suffers from deficiency but an upper level of 150 µg/g should not be exceeded.

4.7.3   ICP-mass spectrometry

The production of positive ions in the plasma allows their identification in ways that have been mentioned above. There can be no spectral interference, but interfering ions may be produced in the plasma; for example, if nitric acid is aspirated $(argon \cdot N)^+$ will be formed having the same mass as ^{54}Fe and sulphuric acid generates SO_2^+ coincident with ^{64}Zn.

The operation of this instrument requires considerable expertise, but as seen from Fig. 4.36 (p. 133) it can detect elements at very low limits. It should also be noted that the older AA spectroscopy technique can be comparable in sensitivity or even superior to ICP-OES: compare Mg, Ag and Pb on the logarithmic scale of this figure.

4.8 Analytical quality assurance

Quality assurance in the analysis of trace pollutants is particularly important in environmental monitoring because wide variations in concentration can be expected within large batches of samples. It is essential that the user of environmental analytical data has confidence in the limits of accuracy of the concentrations found. This is vital if expensive clean-up procedures or legal action appear to be necessitated by the results of environmental monitoring.

Apart from gross errors which occur as a result of instrument failure or loss of sample during treatment, analytical errors can be either systematic or random. Systematic errors are manifested as a consistent bias, which can be determined and a correction applied, but this is not possible for random errors (Fifield, 1995).

Systematic (or determinate) errors are usually caused by the following: (i) sampling and sample preparation, (ii) the analytical method, (iii) faulty instrumentation, and (iv) mistakes made by operators. Systematic errors can be constant or proportional. Constant errors will become increasingly significant as the scale of the determination increases (Fifield, 1995). The identification of these systematic errors is best done by the inclusion of certified reference materials (CRMs) and house reference materials whose composition is accurately known. The accuracy of an analysis is measured by how close within-batch determinations get to the 'true' value of the reference materials. Ideally, the accuracy should be within ± 10%, preferably less than ± 5%.

Random errors can be caused by small variations in a method, such as weighing, measuring volumes, temperature, humidity and time. Some of these errors can cancel each other, others can have either positive or negative effects on the result. In general, so long as a validated analytical method is being used, careful attention to detail and operator experience can help to reduce the magnitude of random errors. The precision (or reproducibility) of an analytical determination provides a measure of the total effect of the random errors in an analytical determination. Precision is determined by the analysis of replicates of samples within batches. Although the cost of analytical determinations is an important consideration, many routine analytical procedures will use duplicates of all samples, and in cases where important decisions need to be made on the basis of analytical data, even triplicates of each sample may be used. However, a minimum of 20% of samples analysed in duplicate is the norm. In general, precision is normally expected to be within ± 10% of the mean value.

It is important to recognize that it is possible to have analytical data which have a high precision (i.e. the values are highly reproducible) but a low accuracy caused by systematic errors. Fortunately, the latter can usually be determined using CRMs and a correction applied. The results of duplicates, CRMs and/or house reference materials analysed under the same conditions as the monitoring samples should be quoted with the results of the environmental samples. Several organizations, including the European Commission Standards, Measurements and Testing Programme, International Atomic Energy Agency, US Bureau of Standards, US Geological Survey, CANMET in Canada, and others offer certified materials, including various types of soils, some amended with sewage sludge, plant and animal tissues, food materials and others. Appropriate CRMs should be selected on the basis of being as close to the monitoring samples as possible in both matrix and concentrations of pollutants. Owing to their cost, it is often not practicable to use samples of these in every batch of analytes and so most laboratories involved in the analysis of large batches use in-house reference materials comprising large bulk samples (tens of kilograms in the case of soils) of thoroughly homogenized material that has been repeatedly analysed (often by other laboratories as well) and calibrated against a relevant CRM. A range of soil CRMs for heavy metal analysis is listed in Alloway (1995).

In addition to the quality criteria of analyses of batches of samples, most laboratories carrying out routine analysis belong to some validation scheme, such as NAMAS (National Measurement Accreditation Service), CON-TEST (Soil Analysis Proficiency Testing Scheme–Laboratory of the Government Chemist, UK), or the VAM (Valid Analytical Measurement) Initiative. These schemes monitor the performance of laboratories through supplying samples for analysis at regular intervals and then reporting back the standard of the results. Although accreditation by one of these quality assurance schemes gives the client more confidence, there is still a need for routine reporting of precision and accuracy.

In many situations, samples with highly anomalous values (greater than mean $\pm 3 \times$ S.D.) are often re-analysed or even the site may be re-sampled in order to be certain about data which can have far-reaching implications.

One important consideration is the data obtained in undergraduate and postgraduate projects. In addition to training in good laboratory practice, it is important that validated analytical procedures are used and that CRMs or house reference materials (cross-checked against CRMs) are included in batches of replicated samples. This analytical quality procedure is essential if the results are to be used in any papers submitted to refereed scientific periodicals, for examination purposes and if items of data are reported to the media or outside organizations. There is a danger of data apparently showing serious levels of pollution being quoted widely and later found to be erroneous. Examples of this include exaggerated values for Cd resulting from failure to use appropriate background correction procedures in AA spectroscopy, or even from errors in calculation of results. This is particularly important where the analyses are being carried out by non-chemists without the supervision of an analytically experienced scientist. Nevertheless, student projects can play a vital role in environmental monitoring because of the diversity of samples taken and the habitats monitored. With appropriate AQA, these data can be given greater credibility both nationally and internationally.

4.9.1   Introduction

**4.9
Environmental
monitoring**

Having discussed some of the methods used for the analysis of organic and inorganic environmental pollutants, it is necessary to consider briefly the rationale for collecting the samples to be analysed. It should be remembered that no matter how accurate and precise the analyses, if the samples are not appropriate the results will be meaningless. Therefore, adequate time should be devoted to preliminary planning and investigation of the areas to be monitored in order to ensure the most cost effective sampling regime. The old adage 'what your purpose is determines the method you choose' is particularly pertinent to environmental monitoring because there are many different ways of monitoring.

In general, monitoring is carried out to determine:

1. The substances entering the environment, the quantities involved, the sources from which they came and their spatial distribution.
2. The effects of these substances on humans, crops, livestock, ecosystems and structures.
3. Any trends in the concentration of these substances over time and the reasons for them.
4. To what extent the inputs, concentrations, effects and trends of pollutants can be modified, by what means and at what cost.
5. To establish baseline concentrations of possible pollutants for comparison with data for other locations and data for the same location after the commencement of polluting activities.
6. To assess the need for legislative controls on emissions of pollutants and ensure compliance with existing regulations.
7. To activate emergency procedures in areas of high risk (e.g. accidental emissions from chemical plants or the accumulation of hazardous levels of pollutants from road traffic).
8. To determine the suitability of water and land resources for various proposed uses.

(adapted from Holdgate (1979) and Hewitt and Allott (1992)).

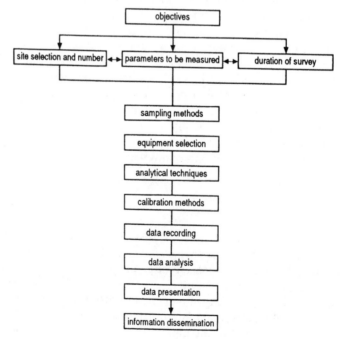

Figure 4.38   Flow chart of the essential components of a monitoring system (reproduced from Hewitt and Allot, 1992).

On the basis of the objectives, a series of decisions needs to be taken about the essential components of a monitoring programme, as shown in the flow chart by Hewitt and Allot (1992) (Fig. 4.38).

Monitoring can be concerned with:

1. sources – investigating the nature and rates of emissions into air or water, or fall-out onto land
2. transport mechanism – concentrations of pollutants, distance transported and whether they have been chemically transformed
3. targets – the amounts and speciation of pollutants reaching or entering individual targets (humans, animals, plants, ecosystems or structures).

4.9.2   Air pollution monitoring

Atmospheric emission (source) monitoring

Emissions sampling from a chimney (or stack) involves the insertion of a sampling probe into the stream of the mixture of gases and particles moving up the chimney (*cf.* pp. 108–9). Isokinetic sampling involves extracting the sample at the same velocity as it is moving in the emissions stream. The sampling of suspended particles will be affected by any changes in velocity, but both sub- and superisokinetic samplers can be used if a bias towards certain particle sizes is desired. Subisokinetic sampling, where the velocity is lower than the stack velocity, is biased towards heavier particles, whereas, superisokinetic sampling at higher velocity tends to draw the mixture into the probe and has a bias towards lighter particles (Laird,1995). Monitoring of mobile emission sources such as vehicles can only be done with any reliability when the vehicle is static but running on a dynamometer, which has flywheels to impose realistic loads on the engine to simulate different driving conditions. PM_{10} emissions (Section 6.1, p. 248) from diesel-powered vehicles are now considered to be very important and receive a lot of attention.

Ambient air monitoring

Apart from sampling emissions, monitoring is also conducted on the concentrations of pollutant gases and particles in the ambient air either in the vicinity of a known source of pollution or at remote sites. Different techniques are required for the various types of pollutant which are to be monitored.

Particulate monitoring.   This can involve:

• total deposit gauges of various types, including isokinetic samplers based on a modified 'Frisbee' design; these can either collect 'total deposits' comprising both solid particles and rainwater containing soluble 'washout' pollutants, or dry deposits (only when an automatically operated cover is used to exclude rain)

- pumped air filters in which ambient air is drawn through filters with various mesh sizes; PM_{10} emissions are monitored using air filters which retain all material $< 10\,\mu m$.

Gas monitoring. This involves:

- pumped air systems, which can involve bubbling the air though Dreschel bottles containing various absorbents to selectively separate/sorb gases, or automated chemical and/or spectrophotometric procedures for continuous sampling and analysis of gases (Table 4.10)
- passive air samplers comprising plastic tubes filled with selective gels to absorb pollutants. These can be also used for some pollutants including NO_2 and hydrocarbons.

Passive air samplers are much less precise than pumped air samplers/analysers but have the advantage of being inexpensive and, therefore, allow a larger number to be used and greater areas to be covered. The tubes need to be left exposed to the air for a week and so can only be used to indicate

Table 4.10 Methods used for the analysis of air pollutants

Pollutant	Analytical method
SO_2	Absorption in H_2O_2 and titration
	Absorption in tetrachloromercurate and spectrophotometry
	Flame photometric analysis
	Gas-phase fluorescence
NO_x	Chemoluminescence with ozone
N_2O	IR spectrometry, gas chromatography-electron capture detector, photoacoustic spectrometry
Total hydrocarbons	Flame ionization analysis
Specific hydrocarbons	Gas chromatography flame ionization detector
CO	Catalytic methanation with
	flame ionisation detector
	electrochemical cell
	non-dispersive IR
Ozone	Chemiluminescent reaction with ethene UV absorption
PAN	Gas chromatography–electron capture detection
Particulates	High volume sampler: filters extracted or digested and analysed for inorganic constituents by AAS, GFAAS, ICP-OES or ICP-MS; organic constituents by GC-ECD/FID or GC-MS
	Total deposit gauges (e.g. isokinetic Frisbee): samples filtered and inorganic constituents determined as above
	Reduction in reflectance of a glass slide (for indication of relative amounts of particulate deposition); the particles can be examined by scanning electron microscope (with electron dispersive X-ray analysis) to identify major constituents

Adapted from Harrison (1990).

average concentrations of pollutants. Passive air samplers are most effective when used in integrated surveys together with pumped air samplers. The most frequently used methods for the analysis of commonly encountered air pollutants are given in Table 4.10 (*cf*. Section 4.4.7).

With air sampling, variations in wind direction need to be taken into account and sampling sites should be on transects along compass bearings where practicable. The data obtained during a monitoring survey of atmospheric emissions may provide comparative values for polluted and non-polluted air because not all samplers will be in the path of the pollutants at any one time. Meteorological data for the area are useful because they provide information on the direction of prevailing winds and the range of weather conditions normally experienced at the site.

4.9.3 Water pollution monitoring

Waters monitored for pollutants include sea water, moving surface waters (streams and rivers), waters at various depths in lakes and pools and groundwater in aquifers. In addition, water used for drinking (potable water) needs to be monitored and will, in most cases, be subject to the strictest regulations for maximum permissible concentrations of pollutants. In addition to chemical pollutants, water can be seriously affected by pathogenic microorganisms, which will seriously impair its suitability for potable use and may also affect aquatic ecosystems.

Water provides a habitat for a very large number of plant, animal and microbial species and these organisms will be directly affected by the composition of the water. Human adults consume an average of 2 l of drinking water/day but normally use a much larger volume for cooking, washing and sanitation purposes. Only the water used for drinking and cooking needs to be of high purity, but in most cases the same domestic supply is used for all purposes and must, therefore, comply with the permissible limiting concentrations. A wide range of analytical methods are used for water pollutants including automated continuous analysis of selected pollutants at industrial plants and some river sampling locations (Table 4.10).

Sampling of rivers and streams

Rivers and streams have water moving down a slope always in the same direction, although the flow will vary as a result of varying precipitation, freeze–thaw and runoff conditions. Moving water monitoring should include an adequate number of sampling points above known/suspected input points of pollutants (e.g. outfall pipes) to provide reliable control data. Sampling locations below the outfall should allow for both the mixing of the pollutants with the river water and also the effects of tributary drainage downstream. If this tributary is heavily polluted, it could lead to confusion about releases from the source under investigation. It is important to maintain a constant depth of sampling for surface waters and specialist equipment is available to collect samples of water at preselected depths. Care should be taken in urban and rural areas where streams and canals may be infected with pathogenic *Leptospirosis* bacteria. Canals are a special case of waterway since in many cases the water is not continuously moving but only moves when a lock system is used. This tends to restrict the dispersion of pollutants and so the water and surface sediment at any particular site will only reflect localized pollution. However, where a river has been canalized, the same principles as rivers will apply although the flow will have been modified by locks and constructed banks. Heavy commercial use by barges together with industries located along the banks will result in large canalized rivers being more heavily polluted than rivers not used for commercial navigation.

Sampling lakes and ponds

Unlike rivers, lakes and ponds do not have a continuous flow in the same direction. Water in lakes is moved by the wind and this can lead to varying circulation patterns (pp. 37 and 38); these in turn affect the balance of oxidation and reduction. It must be remembered that lakes can vary in size over a vast range, from the largest (Lake Superior, USA and Lake Baikal, Siberia) to small water-filled quarries. Reservoirs are lakes which have been constructed for potable water storage and so the composition of the water is of prime importance. Sampling of lakes needs to take wind-induced currents into account and also to include samples from the epilimnion and the hypolimnion. Fixed-depth sampling equipment will be necessary for this. Some lake waters can contain high levels of blue-green algae, which can cause a severe neurotoxic condition in humans and other animals and care should be taken by samplers.

Stream and lake sediment sampling

The mineral and organic particulate material on the beds of streams/rivers and lakes can be an important sink for pollutants which is less prone to rapid changes in composition than the water above it. Therefore, sediment samples can usually provide a useful guide to the pollution history of the water body. From a geochemical survey point of view, low-order streams (tributary drainage) provide a more useful indication of regional geochemical variations than rivers. Stream sediment samples can often be collected by hand from shallow sections of the stream using either a trowel or hands to lift the sediment. Care should be taken not to fall into deep pools or to be washed down river under fast-flowing conditions. Some streams in urban and rural areas may contain dangerous levels of pathogenic bacteria *Leptospirosis* and samplers should be aware of this danger. Sediment can usually be found in the lee (downstream 'shadow') of boulders/rocks on the stream bed. Contamination by runoff from roads crossing or adjacent to streams and from outfalls of various types (including field drainage) should be taken into account at both the sampling and data interpretation stages. Stream sediment sampling and analysis has proved to be a very useful geochemical reconnaissance tool in mineral prospecting and studies of heavy metal pollution (Thornton, 1983).

Lake and large river sediment sampling requires the use of a boat and specialized grab or sediment core sampling equipment. Grab samples can be taken more rapidly and allow larger areas of lake or stream bed to be sampled; however, the technique is less precise than coring where either extrusion or cutting a plastic-lined core down the middle allows variations with depth (and redox conditions) to be investigated. It is important to be able to fix accurately the location of the sediment sampling point and this requires the use of navigational instruments, or more recently, global positioning satellite systems (GPS). If it is intended to study the dynamics of pollutants in sediments, it is necessary to refrigerate the core immediately after collection and to cut and examine/subsample it under an atmosphere of nitrogen in a special glass-fronted cabinet fitted with gloves. These precautions are to preserve the reducing conditions found under the thin oxidized layer of surface sediment.

Groundwater sampling

Since, by definition, groundwater is not accessible at the surface, it needs to be studied either in the water-bearing geological strata (aquifer) where it occurs or at the springs where it emerges at the surface (Section 2.5.2). In some cases, boreholes are specially drilled for the purpose of monitoring groundwater, but frequently it is possible to sample the water in existing wells to obtain a reasonable coverage of the groundwater in major aquifers near the surface. Deep aquifers will need a special programme of investigation if not already used for water supply. The major pollution problems in groundwater are leakage from underground pipelines and storage tanks (LUSTs) containing a wide range of chemicals, including hydrocarbon fuels and organic solvents (NAPLs), leachates from landfills, which can include many different aqueous solutes, organic compounds and microorganisms, and pesticides. Landfills in current use are required to have a series of boreholes around them for monitoring purposes, but old landfills and derelict industrial sites may need to have boreholes specially drilled. These will need to be lined and capped to the appropriate specifications to prevent seepage of surface contaminants down the sides of the borehole.

Monitoring of drinking water

In many cases, drinking water can be sampled from domestic taps, so long as the water is run for a sufficient period of time to allow possible products of the corrosion of the pipe and tap to be flushed away. The 'first draw' water, when the tap is turned on is sometimes sampled intentionally to assess the contamination from the pipes and tap (plumbing system) As a result of the practice of sterilization through the chlorination of water, the composition of the water will often vary along the length of the distribution network from the treatment works where the water is chlorinated to the domestic consumer. Chlorinated organic compounds can be formed, such as trihalomethane, and these could have a potentially adverse effect on health. The current EC, British, US, Canadian and WHO standards for inorganic drinking water are given in Table 5.20 in Section 5.3.5. The WHO standards for organic pollutants in drinking water are given in Table 3.10. Accidental contamination of domestic water supplies is rare but can sometimes occur. In 1988 the domestic water supply around Camelford in North Cornwall (UK) was accidentally contaminated with a very high concentration of acidic aluminium sulphate (normally used as a flocculating agent). This caused a wide range of traumatic effects in people drinking and washing in the water and there is still uncertainty about the possibility of long-term health effects. When the problem was identified, the water was diverted into the local river and this caused the death of a large number of fish and other aquatic organisms.

In parts of the UK and other countries, many old houses still have drinking water pipes made of Pb and this results in the water containing concentrations of Pb above the maximum safe level of 0.01 mg/l Pb (WHO, 1993) and even above the British statutory limit of 0.05 mg/l Pb. These pipes will be replaced in due course but it is a prohibitively expensive exercise to replace all the existing domestic water pipes made of Pb with either Cu or plastic pipes, which would prevent the elevated intake of potentially toxic Pb. This problem is exacerbated in upland areas where the water has a higher concentration of dissolved organic compounds (DOC) which increase plumbosolvency.

4.9.4   Soil pollution monitoring

Agricultural land

As discussed in Section 2.6.1, the soil tends to be a sink for most pollutants and also acts as a filter which protects the groundwater from pollution. In view of the sorptive properties of soils, their composition will not be subject to the temporal fluctuations seen in air and water. As a consequence, soils are often surveyed for their composition and are not regularly monitored.

The timing of soil sampling is less critical than that of the fluid media. This enables agricultural soils to be sampled at a time which is convenient for the farmer. Nevertheless, it is still essential to prevent contamination of samples in the same way that it is for air and water samples. However, soils are inherently heterogeneous with regard to many properties and pollutants. In the case of metals, the rock-forming minerals in the soil parent material vary considerably in metal composition and so even relatively unpolluted soils differ spatially in metal composition (see Section 5.3.1, p. 190). However, most soils in the world have been polluted, at least to a slight extent, from atmospherically deposited pollutants (both inorganic and organic), from fertilizers, agrichemicals and manures. As a result, contemporary monitoring of inputs of pollutants is complicated by those extraneous substances which are already present. Given this variability in the background soil contents of a wide range of pollutants, it is essential to sample a sufficiently large number of sites to determine local background concentrations of pollutants for comparision with the more recent or contemporaneously polluted sites. For example, if monitoring is carried out for TCDD in soils near to an incinerator, the soil content of the pollutant may already have been significantly high because of the use of herbicides with traces of the pollutant or the disposal of sewage sludge containing traces of TCDD. In the past, background monitoring has not always been done because workers have relied on comparisons of the data for suspected polluted sites with national or international average values. If all the soils in an area had previously been polluted or the soil parent material was naturally (geochemically) enriched in some metals (for example Cd, As, Pb, Cu, Zn, Mo, and V in marine black shales) then it is not possible to tell whether the suspect soil is polluted or just naturally enriched. Although this may not be pollution in the sense of Holdgate's definition (Chapter 1), the soil could still contain anomalously high levels of heavy metals which may be taken up by food plants and pass along the food chain and have similar effects as the same metals arising from anthropogenic sources (pollution). Nevertheless, the distinction of the source may be important for legal reasons if action is being taken against an alleged polluter.

Soil sampling for atmospherically deposited pollutants will normally be based on shallow sampling (0–2.5 or 0–5 cm) in topsoils. Internationally applied materials containing pollutants (sewage sludges, fertilizers and manures) which were subsequently ploughed into the soil need to be studied in samples of both the topsoil (normally 0–15 cm, but sometimes 0–10 cm) and samples from deeper in the soil profile (15–30 cm or 30–45 cm). For normal agricultural testing purposes, a composite sample of 25 auger cores of soil in a W-pattern covering an area of about 5 ha are collected and bulked together to give one composite sample. Patches of different soils, such as distinctly wet areas or where the crop shows distinct differences in colour or growth should be sampled separately. The taking of 25 auger cores helps to overcome the problems of soil heterogeneity. If the bulk samples are too large for logistic reasons, they should be thoroughly mixed (in a clean plastic bucket) and systematically split to obtain smaller representative samples. Soil sample collection, treatment and analysis for heavy metals is comprehensively covered by Ure (1995).

Domestic gardens and allotments

Garden soils are frequently used for growing vegetables and children also play on and with the soil and are, therefore, at risk of exposure to soil pollutants. Gardens and allotment plots are usually relatively small in area but are likely to have been subjected to more concentrated pollution than most agricultural land. Sampling therefore needs to take into account the possible sources of pollution and the parts of the garden used for growing food crops and for children to play on. All of the garden (and the indoor environment) will have been subject to atmospheric pollution and the deposition of particulate material. The extent to which this has occurred will depend on the proximity of major roads and other sources of pollution, including automobile repair garages and petrol filling stations, railway lines, scrap yards and industries of various types. The burial of metallic scrap in gardens, disposal of coal fire and bonfire ash, disposal of used car engine oil, materials from hobbies and even home-based manufacturing, flakes of Pb-rich paint from windows and other painted surfaces, Pb in airgun pellets and lost toy soldiers and possible excessive use of garden chemicals will all have contributed to the pollutant load of domestic garden soils. Thornton *et al.* (1983) found concentrations of $< 14\,125\,\text{mg/kg Pb}$, $< 16\,800\,\text{mg/kg Cu}$, $< 17\,\text{mg/kg Cd}$ and $< 14\,568\,\text{mg/kg Zn}$ in samples from 3550 urban gardens in the UK. Allotments are less likely to have been subjected to such a wide range of pollutants, except atmospheric pollution, garden chemicals and bonfire ash. However, one of the major problems with many allotments is that they have sometimes been established on previously contaminated sites which had not been built upon so were convenient areas of open land near to housing in urban areas.

Sampling of garden and allotment soils cannot be based on the system described above for farm land but all samples should comprise at least five samples (0–15 cm and possibly 15–30 cm) taken at random from various parts of the garden where pollution would result in the greatest exposure to people.

Contaminated or derelict land

Analytical surveys of contaminated or derelict land differ from many other types of environmental monitoring because they are usually carried out on a one-off basis and are intended to determine whether any pollutants are present in significantly high concentrations which will need special attention in the development of the site or which will prevent the re-use of the site for sensitive uses, such as housing and/or crop growing. There are many cases in the literature of housing being built on hazardous waste landfill sites that had not been thoroughly investigated before development (Section 8.6). Site surveys are usually based on a sampling grid going down several metres to the undisturbed subsoil. If the buildings have already been cleared, the sample grid should be selected using any site plans or any other information available to ensure that the areas most likely to be severely polluted (such as waste lagoons, chemicals, etc.) are included in the survey (Finnecy and Pearce, 1986; Alloway, 1992).

Soil samples are usually air-dried, disaggregated, passed through a nylon sieve with 2 mm apertures and then thoroughly homogenized ready for analysis. All soil analyses require the pollutants to be extracted with an appropriate solvent in the case of organic compounds and with chelating agents, acids or salt solutions for extractable metals. For total metal analysis, it is a common practice to grind subsamples of the < 2 mm fraction to a fine powder in a ball mill with agate balls (such as a Fritsch Pulverisette) to facilitate digestion of the soil in boiling concentrated acid. True total analysis requires the use of hydrofluoric acid to break down the silicate minerals, but in many circumstances 'pseudototal' concentrations obtained by boiling concentrated nitric acid, nitric/hydrochloric acids (aqua regia) or nitric/perchloric, are acceptable since they represent the fraction which is potentially bioavailable. Perchloric acid is a powerful oxidizing agent and can cause explosions with samples containing organic matter and so should be treated with special care or avoided. The final extracts and acid digests are analysed for metals by AA spectrophotometry or inductively coupled plasma atomic emission spectrometry (ICP-AES). Organic pollutants are determined by the chromatographic methods described in Sections 4.1–4.4.

4.9.5   Plant sampling and analysis

Terrestrial plants provide a good indication of the bioavailability of a pollutant in the soil and also an indication of the potential hazard to eco-systems and humans and livestock consuming plants growing on contaminated soil. However, it is important to avoid contamination of samples from atmospheric deposition and soil splash, since these will give misleading results with regard to the soil–plant transfer of pollutants. Nevertheless, the pollutants on the plant surface can still constitute an important source of potentially harmful substances to the consumer.

Where possible, plant sampling should be based on the same tissue in the same position on the plant (e.g. flag leaf in cereals) at the same height above ground and from plants of the same species at the same stage of growth. In practice, this is not always possible but should be strived for wherever possible to exclude several potential complicating factors in the interpretation of the plant analysis data. Leaf samples should be washed with deionized distilled water to remove particulates deposited from the atmosphere and soil splash and then the leaves should be dried in a forced draught oven at 60°C until they reach constant weight. The dried material should be ground to a powder in a mill with no contaminating metal parts in contact with the sample. Analysis of the plant material for metals usually involves either digesting by refluxing in boiling concentrated acids such as nitric acid or nitric/perchloric acids or samples should be ignited in a muffle furnace to destroy the organic matter. Special microwave heating apparatus is being used increasingly for acid digestions and this has increased the speed and efficiency of the operation. The ignition of plant material can be incomplete in some parts of the muffle furnace if the supply of air is insufficient. Some organic pollutant determinations involve leaching macerated tissue with organic solvents and further treatments prior to chromatographic analysis (Sections 4.1–4.4).

4.9.6   Sampling and analysis of animals and animal tissue

Ecosystem target monitoring may require certain animals to be sampled and parts of them analysed especially in food-chain studies. Various methods are available for soil invertebrates, including pit traps and digging out squares of soil and removing all the earthworms. Care needs to be taken to avoid soil contamination of the animal material and earthworms are often fed on clean material to purge the soil from their gut before analysis. Fish and other aquatic animals are usually caught in special nets but shell fish may, in some cases, be dug up from sediments at low tide.

Mammals are sometimes monitored and this can sometimes be done postmortem on animal corpses found in the study area, but trapping, blood sampling and other tests may be required. Experiments on animals are tightly regulated in most countries and appropriate permits are required. This necessitates having appropriately qualified staff to carry out specialist tasks on animals. Most animal tissue analysis involves destruction of the organic matter usually by digestion in strong acids (as outlined above for plant material). Care needs to be taken if perchloric acid is used for digesting tissue because the risk of explosion is greater with lipid-rich material than with soil organic matter.

4.9.7   The use of biological indicators in environmental monitoring

The ecotoxicological effects of pollution can only be assessed by studying the ecosystems affected. In view of the diversity of species normally encountered, most biomonitoring methods tend to concentrate on indicator species which provide an indication of the effects and extent of pollution. Manly (1995) listed the types of effect which can provide an indication of pollution as:

- changes in species composition and dominance in ecological communities
- changes in species diversity in communities
- increased mortality rates in populations (especially in sensitive early development stages)
- physiological and behavioural changes in individuals
- morphological and histological aberrations in individuals
- the build-up of pollutants or their metabolites in the tissues of individuals.

An interesting example of morphological and histological changes was provided by a cultivar of wheat which was found to be intoxicated by a residue of tar oils used as an acaricidal winter bark dressing on fruit trees.

Table 4.11   Principal methods employed for biomonitoring

Type of surveillance	Major organisms used	Principal pollutants assessed	Advantages	Disadvantages
Community structure studies	Invertebrates, macrophytes	Organic and toxic wastes,[a] nutrient enrichment	Easy to use, low cost, no specialist equipment or knowledge required	Some specialist knowledge may be required, localized use, non-specific
Bioindicators	Invertebrates, macrophytes, algae, lichens	Organic wastes, nutrient enrichment, acidification, toxic gases	Easy to use, low cost, no special equipment required	Some specialist knowledge may be required localized use, non-specific
Microbiological methods	Bacteria	Organic and faecal matter	Relatively low cost, directly relevant to human health	Some specialist equipment and knowledge required
Bioaccumulators	Macrophytes, macro-invertebrates, vertebrates	Toxic wastes,[a] radionuclides	Indicative of bioavailability of pollutants, relevant to human health	Time-consuming expensive equipment and trained personnel required
Bioassays	Microorganims, macrophytes, algae, invertebrates, lower vertebrates	Organic matter, toxic gases, toxic wastes[a]	Rapid results, relatively low cost, some continuous monitoring possible	Difficulties in extrapolating laboratory results into field situation

[a] For example, heavy metals, pesticides, fossil hydrocarbons, PCBs.

Seedlings of the sensitive cultivar showed very distinct root abnormalities within a few days when placed in a water extract of soils contaminated with tar oil residues. The characterization of the actual pollutant proved very difficult owing to the large number of different chemicals present in the 'tar oil' distillation fraction; however, the bioassay was very rapid and sensitive and very much less expensive (B. J. Alloway, unpublished data). An example of pollutant build-up in animal tissues is provided by the 'Mussel Watch Programme' in the USA, where regular sampling and analysis of mussels provides a good indication of general levels of heavy metals, such as Pb, in the environment (p. 203).

The principal methods used in biomonitoring are summarized in Table 4.11 (from Manly, 1995) and readers are referred to this source for a more detailed coverage of the broad topic of biological monitoring.

Processing and interpretation of environmental monitoring data

Interpretation of monitoring data often necessitates the plotting of isopleths of pollutants onto maps of the area affected, analysis of the variance of data for control and affected areas and various types of regression analysis to evaluate apparent relationships between parameters. Useful references on data handling include Black (1991), Rowell (1994), and Webster and Oliver (1990).

The reader is referred to Hewitt and Allot (1992) for more detailed coverage of the theory of environmental monitoring and Hewitt (1991) and Fifield and Haines (1995) for additional information on methods of analysis of environmental samples.

References

Ahuja, S. (1992) *Trace and Ultratrace Analysis by HPLC*. Wiley, New York.

Alloway, B. J. (ed) (1995) *Heavy Metals in Soils* (2nd edn). Blackie A and P, Glasgow.

Alloway, B. J. (1992) In *Understanding Our Environment* (2nd edn), Ch. 5. (ed. R. M. Harrison). Royal Society of Chemistry, Cambridge.

Black, S. C. (1991) Data analysis and presentation. In *Instrumental Analysis of Pollutants*, pp. 335–355 (ed. C. N. Hewitt). Elsevier Applied Science, London.

Braithwaite, A. and Smith, F. J. (1990) *Chromatographic Methods* (4th edn). Chapman and Hall, London.

Braithwaite, A. and Smith, F. J. (1990) Concentration and determination of trace amounts of chlorinated pesticides in aqueous samples. *Chromatrographia*, **30**, 129–134.

CHROMPAK Ltd (1992) Hydrocarbon smog, Bilthoven air, May 17, 1990. *J. Chromatog. Sci.*, **30**, 40.

Conway, G. R. and Petty, J. N. (1991) *Unwelcome Harvest*. Earthscan, London.

Ebdon, L. (1982) *An Introduction to Atomic Absorption Spectroscopy*. Heyden, London.

Fifield, F. W. (1995) In *Environmental Analytical Chemistry*, Ch. 2. (eds F. W. Fifield and P. J. Haines). Blackie A and P, Glasgow.

Fifield, F. W. and Haines, P. J. (1995) *Environmental Analytical Chemistry*. Blackie, London.

Finnecy, E. E. and Pearce, K. W. (1986) In *Understanding Our Environment*, Ch. 4. (ed. R. E. Hester). Royal Society of Chemistry, London.

Gopalan, R. (1977) Ph.D. thesis, University of London.

Harrison, R. M. (ed.) (1990) In *Pollution: Causes, Effects and Control* (2nd edn), Ch. 7. Royal Society of Chemistry, Cambridge.

Hewitt, C. N. (ed) (1991) *Instrumental Analysis of Pollutants*. Elsevier, London.

Hewitt, C. N. and Allot, R. (1992) In *Understanding Our Environment* (2nd edn), Ch. 7. (ed. R. M. Harrison). Royal Society of Chemistry, Cambridge.

Holdgate, M. W. (1979) *A Perspective of Environmental Pollution*. Cambridge University Press, Cambridge.

Kuwata, K., Uebori, M. and Yamazaki, Y. (1980) Determination of phenol in polluted air as *p*-nitrobenzenazophenol derivative by reversed phase HPLC. *Anal. Chem.*, **52**, 857–859.

Laird, C. K. (1995) In *Environmental Analytical Chemistry*, Ch. 14. (eds F. W. Fifield and P. J. Haines). Blackie A and P, Glasgow.

Lajunen, L. H. J. (1992) *Spectrochemical Analysis by Atomic Absorption and Emission*. Royal Society of Chemistry, Cambridge.

Lane, D. A. (1982) Mobile mass spectrometry: a new technique for mobile environmental analysis. *Environ. Sci. Technol.*, **16**, 45A–49A.

Manly, R. (1995) In *Environmental Analytical Chemistry*, Ch.12. (eds F. W. Fifield and P. J. Haines), pp. 249–279. Blackie A and P, Glasgow.

Possanzini, M. and Di Palo, V. (1990) Improved HPLC determination of aliphatic amines in air by diffusion and derivatisation techniques. *Chromatographia*, **29**, 152–154.

Rastogi, S. C. (1990) A routine GC method for the analysis of CFCs in aerosol cans. *Chromatographia*, **29**, 13–15.

Renberg, L. and Lindstrom, K. (1981) Reserved phase trace enrichment of chlorinated phenols in water. *J. Chromatog.*, **214**, 327–334.

Rowell, D. L. (1994) *Soil Science: Methods and applications*. Longman Scientific & Technical, Harlow. Essex.

Slavin, W. (1992) A comparison of atomic spectroscopic analytical techniques. *Spec. Intern.*, **4**, 22–27.

Stahl, E. (1969) *Thin Layer Chromatography, a Laboratory Handbook*, p. 642. Allen & Unwin, London.

Suelter, C. H. and Watson, J. T. (eds) (1991) Biomedical applications of mass spectrometry. In *Methods of Biochemical Analysis*. Wiley, New York.

SUPELCO Ltd (1990) Temperature programmed separation of mixed hydrocarbons. *J. Chromatog. Sci.*, **28**, GC116.

Takeuchi, T. and Ishii, D. (1981) High performance micro-packed flexible columns in liquid chromatography. *J. Chromatog.*, **213**, 25–32.

Thornton, I., Culbard, E. and Moorcroft, S. *et al.* (1983) *Environ. Technol. Lett.*, 137.

Ure, A. M. (1990) In *Heavy Metals in Soils*, Ch. 4. (ed. B. J. Alloway). Blackie A and P, Glasgow.

Voyksner, R. D. (1994) Atmospheric pressure ionisation LC/MS. *Environ. Sci. Technol*, **28**, 118A–127A.

WHO (1993) Guidelines for Drinking-water Quality (2nd edn). WHO, Geneva.

Further reading

Holland, M. R., Smith, B. L. and Perrett, D. (1987) *A Guide to GLC and HPLC*. Assoc. Clin. Biochem., London.

Fifield, F. W. and Haines, P. J. (1995) *Environmental Analytical Chemistry*, various chapters. Blackie A and P, Glasgow.

Karasek, F. W., Hutzinger, O. and Safe, S. (1985) *Mass Spectrometry in Environmental Science*, Plenum, New York.

Sherma, J. and Fried, B. (1990) *Handbook of Thin Layer Chromatography*. Dekker, New York.

Thornton, I. (ed) (1983) *Applied Environmental Geochemistry*. Academic Press, London.

Webster, R. and Oliver, M. A. (1990) *Statistical Methods in Soil and Land Resource Survey*. Oxford University Press, Oxford.

Part Two
The Pollutants

Inorganic pollutants 5

5.1.1 Historical introduction

Ozone was first identified in 1840 by Schonbein who described its formation by electrolysis and also when an electric discharge passes through the oxygen of the air. Its peculiar odour is detectable on the London underground and in rooms with Xerox-type photocopiers.

5.1.2 Formation

Electrical discharge. Laboratory ozonizers depend on energizing oxygen by passage through a silent discharge of some 10 000 V, which at 0°C produces a gas stream of about 4% ozone by volume. This low conversion results from the ready thermal decomposition of the initial product:

$$3O_2 \xrightarrow{\Delta} 2O_3 \tag{5.1}$$

If air is passed through an ozonizer, then some combination of nitrogen and oxygen occurs to give NO, which is more stable to heat than ozone. This reaction occurs during thunderstorms and most significantly in the internal combustion engine.

$$N_2 + O_2 \longrightarrow 2NO \tag{5.2}$$

5.1.3 Physical properties and structure

Ozone has a melting point of $-93°C$, a boiling point of $-112°C$ and a density of 1.6 relative to air ($=1.0$). The bonds linking the oxygen atoms in ozone are 0.128 nm in length and they are disposed at an angle of 116°. The principal structure (**1**) is a charge separated species which arises from the

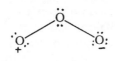

1

donation of a shared electron pair from one terminal bond to the other (*Advances in Chemistry*, 1959).

Absorption of UV light.   Oxygen absorbs visible light and this corresponds to dark Fraunhofer lines in the absorption spectrum of sunlight, notably at 759 nm (extreme red) and 687 nm (red). Activation at these low quantum levels does not lead to chemical change only to dispersal of the energy as heat. However, absorption of higher energy radiation at shorter wavelength, typically at 200 nm, leads to cleavage of the molecule and the formation of very reactive atomic oxygen (Equation 5.3). When atomic and molecular oxygen collide, ozone is formed and survives the collision provided that a third inert entity (*M*), normally nitrogen present in excess, is available to take up excess energy.

$$O_2 \xrightarrow{hv} [O] + [O] \tag{5.3}$$

$$[O] + O_2 + M \longrightarrow O_3 + M \tag{5.4}$$

$$O_3 \xrightarrow{hv} O_2 + O \tag{5.5}$$

The activated forms atomic oxygen and ozone are sometimes described as 'odd oxygen'. The principal UV absorption of ozone occurs in the range 200–300 nm and irradiation with light of this wavelength will lead to reversal of ozone formed, as in Equation 5.3. Ozone is detectable by absorption spectroscopy at levels down to 0.01 ppm.

5.1.4   The ozone layer

Figure 5.1 shows the distribution of atomic and molecular oxygen and of ozone as a function of distance from the earth's surface. In the distant stratosphere oxygen is at a low concentration and the rate of formation of ozone, shown in Equation 5.4, is reduced; at the same time, the radiation becomes more intense, which enhances the decomposition of ozone according to Equation 5.5. As a result atomic oxygen is here the dominant form of odd oxygen.

At about 60 km from the earth's surface, increasing concentrations of molecular oxygen and nitrogen favour ozone formation (Equation 5.4), while a decrease in the incidence of higher energy light disfavours formation of atomic oxygen. As lower levels are reached, ozone becomes the only form of odd oxygen and it forms a layer in the region of 35 km where it accumulates at a concentration of about 1% of the atmosphere. The formation of this layer of ozone extends the absorption of UV light to include the range 200–300 nm, and these frequencies are filtered from the sun which falls on the earth. Below the ozone layer in the troposphere the undisturbed system is protected from high-frequency light and

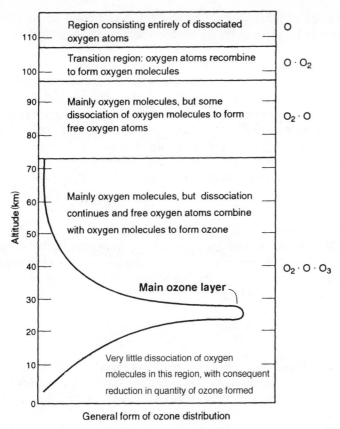

Figure 5.1 Distribution of ozone and its precursors in the stratosphere (Nebel, C., *Encyclopedia of Chemical Technology*, (3rd edn), © 1981, reprinted by permission of John Wiley & Sons Inc.).

in its absence the formation of atomic oxygen and hence of ozone is minimal.

5.1.5   Factors which disturb the natural environment

Depletion of the ozone layer. Measurement of ozone concentrations depends on the absorption of UV light already referred to and also to the typical absorption or emission of IR light. Analysis is complicated by natural seasonal variations and also by changes in the solar emission cycle and by stratospheric winds. Observations have been going on since the early 1970s; however, a combination of ground-based and satellite-based instruments has provided the most telling data of weakening of the layer.

A large body of evidence includes a spring depletion of up to 30% above Antarctica, substantiated by an observed sequential October decline in the 7 years to 1987.

Continued depletion of the ozone layer must have serious consequences for the earth's ecology since it would lead directly to an increased exposure to light in the 200–300 nm range. This radiation is toxic to unicellular organisms and to the surface cells of animals and higher plants. An increase in *treatable* human skin cancers has already been confirmed, although it is not yet clear whether an increase in the more dangerous melanoma arises in the same way, since there are other environmental factors and melanomas usually occur in parts of the body not directly exposed to sunlight.

Aviation. In addition to contributing to the burden of hydrocarbons, supersonic transports, such as Concorde, operate at a height sufficient to interfere with the protective ozone layer. Nitrogen and oxygen combine to produce NO at the high temperature of the engines, the ozone level is then reduced by the combination:

$$NO + O_3 \longrightarrow NO_2 + O_2 \tag{5.6}$$

Release of chlorofluorocarbons. These substances were first marketed by General Motors Laboratories in 1930 as they were non-toxic, non-flammable and could replace SO_2, NH_3 and carbon tetrachloride in refrigerators. On the evidence available at the time, they appeared ideally friendly to the environment since they were stable to all the reactions which take place in surface air pollution (Section 2.4.3, p. 34) and are unaffected by normal daylight radiation. They came into increasing use as commercial refrigeration became widespread and also found applications as industrial solvents and blowing agents (Section 6.3.3).

The two substances of significance are trichlorofluoromethane or CFC-11 and dichlorodifluoromethane or CFC-12, where the units refer to the number of fluorine atoms and the ten to the single carbon atom. They tend to persist in the troposphere on account of their chemical stability and because they are volatile by choice, insoluble in water and are not removed by rainout. It is now realized that these properties make them undesirable, as 'greenhouse' gases; even more seriously it was shown by Molina and Rowland in 1974 that they diffused upwards into the stratosphere and were there exposed to high-frequency radiation which caused them to react and destroy ozone (see Equations 5.7–5.9). At that time, production of CFC-11 was 0.3 million tons/year and of CFC-12 0.5 million tons/year and it was shown that stratospheric levels then correlated with the total produced up to that time. In subsequent years, growth in the USA was almost 9% per year.

5.1.6   Chemistry of stratospheric CFC

In contrast to their stability in the troposphere the higher energy radiation at 20–40 km causes photolysis (Wayne, 1991):

$$CFCl_3 \xrightarrow{hv} \dot{C}FCl_2 + Cl^{\bullet} \qquad (5.7)$$

and

$$CF_2Cl_2 \xrightarrow{hv} \dot{C}F_2Cl + Cl^{\bullet} \qquad (5.8)$$

In the Earth's atmosphere, chlorine atoms produced in this way react to destroy ozone in a manner which is analogous to the action of N_2O (Equation 5.6). However, continuing reaction of the chlorine monoxide radicals formed in Equation 5.9. releases chlorine atoms (Equation 5.10) which re-enter the cycle in a chain reaction which destroys very many 'odd oxygen' molecules.

$$Cl^{\bullet} + O_3 \longrightarrow ClO^{\bullet} + O_2 \qquad (5.9)$$
$$ClO^{\bullet} + [O] \longrightarrow Cl^{\bullet} + O_2 \qquad (5.10)$$

This exact sequence is not followed in stratospheric ice-clouds because the concentration of oxygen atoms is too low for reaction 5.10 to be significant. Here an alternative chain sequence is important in the destruction of ozone, it involves the chlorine monoxide radical as its dimer:

$$ClO^{\bullet} + ClO^{\bullet} + M \longrightarrow Cl-O-O-Cl + M \qquad (5.11)$$

This initial step, which is favoured by the low temperature in these clouds, is followed in sunlight by photolytic cleavage:

$$Cl-O-O-Cl \xrightarrow{hv} Cl-O-O^{\bullet} + Cl^{\bullet} \qquad (5.12)$$

and

$$Cl-O-O^{\bullet} + M \longrightarrow Cl^{\bullet} + O_2 + M \qquad (5.13)$$
$$2Cl^{\bullet} + 2O_3 \longrightarrow 2ClO^{\bullet} + 2O_2 \qquad (5.14 \; cf. \; 5.9 \; above)$$

By addition of Equations 5.11–5.14, the net reaction is:

$$2O_3 \xrightarrow{hv} 3O_2 \qquad (5.15)$$

where chlorine atoms and chlorine monoxide act overall as catalysts.

Nomenclature.   For brevity a system is used which is based on the form CFC-XYZ where X = number of carbon atoms minus 1 (omitted if $X = 0$); Y = number of hydrogen atoms plus 1; Z = number of fluorine atoms. So that $CFCl_3$ becomes CFC-11 and CF_2Cl_2 is CFC-12.

Ozone depletion potential (ODP).   This is based on model calculations made at the University of Oslo and CFC-11, the most damaging chlorofluorocarbon, is assigned an ODP of 1.0. On this scale CFC-12 has an ODP of 0.9. The potential for depletion will be reduced if a substance is less persistent: CFC-11 survives the decade it takes to diffuse into the stratosphere and persists for many years more; however, research is going on to develop alternatives whose concentration is reduced by reaction in air at lower levels. These compounds typically include a reactive C–H bond and are known as HCFCs, they decompose in the troposphere under attack by species such as the $^\bullet$OH radical. This initiates change by the abstraction of H$^\bullet$ from the C–H bond. The reaction type is similar to that discussed in Section 6.2.4, p. 259.

Examples of suitable compounds are: dichlorotrifluoroethane CF_3CHCl_2 (HCFC-123) for blowing foam (ODP 0.013); tetrafluoroethane CF_3CH_2F (HCFC-134) as a refrigerant (ODP zero); and 1,2-dichloro-1,1-difluoro-ethane CH_2Cl-CF_2Cl (HCFC-132) as solvent in the electronics industry.

Further examples of the environmental exposure of these compounds are given in the section on solvents (Sections 6.3.3–6.3.5).

5.1.7   Control measures (American Conference of Government Industrial Hygienists, 1990)

Legislation was enacted in the USA to restrict the use of CFCs as early as 1975 and worldwide production declined through the 1980s. In 1987, an International Protocol was agreed at Montreal whereby the signatories agreed to halve their current production by 1999. In recent years, the public and governments have become increasingly concerned. An international work group reported 18% depletion of the ozone layer over Europe in the winter of 1991/2 and work in the USA has confirmed this trend. An increase in skin cancer and eye cataracts has been predicted and in the UK the Health Education Authority has issued a warning about over-exposure to the sun.

An international meeting at Copenhagen in November 1992 took account of the growing problem and agreed for the most part to phase out CFCs altogether by 1995. Desirable though this is, it should be noted that considerable usage continues in countries outside the agreement and some manufacturers have been accused of selling stocks to these consumers.

5.1.8   Ozone in the troposphere

The filtering action of the ozone layer removes the shorter wavelength radiation with sufficient energy to produce odd oxygen (Section 5.1.4). Under natural conditions, small amounts of ozone enter by transfer from the stratosphere. However, tests in pollution zones reveal substantial levels, which in sunlit urban areas can rise to 1 ppm; typical peak urban concentrations lie in the range 0.01–0.04 ppm (200–800 $\mu g/m^3$). In these areas nitrogen is fixed as NO in the internal combustion engine where oxygen is present in limited excess (Equation 5.2). In daylight this initial product reacts with atomic oxygen, or other oxidizing agent present in low concentration in the natural environment, to give the peroxide:

$$NO + O \longrightarrow NO_2 \qquad (5.16)$$

This is the primary source of ozone because it can be photolysed by available blue light of wavelength near 400 nm. So through the morning, sunlight initiates decomposition of the peroxide to give the lower oxide and releases atomic oxygen:

$$NO_2 \longrightarrow NO + [O] \qquad (5.17)$$

This then combines with molecular oxygen to produce ozone.

In the absence of other agents, oxygen and NO_x would achieve a steady state including a low level of ozone, but Equation 5.17 is driven to the right by the removal of NO through its reaction with other pollutants, notably CO and unburnt hydrocarbons (p. 258). As a result, the concentration of atomic oxygen and hence of ozone is further increased.

In this context, it must be stressed that the environment is not compartmentalized and although use of HCFCs reduces the damage to the ozone layer they act in the troposphere as greenhouse gases (p. 167).

Methyl bromide is a substance of current concern not so much for its action in the troposphere, but from the realization that its survival in low concentrations in the stratosphere releases very damaging bromine atoms. About 50% of methyl bromide comes from natural marine sources, but the rest is artificially made because its low boiling point of 5°C has led to widespread use as a space and soil fumigant. The quality and variety of foodstuffs available in advanced countries depend on the pesticidal action of this compound. It is used on food crops, coffee, cocoa, forage, timber, tobacco, etc., besides the protection of foods and grain in storage and during transportation in ships, trains and trucks. There is no single substitute, although ethyl ether may be used as a fumigant, Malathion (p. 316) for stored products and chloropicrin for soils.

5.1.9   Diurnal variations of ozone levels

In the stratosphere, little change occurs during the dark hours since, although production ceases, the concentration of atomic oxygen falls away and ozone is no longer lost by the reaction:

$$O + O_3 \longrightarrow 2O_2 \qquad (5.18)$$

In contrast, levels in the troposphere decline after dark because ozone

Table 5.1 Air quality health standards for ozone

	Ozone levels (ppb)		
	8 h avg.[a]	1 h avg.	Daily mean
European Commission Directive 92/73/EEC	50–60	76–100	32
US EPA	–	120	–
WHO	33(24 h)	100	–

[a] Obtained from the hourly average by taking them in groups of eight reported in sequence, e.g. 00.00–07.59/ 01.00–08.59/etc.

continues to react with other residual pollutants such as hydrocarbons. A typical diurnal cycle is shown in Figure 6.5 (p. 260).

Ozone differs from molecular oxygen in that it is a polarized molecule (see **1**) and hence is 12 times as soluble in water as oxygen under comparable conditions of temperature and pressure: ozone is, therefore, scrubbed from polluted air by rainfall. Uptake of ozone from ozonizers may produce aqueous solutions with concentrations of the order of 40 mg/l and these are useful in destroying the last traces of pollutants in waste waters (Section 8.4).

5.1.10   Toxicity and control

The primary site of injury by ozone to humans is the lung, which may become congested with swelling of the tissue (oedema) and possible haemorrhage. In humans exposure for 2 hours to concentrations of 1.5 ppm is regarded as dangerous. Workers particularly at risk are welders using the shielded arc processes. A threshold limit value (TLV) of 0.1 ppm is being considered in Europe and the USA (Table 5.1).

Where NO emissions are locally high they reduce ozone through the reaction:

$$NO + O_3 \longrightarrow NO_2 + O_2 \tag{5.19}$$

and in consequence ozone levels tend to be higher in rural areas (Bower *et al.*, 1995). In Jan/Dec 1992, the EC daily mean (32 ppb) was exceeded on 111 days at Great Dun Fell (Cumbria) and on 104 days at Lullington Heath on the Sussex coast (Warren Spring Laboratory, 1992).

5.2
Oxides of carbon, nitrogen and sulphur

The problems of growth associated with demand for energy and supply by combustion of fossil fuels were stressed in Chapter 1. This section covers the chemical and physical properties of the pollutants which are produced. The composition of clean air is given is Table 5.2.

Table 5.2 The composition of clean air at STP

Gas	Volume %	Mass t
Dry air	100	5.12×10^{15}
N_2	78.1	3.87×10^{15}
O_2	20.9	1.18×10^{15}
Ar	0.93	6.59×10^{13}
CO_2	0.0315	2.46×10^{12}
Ne	1.82×10^{-3}	6.48×10^{10}
He	5.24×10^{-4}	3.71×10^{9}
Kr	1.14×10^{-4}	1.69×10^{10}
Xe	8.7×10^{-6}	2.02×10^{9}
CH_4	1.5×10^{-4}	4.3×10^{9}
H_2	5.0×10^{-5}	1.8×10^{8}
CO	1.2×10^{-5}	5.9×10^{8}
O_3	varies	$ca. 3.3 \times 10^{9}$
N_2O	3×10^{-5}	2.3×10^{9}
NO_2	1×10^{-7}	8.0×10^{6}
NH_3	1×10^{-6}	3.0×10^{7}
SO_2	2×10^{-8}	2.0×10^{6}
H_2S	2×10^{-8}	1.0×10^{6}

From Greenwood, N. N. and Earnshaw, A. (1984) *Chemistry of the Elements*, p. 701. Pergamon, Oxford.

5.2.1 Carbon dioxide

Variation of yield with source. It is shown in Section 6.2 that most coals contain 85–90% of carbon, almost all of which is released as CO_2 on combustion. This must be compared with a typical fuel oil whose composition will be close to that of a saturated aliphatic hydrocarbon, i.e. $C_nH_{(2n+2)}$. Of this group, methane (CH_4) has the lowest carbon content of 75%, which rises on ascending the series to 84% for octane (C_8H_{18}). Hence oil, and especially natural gas, contribute less per energy unit to the CO_2 burden than does combustion of a representative coal (Table 5.3).

While the more volatile petroleum fractions, including gasoline, are to be preferred for the control of CO_2 emissions, the release of oxidized minor components, especially nitrogen and sulphur, must also be considered. The consequences of NO_x and SO_2 release are discussed below.

The observed increase in CO_2 levels. The natural level of 0.03% corresponds to 275 ppm and correlates with that found in ice cores pre-dating the Industrial Revolution and also with a summary of earlier data made by Callender in 1958. The subsequent upward trend (Fig. 5.2) has been established remote from local discharges, at the Mauna Loa observatory, following an initiative by Keeling.

These observations are supported by others made at the South Pole and in Alaska at Point Barrow, also by samples taken during aircraft flights. From Fig. 5.2, it can be seen that there is a seasonal downward 'wobble' every year caused by the uptake of CO_2 during photosynthesis by deciduous trees and

Table 5.3   Relative CO_2 production/unit
of energy[a] from various fuels

Oil	2.0
Gas	1.45
Coals	2.5

[a] The quad = 1015 BTU or 1.055 kJ.

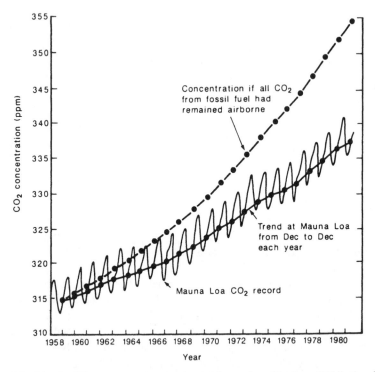

Figure 5.2   Rise in CO_2 concentration in air at Mauna Loa (Keeling, (1989) *Geophysics Monograph*, **55**, 165–236, copyright by the American Geophysical Union).

other plants. It is also apparent that only part of the released CO_2 remains in the atmosphere. This is to be expected since it is soluble in water to the extent of 3.3 g/l at 0°C and 1.7 g/l at 20°C. Ocean sediments arise from its solution in sea water but the surface contact with natural water is limited and only moderates the rise in concentration.

Ice cores.   The solubility of CO_2 in water ensures that equilibrium is established through time between the levels in air and water. Hence from a knowledge of the level in polar ice of known date, it is possible to evaluate the level in air in equilibrium with it at the same period. It has been shown that in the years preceding the Mauna Loa analysis CO_2 ranged from an unperturbed level of 270 ppm to 320 ppm. Figure 5.3 shows the rise in emission from individual fuels and Fig. 5.4 predicts the relative contributions for the year 2025 from major consumer countries.

Wood is an important energy source in Third World communities. For example in India a village will consume about 3800×10^9 joules annually and of this almost 90% comes from firewood. Taking India as a whole, wood and charcoal account for one-third of its energy demand but this is small on a global scale since the inhabitants consume only one-twentieth of that consumed by inhabitants of developed countries.

By reference to Fig. 5.4 and Fig. 1.3 (p. 12) it is seen that as heavily populated Third World countries such as India and China strive to raise their living standard, residual CO_2 levels must rise at a greater rate than before. At present these countries account for 80% of the world's population and consume only 25% of commercial energy. Their demand is expected to double during 1990–2010 and redouble during 2010–2030. After that time if 10 billion people are to meet the desired 60% cut in present levels per capita, emission must fall to one-fifth, i.e. to 0.2 tC/year.

The greenhouse effect (Dickinson and Cicerone, 1986).   Solar energy falling on the earth's surface is absorbed and transferred to the atmosphere by

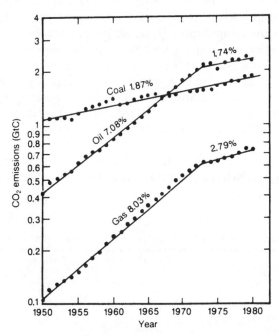

Figure 5.3 Annual emissions of CO_2 by fuel type with indicated growth rates (Rotty, R.M. (1983) *J. Geophys. Res.*, **88**, 1301, copyright by the American Geophysical Union).

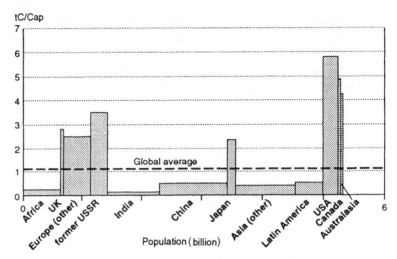

Figure 5.4   Relative carbon emissions per capita and population 1989 (after Grubb, 1991 and United Nations, 1994).

evaporation and as heat flux, including IR radiation. At present the energy fall at the surface is about 157 W/m^2; rather more than twice this amount enters the stratosphere but is depleted by reflection from clouds and dust and also by absorption by clouds, ozone and water vapour. The IR energy absorbed by these components is governed by the intensity of IR emission at the earth's surface and in the lower levels. The emission of energy from the upper levels is reduced as the temperature of the troposphere falls at about 5°C/km of altitude, leading to a net energy gain and surface warming.

The activities of humans have introduced additional absorptive molecules, so trapping more heat and disturbing the natural equilibrium (Wayne, 1991). The principal components and their efficiencies are shown in Table 5.4.

Figure 5.5 shows the span of the the IR spectrum in the frequency range 500–4000 cm^{-1}, which is equivalent to a wavelength range of 20–2.5 μm.

Table 5.4   Trapping of IR radiation by trace constituents (1985)

	Concentration (ppb)	Trapping (W/m^3)
CO_2	345×10^3	c. 50
Methane	1.7×10^3	1.7
Ozone	10–100	1.3
N_2O	304	1.3
CFC-11	0.22	1.06
CFC-12	0.38	0.12

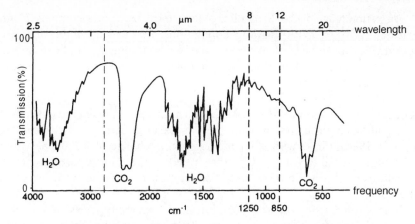

Figure 5.5   Background IR absorption by water vapour and CO_2.

The escape of IR energy is blocked by the absorption of water vapour and CO_2 save for a narrow window near $3.5\,\mu m$ $(2800\,cm^{-1})$ and a broad window in the range $8-12\,\mu m$ $(1250-850\,cm^{-1})$. It can now be seen why the effect of polluting chemical species is so much greater than is immediately suggested by their relatively low concentration compared with CO_2 (Table 5.4): this is because they have appreciable absorption within the natural escape window. Ozone absorbs strongly at $9.6\,\mu m$ and, on a per-molecule basis, ozone is 30 times as absorbant as CO_2. The release of ozone and CFCs has been mentioned above; methane is a significant and growing contributor (Section 6.2.3) and the contribution of NO_x is discussed later in this section.

Release of greenhouse gases from soil.   The carbon content of soils ranges from less than 1% in sandy soils to over 40% in peats. Annual CO_2 release varies correspondingly between $80-550\,g$ carbon/m^2. The act of draining soils leads to an increase estimated worldwide, at $150-180\,Mt/a$ (*cf.* Fig. 5.3). Cultivation of organic soils increases the aeration with an estimated global release of $30\,Mt$; soil carbon is also lost whenever forest and grassland are converted to arable (Armstrong-Brown *et al.*, 1995).

Growing biofuel crops on surplus agricultural land helps reduce emissions and burning biofuel returns CO_2 to the atmosphere, whereas fossil fuel burning leads to a net addition to the carbon cycle.

Methane is released directly through rice stems in flooded paddy fields; this may be limited by siting them as far as possible on mineral soils. In temperate regions, landfill sites are a major source and it has been estimated that $20-50\,Mt$ of methane are released annually in biogas, an equimixture of methane and CO_2. Retention of NH_4^+ in soil reduces emission as it inhibits methane oxidation. The digestive systems of termites produce significant amounts (*cf.* Section 6.2.3).

An estimated 0.5–2.0% of nitrogenous fertilizers are metabolized and released as N_2O. With growing populations, this will inevitably increase worldwide, but some control can be exercised by timing the application, e.g. by avoiding autumnal distribution, when plant uptake is low and maximum losses occur by leaching.

Effect of greenhouse warming.   A fundamental uncertainty is the past effect of variations in the emissions from the sun, the brightness of which varies by up to 0.1% in phase with magnetic activity. A change of 0.2–0.5% would account for a fall of 0.4–0.6°C in the Little Ice Age. Future forecasters will benefit from observations by NASA, which now records the sun's energy to 0.01%.

It is known from the study of ice cores that CO_2 levels in the past interglacial age were similar to those of today and the conditions on other planets correlate with their surface temperatures. Mars has a thin atmosphere composed mainly of CO_2, a surface less than one-hundredth that of earth and a surface temperature of 223 K. Venus has an atmosphere which is 96% CO_2 at a surface pressure 100 times that of earth and a surface temperature of 732 K (Wayne, 1991).

The earth's surface temperature has risen by 0.5°C over the last century and predictions depend on forecasting the likely rate of change in the concentration of greenhouse gases. The earth's atmosphere contains 2.6×10^{15} kg CO_2 to which human activities added 5×10^{12} kg in 1985 and the current rate of addition is 1.9×10^{13} kg/year. Fossil fuel usage over the past century increased at 4.5%/year, a doubling time of 16 years, leading to the present level of 350 µg/g. If the future rate of usage increases at 4% then the natural level will be doubled to 600 µg/g by the year 2030 (Fig. 5.6). With growth nearer to the present rate of 2%, this point will not be reached until 2050. These predicted dates will vary somewhat depending on the fraction of emitted CO_2 which persists in the atmosphere: at present this is almost one-half.

Computer models which have been validated against past variations indicate a temperature rise of 3°C on doubling present CO_2 levels. This must be qualified as there are both positive feedback mechanisms which enhance the effect and negative feedback which reduces it. One example of negative feedback is the reduction of incident radiation as a result of increased cloud cover as oceans warm up. An example of positive feedback is the reduced reflectivity (albedo) caused by the shrinkage of polar ice. Warming is likely to be faster in the northern hemisphere with its higher level of industrial activity and smaller ocean area. At the second World Climate Conference at Geneva in 1990 a forecast was made of 2–5°C warming by the end of the twenty-first century.

Aside from CO_2 emissions, it must be remembered that increases are to be expected in the concentration of other airborne pollutants, especially those listed in Table 5.4, which introduces further uncertainty into these predictions.

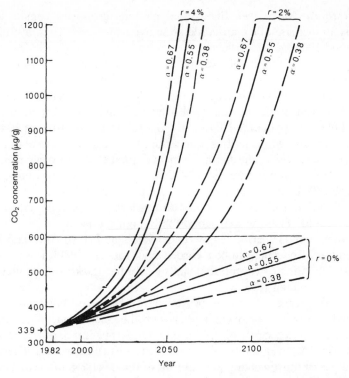

Figure 5.6   Predicted CO_2 levels for zero, 2% and 4% growth (from Roberts *et al.*, 1990).

Consequences of greenhouse warming.   Some events which may arise from warming have already been noted:

1. Persistent water shortages in southern England – construction of a desalination plant is being considered by Southern Water.
2. The six warmest years since records were kept in the UK all fell in the 1980s.
3. Evidence of warmer water off the Cornish coast comes from the appearance there of the tropical triggerfish and also of red mullet and the marbled electric ray.

Apart from variations in the sun's emissions, another phenomenon which may account in part for these observations is 'El Niño', a periodical warming of the Pacific west of Peru, which occurred twice in the 1980s. Evidence has recently been obtained of fossilized beech leaves near the South Pole, indicating a dramatic warming in the recent geological past.

The effect on sea levels.   In 1985 responsible predictions based on computer models were of about 1 m rise by 2050; these have since been revised downward. In May 1990, the United Nations group on climate change forecast an increase of 6 cm/decade or an average rise of 20 cm by 2030 and 60 cm by 2100.

Levels rise as a result of expansion of the water body as temperatures increase and also of its enhancement by melting South polar ice – melting of North polar ice will not directly change sea level as it is already afloat (a full glass of an iced drink does not spill over if left at room temperature). The North Pole is more vulnerable than the South, which was formed 10 million years earlier, and Arctic ice has been reduced in thickness from about 7 m in 1976 to 4 m in 1987. Hermann Flohn has predicted complete melting of Arctic ice on a temperature rise of 4°C and this is equivalent to a localized increase of 38°C in air temperature at the pole, presently −34°C.

The Antarctic ice has contributed to sea levels as average temperatures have increased. There was an abnormally high release of icebergs in 1930 and another similar period is now underway. The British Antarctic Survey assessed the 24th iceberg to be recorded since the mid-1980s; it is moving at 3 miles/day and is of exceptional size at 50 miles (80 km) in diameter and weighing 10^3 billion tons.

The consequences of melting of ice will not be spread uniformly and the predicted average rise of 20 cm in sea level will be seen as one of about 35 cm in Europe. This is because of the more rapid disappearance of Arctic ice and its replacement by a body of darker heat-absorbing water. There exist possibilities of drastic changes as a result of the diversion of ocean currents. Warm Gulf Stream waters could take another course to the detriment of the climate of the British Isles and Europe. The direct threat will be to communities on low-lying islands such as the Maldives, 3.66 m above the level of the Indian ocean. In the South Pacific, Tonga, Kiribati and Tuvalu are at risk and observation posts have been established there by the Australian Government.

The backing up of rivers will lead to flooding. The most serious risk is to Bangladesh, in the delta of the Brahmaputra. In England, the threat is to the fenlands and the estuaries of the Thames, Humber and Severn. In Europe, rivers at risk include the Garonne, Loire, Seine, Rhine and the Elbe.

The effect on agriculture.   Changes in the Arctic referred to above will lead to a progressive shift in the meteorological equator, at present at 6°N to within 10–12°N. Although this change may be long delayed, the consequences are so serious as to demand our attention.

In the UK, the south east will become drier and growth of arable crops would tend towards the west. It is also likely that industry would move to the north-west to secure water supplies. Typical French crops such as maize and sunflowers would grow in the south and the malarial mosquito could breed in these areas of England.

The African desert would move northward and areas of Spain, Italy and Greece could become deserts. Similarly, with reduction of winter rainfall,

California would come to resemble Mexican desert and the change would also be seen in the Punjab of India.

Control measures. In a review by the Brookings Institute (Epstein and Gupta, 1990) it was proposed that control of emissions could be achieved by allocating emission levels based internationally on previous consumption. These would be set so that the heavy polluters (e.g. USA, USSR, Japan, UK, Germany) would be restricted and countries in surplus (e.g. India, China, Indonesia, Bangladesh) would be allowed growth. A nation wishing to exceed its quota could only do so by purchasing part of that allocated and unused by another nation with an emission balance in hand. Such a measure would discourage the use of coal, which produces more CO_2/energy unit than oil.

The United Nations panel (Houghton *et al.*, 1990) considers that in order to limit temperature rise to within 0.1–$0.2°C$/decade the global emissions of CO_2, CFCs and NO_x (Sections 5.1.6, 5.2.2) must be reduced by 60% and those of methane by 20%. Set against a history of economic growth and demand for energy by a burgeoning population, this is an improbable scenario. However, some desirable measures may be summarized:

1. Energy conservation in building design. The Confederation of British Industry (CBI) has stated that 20% of energy was wasted out of an investment of £38 billion in fossil fuels.
2. Stricter control of vehicle emissions, more efficient use of fuels through speed limitation, greater use of public transport and introduction of the electric car in cities.
3. Overall increase in forestation. This is likely to make only a limited contribution as an increase of about 10% of all forests over and above felling would be required to take up half of the present CO_2 production of 2.5 Gt carbon/year. The value of this area of suitable land and the demand for its use in food production would compete with planting proposals. Trees planted in areas at the borderline of fertility could become unhealthy if soil warming occurred.
4. Encouraging uptake by algal growth. The National Research Council advised US Congress that dumping iron-rich residues in the Antarctic ocean and off Alaska would make up for an existing growth deficiency in these waters.
5. Taxing inefficient sources such as coal and giving preference to natural gas (p. 165).
6. Removal of CO_2 at power stations by scrubbing emitted gas. To be effective, it would be necessary to dispose of the trapped gas out of contact with the atmosphere. Burial in the deep sea after entrapment by and subsequent release from ethanolamine has been proposed. Such processes are themselves energy intensive and could even double the cost of electricity.

7. Wind, wave, hydro-, solar and nuclear power are preferable to option 6. An encouraging report is the construction of a cell which harnesses sunlight with an efficiency of 10%. This is higher than sugar cane, which at 7% is the most efficient of the plants. The use of nuclear power (Section 5.5) is attended by unsatisfactory economics and the risk of exposure to radioactivity.

Monitoring. The UN panel on climate change has been mentioned; 25 countries are represented on it and it last reported in 1996. In the UK, contributions are made by the Centre for Agricultural Strategy at Reading and by the Institute for Terrestrial Ecology at Abbots Ripton.

The Scripps Institute at La Jolla is attempting to detect warming by placing a source of sound at Heard Island in the southern Indian ocean. Using the boundary between warm and cold water as a wave guide, the time taken to reach remote receiver stations can be measured and will increase perceptibly if ocean warming occurs.

The International Panel for Climate Change has recently accepted the reality of greenhouse warming, but a contrary view has been expressed by some members of the European Science and Energy Forum (Emsley, 1995).

5.2.2   Oxides of nitrogen

Sources. Nitrous oxide (N_2O) is a stable long-lived gas formed naturally by blue-green algae and by *Rhizobium* bacteria active in the nodules of peas, beans and other legumes. The internal combustion engine, power stations and nitrate fertilizers now produce almost half the amounts arising from the natural sources.

Vehicles are the principal source of NO and of its oxidation product NO_2. The brown peroxide dimerizes to the yellow tetroxide (N_2O_4) which predominates in highly polluted situations. The analysis of these gases is complicated by their interconversion both in the environment and during chromatography, and they are described collectively as NO_x.

Levels in air. In 1987 the average concentration of N_2O was 307 ppb and this is increasing by 0.2% annually. Figure 5.7 shows growing contributions to NO_x levels from various anthropogenic sources in the USA, rising in 1985 to 22×10^6 tons.

The levels in other communities can be estimated by scaling these figures in the ratio of their GDP to that of the USA (p. 12).

Effect on the environment. In addition to the greenhouse effect, NO_x contributes to air pollution in three ways:

1. depletion of the ozone layer
2. production of acid rain (p. 175)
3. general air pollution.

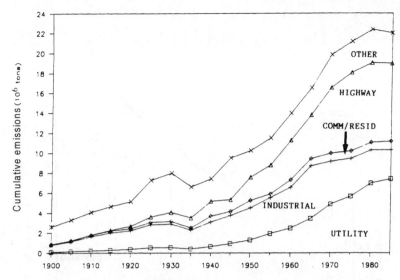

Figure 5.7 NO_x emissions by categories in the USA (COMM/RESID is residential). (From Pahl, D. A., Zimmerman, D. and Ryan, R. (1990) In *Emissions from Combustion Processes* (ed. R. K. Clement and R. Kagel). Lewis, Boca Raton, FL, with permission.

Depletion of the ozone layer.   Nitrous oxide has a lifetime of about 100 years owing to its low reactivity, and once formed naturally or by reduction of surface nitrate it can survive to reach the stratosphere. There it reacts with excited oxygen:

$$N_2O + O \longrightarrow 2NO \qquad (5.20)$$

This reaction reduces ozone formation (*cf.* Equation 5.4) and the two molecules of NO formed lead to ozone decomposition (*cf.* Equation 5.6). This link between excessive use of nitrate fertilizers and depletion of stratospheric ozone could hardly have been predicted.

Production of acid rain (cf. Section 5.2.3).   Although oxygen is itself a powerful oxidizing agent, it fortunately does not readily combine with nitrogen since the formation of NO by this route (Equation 5.21) has a free energy of 87 kJ/mol. However, the necessary input of energy may be obtained from lightning discharges and from the internal combustion engine.

$$N_2 + O_2 \longrightarrow 2NO \qquad (5.21)$$

Further oxidation of NO by ozone leads to NO_2 which may then react with hydroxyl or other free radicals to form nitric acid:

$$^{\bullet}OH + NO_2 + M \longrightarrow HNO_3 + M \qquad (5.22)$$

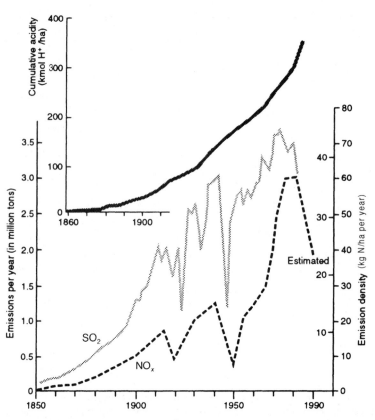

Figure 5.8   Annual emissions of SO₂ and NOₓ in West Germany (reprinted with permission from Ulrich, B., *Environ. Sci. Technol.*, **24**, 439, © 1990 American Chemical Society; Social Trends, 1991).

This product then falls as rainout, contributing significantly to forest decline and with an adverse effect on fisheries, crops and buildings.

Experience in West Germany shows that emissions of NO_x and SO_2 are comparable (Fig. 5.8).

The steep rise in NO_x in the years following 1950 are attributable to an increase in vehicle emissions. In the USA, over 40% of the NO_x burden arises from vehicle and there it is at a maximum relative to SO_2, which is derived mainly from other sources (*cf.* Fig. 5.8). In 1987 in the USA, releases of SO_2 and NO_x were comparable at about 20 Tg, six times the West German output. In the UK in 1980, the burden of NO_x from transport was 28% with the major contribution of 46% from power stations. Since the internal combustion engine accounts for only about 3% of released SO_2, a comparison of vehicle usage indicates where measures for control of NO_x are best directed (Table 5.5).

Table 5.5 Transport patterns in some major cities (1980) (Newman and Kenworthy, 1989)

	Petrol use (10^3 MJ/head)	Vehicles/1000 population	Proportion public transport (%)	Proportion private transport (%)
Los Angeles	58	670	7.5	88
New York	44	460	28	64
London	12	355	39	38
Munich	12	400	42	38
Paris	14	385	40	36
Amsterdam	9	340	14	58
Melbourne	29	530	21	74
Moscow	0.4	40	74	2

A very high level of private vehicle usage is observed in Los Angeles, consistent with the present social threat this represents. The major European cities have a similar pattern of use. Summation of the last two columns gives a transport shortfall for each city which is accounted for by cyclists and pedestrians; this is still significant in Europe and in Amsterdam amounts to 28%. The record for Moscow shows that a major city can function with minimal use of private transport.

NO_x and air pollution.   The critical role of NO_2 as a trigger for tropospheric air pollution has been discussed (Section 5.1.8) together with the consequent production of ozone. It also reacts to form PAN as part of a complex web of reactions that involve methane and other airborne chemicals. A discussion of this chemistry is deferred until the issues arising from the release of organics has been addressed in Section 6.2.

Control of NO_x emissions

Emissions from vehicles.   It is evident from Table 5.5 that much can be done through economies of consumption of motor fuel and by suitably modifying existing engines. Other possible measures include:

1. Keeping the compression ratio below the usual 10:1. This has the effect of reducing the combustion temperature, which is unfavourable to the initial fixation reaction (Equation 5.21) and hence reduces the yield of NO.
2. Enriching the mixture of fuel to bring the post-combustion level as low as possible. However, this is energy inefficient and leads to an increase in the release of unburnt fuel.
3. Unburnt fuel and NO can be removed from the exhaust using a catalytic converter – the option adopted in North America and some European countries. The action takes place in two stages. In the first NO_x is

Table 5.6 Percentage compliance of petrol car emissions with EC standards (Dunne, 1990)

	Pre-1992 (A)		Post-1992 (B)	
	CO	THC + NO$_x$	CO	THC + NO$_x$
EC standard	70%	24%	22%	6%
Uncatalysed	165	81	525	334
Catalysed	10	12	32	47

THC = total hydrocarbons.

reduced to N$_2$ using unburnt fuel and CO as reducing agents, and in the second stage the remaining CO and fuel are oxidized by injection of air at 400°C. The catalysts are platinum or alloys of platinum/rhodium, which are poisoned by leaded fuels. The SO$_2$ in the emissions is converted to SO$_3$ and thence to sulphuric acid (p. 180).

4. Because of concern at the high emissions of NO$_x$ from cars, all those sold in the EC must be fitted with catalytic converters. Table 5.6 shows the relative efficiency of comparable groups of cars, with average engine capacity of 1.7 litres, relative to the pre-1992 EC standard (A) and to the post-1992 standard (B).

It is seen that although THC + NO$_x$ is at present within limits, incomplete combustion takes the CO level outside. Furthermore, unless changes are made, CO will be five times the future standard and THC + NO$_x$ over three times that permitted. Control of vehicular pollutants is discussed further on pp. 256–8.

The diesel engine. The use of diesel engines has been favoured because of concerns about greenhouse warming, as they produce less CO$_2$/mile travelled than do petrol engines (Royal Commission, 1991). However, this favourable factor is counteracted by a relatively greater output of SO$_2$ and NO$_x$ (Table 5.7) and recent awareness of their particulate emissions (Section 6.1).

In the UK in 1989 there were 567 000 diesel goods vehicles and 103 000 diesel transport vehicles; their contribution to NO$_x$ emissions is shown in Table 5.8.

Table 5.7 Typical sulphur content of petroleum fractions

	Percentage S	Quantity used[a]
Motor spirit	0.1	24
Diesel fuel	0.3	10.6
Gas oil (domestic)	0.7	8
Fuel oil (power station)	2.0	6.5
Heavy fuel oil (industry)	> 3.5[b]	11

[a] 1990 in the UK, 10^6 tonnes.
[b] 100 tonnes yield 7 tonnes SO$_2$.

Table 5.8 NO_x emissions in the UK (Royal Commission, 1991)

Source	Yield (m tonnes)	Percentage
All vehicles	1.3	48
Petrol-powered vehicles	0.74	27
Diesel vehicles	0.56	21
Power stations	0.78	29
Manufacturing industry	0.27	10
All other	0.36	13

The numbers of diesel passenger cars has risen in the UK from 5% of cars in 1986 to 19% in 1993. The output of NO_x from these vehicles is at an optimum of 0.5 g/km travelled at speeds between 40 and 90 kph, but rises to about 0.9 g/km outside these limits.

Emissions from power stations. Nitrogen compounds are present at low concentrations in the fuels and give rise to some NO_x on combustion, but the principal source is fixation of nitrogen by reaction of air in the furnace. Fluidized bed combustion is becoming the standard practice in order to contain emissions of SO_2 (p. 189). These operate at lower temperatures, which, as seen above, will also reduce the yield of NO_x.

5.2.3 Oxides of sulphur

Sources

Coal. The content of sulphur in coal (Table 6.1) rises to an extreme of 13% in Czechoslovakian supplies but normally varies between 0.5 and 4.0% with an average amount of about 1.3%; of this about 40% occurs as iron pyrites and the rest is organically combined. There may be significant minor inclusions of chlorine in the range 0.1–0.7% and traces of fluorine, phosphorus, lead and arsenic.

The biogenetic degradation of natural peptides during the formation of petroleum leads to the inclusion of nitrogen and sulphur compounds. Some of the typical organic components of petroleum, including those containing nitrogen and sulphur, are shown in Fig. 6.3 (p. 254); the sulphur content in various fuels is given in Table 5.7. It is also possible that some petroleums are of abiogenic origin and originate in the polymerization of methane, which is commonly trapped in sediments and elsewhere in the earth's crust.

Formation and fate. The worldwide consumption of oil in 1989 was 3.0 billion tonnes, with coal consumption at 2.7 billion tonnes oil equivalent (Sprague, 1990). The former was subdivided into low sulphur oil and gasoline (0.5% S), and fuel oil plus heavy oil (2.5% S) yielding some 28 million tonnes of SO_2 in that year. The coal consumed (1.3% S) afforded

39 million tonnes SO_2. In contrast to CO_2, SO_2 is highly water soluble (23 g/100 g at 0°C; 11 g/100 g at 20°C and NTP) and has a lifetime of only a few hours in the atmosphere before it dissolves in surface water. Its release from vehicles is small compared with that of NO_x but power stations are again implicated, being responsible for 60% of the SO_2, twice the quantity produced by industry including the refineries.

Unlike nitrogen, sulphur reacts readily with oxygen during combustion:

$$S + O_2 \longrightarrow SO_2 \tag{5.23}$$

$$SO_2 + \tfrac{1}{2}O_2 \longrightarrow SO_3 \tag{5.24}$$

The oxidation to SO_3 (Equation 5.24) does not occur readily in dry air but proceeds when catalysed by platinum, vanadium or sunlight; ozone is also an effective oxidant (Wayne, 1991). Sulphur dioxide forms sulphuric acid in its reaction with NO:

$$H_2O + 2NO + SO_2 \longrightarrow H_2SO_4 + N_2O \tag{5.25}$$

but in polluted air the principal route involves the $^{\bullet}OH$ radical. In the presence of water and oxygen this can be represented as:

$$^{\bullet}OH + SO_2 + (O_2,H_2O) \longrightarrow H_2SO_4 + {}^{\bullet}OOH \tag{5.26}$$

or more particularly as the initial formation of the bisulphite radical:

$$^{\bullet}OH + SO_2 + M \longrightarrow HO.SO_2^{\bullet} + M \tag{5.27}$$

followed by its reaction with water and oxygen:

$$\tag{5.28}$$

Effect on the environment.   Sulphur dioxide and the sulphuric acid formed from it have four adverse effects:

1. toxicity to humans
2. acidification of lakes and surface waters
3. damage to trees and crops
4. damage to buildings.

Toxicity to humans (Park, 1987).   The detection limit of SO_2 by humans is about $0.5 \mu g/g$ and at levels below $1 \mu g/g$ it has no obvious effect but breathing difficulties are experienced above $1.5 \mu g/g$. Exposure at $200 \mu g/g$ for 1 min causes great discomfort, while the limit for prolonged exposure is $5 \mu g/g$. The gas exerts its worst effects on asthmatics and bronchitics, who are at risk if exposed for 1 day at the detection limit.

Inhalation of SO_2 was established as the cause of death in a number of incidents:

1930 In the Meuse valley, levels over 10 ppm built up during a thermal inversion with input from power stations, iron and steel works and other industries. Sixty people died on two December days from heart failure linked to respiratory disorder.
1948 Under similar conditions at Donora near Pittsburg over 10 000 people were affected during 5 days in October and 20 deaths were attributed to stress from SO_2.
1952 One of the authors experienced the London 'smog' between 5–9 December, which was the worst of a recurrent series of these incidents, when to walk outdoors was to travel the unknown. The conditions have been recorded in fiction by Dickens and Conan Doyle. In the months following this worst episode, when SO_2 levels reached $0.7 \mu g/g$, 4000 deaths from lung and heart disease above the normal expectancy were recorded. This incident prompted the enactment of the Clean Air Acts of 1956 and 1968 (*cf.* Section 1.3).

Acidification of lakes and surface waters.   A measure of acidity is the pH value, which is defined as $\log_{10} 1/[H^+]$. The relatively weak carboxylic acids have dissociation constants (K_a) close to 10^{-5}; for acetic acid the figure is $10^{-4.76}$:

$$K_a = [H^+][AcO^-]/[AcOH] = 10^{-4.76} \qquad (5.29)$$

About 1% of the acetic acid in a 0.1 molar solution is dissociated. From Equation 5.29, $c = [AcOH]$ and $[H^+] = [AcO^-]$, hence $[H^+]^2 = K_a \times c = 10^{-4.76} \times 10^{-1} = 10^{-5.76}$. Therefore $[H^+] = 10^{-2.88}$ and the pH of a 0.1 molar solution of acetic acid is 2.88.

Decrease of pH can be prevented by buffer action, which is best illustrated by a solution containing a salt of a weak base, for example sodium acetate,

Table 5.9 Some typical pH values

Strong alkali	14.0	Pure rain	5.7
Sea water	8.5	Acid rain	4.0–4.4
Blood	8.0	Wine	4.0
Conductivity water	7.0	Lemon juice	2.3

which captures added protons to release the largely undissociated weak acetic acid:

$$Na^+AcO^- + H^+ \longrightarrow Na^+ + AcOH \qquad (5.30)$$

For strong acids the dissociation is essentially complete so that $[H^+] = 1.0$ and the pH $= \log 1.0 = 0$. Some typical pH values are shown in Table 5.9.

Many scientists would accept a value less than 5.7 for unpolluted rain, but few would accept the figure of 5.0 proposed by the former CEGB, which reveals a vested interest. Because the pH scale is logarithmic, it tends to conceal the fact that rain of pH 5.0 has a hydrogen ion concentration five times (log 0.7) that of rain of pH 5.7.

Low-risk areas are those remote from industrial sources or those which have alkaline soils which 'buffer' and sustain a higher soil pH. Acid lakes can be improved by seeding them with lime, but this requires renewal and is expensive.

High-risk areas are those which are close to industrial sources or which are exposed to transported acids, and the worst consequence appear when the fallout occurs onto thin soils lying on granitic bedrock. Surface waters on granites are not buffered and reveal the trend before soils and plants. A notable consequence is the decline in fish populations where the critical pH is 6.0; Fig. 5.9 shows changes in the acidity of lakes in Sweden and New York State during 1930–75.

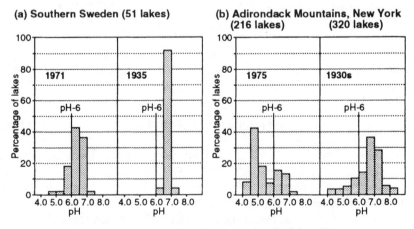

Figure 5.9 Acidification of lakes, 1930–75 (Wright, 1976).

Table 5.10   Changes consequent on a fall in the pH of surface waters

pH 6.0	Number and variety of species begin to decline
pH < 5.8	Green algae and diatoms disappear; more light penetrates, the water is clearer
pH < 5.5	Growth of sphagnum moss and filamentous algae is enhanced; matting on lake floor; oxygen access reduced with accumulation of organic matter; bacteria no longer survive

Natural buffers in rivers include phosphate, amino acids, bicarbonate and organic matter and their pH depends on an input from soil and rocks in the catchment, rather than from direct rainfall on the relatively small area of the river itself. The pH can be affected by changes in land use. A sequence of changes is identified in Table 5.10.

It is now established beyond doubt that atmospheric pollution is linked to the acidification of surface waters. A significant observation is that of Battarbee *et al.* (1985) who showed that trace metals unique to anthropogenic emissions correlated with surface acidity.

Effect on fish stocks.   Roach, trout and perch cannot tolerate acid waters and many upland lakes such as those monitored for Fig. 5.9 have no fish left in them. Depletion of fish populations has been recorded in Scotland, notably in Galloway, and the headwaters of many rivers elsewhere show signs of a decline.

In acid waters, the solubilization of heavy metals (Table 5.17) such as manganese, cadmium, mercury and, especially aluminium is the principal cause of fish death. The last is liberated from clays at pH 4.0 and is toxic to fish even when the pH of the water is not at a harmful level.

Aluminium interferes with reproduction and also damages the gills, while precipitation of aluminium phosphate reduces the primary production of plants and phytoplankton so limiting the food supply.

Effect on soils and vegetation.   Soils usually contain buffering components and hence are much more acid tolerant than surface waters and the practice of liming will give additional protection. A better guide than pH is the *buffer capacity* which is defined as: the number of mols of H^+ or OH^- required to lower or raise the pH of 1 kg of soil by 1 pH unit.

Tree growth leads to soil acidification as Ca^{2+} and Mg^{2+} are taken up and, in order to maintain a charge balance, H^+ is released. Acid soils may become impoverished through the leaching of K^+, Ca^{2+} and Mg^{2+} and by inhibition of the natural decay of litter, leading in turn to increased ammonia and reduced nitrate levels. When soil bacteria are active in converting NH_4^+ to NO_3^-, H^+ is released and represents a source of acidity not indicated by pH measurements; this is an adverse effect from intensive agriculture, where it is common practice to apply ammonium nitrate at 150 kg of nitrogen/ha. The dieback of ash trees post-1950 has been attributed to this practice.

Aluminium is found in rain at about 0.012 mg/l and may reach a level of 0.45 mg/l in soil solution and 0.14 mg/l in streams. Relatively little sulphate deposited on soil reaches streams, but at pH < 4.0 Al^{3+} is mobilized and its ability to adsorb SO_4^{2-} then decreases. The role of Al^{3+} has been disputed but Ulrich (1980) has established that it has a direct effect in damaging fine root hairs and so reducing the uptake of nutrients.

Vegetation damage was first observed in 'desert' areas close to point sources such as smelters, where pine trees are often absent within a radius of 20 km. Lichen deserts also point to acid pollution; they are especially susceptible as they have no roots and depend on nutrients in rain. Trees and higher plants are affected through (i) denial of nutrients to the roots and (ii) damage to exposed foliage.

The 1970s saw evidence of inexplicable damage to coniferous and deciduous trees in the Black Forest of West Germany ('Waldsterben'), which was exposed early on owing to its location close to major industrial output, and in 1980 Ulrich attributed the wasting of its trees to acid rain and defined an unnatural sequence:

1. deposition of nitrate and NO_x in soil accelerates growth
2. the pH falls and aluminium is mobilized
3. at pH 4.2 destruction of the roots starts the final deterioration.

Regeneration is difficult on inhospitable acid soils and by 1984 extensive damage to spruce and silver fir to a value of 10 billion DM was reported. Wasting of trees was subsequently identified in the rest of Europe, USA and Canada, which has an extensive forest industry.

Table 5.11 (Innes, 1987) shows the heavy damage to both coniferous and deciduous trees in West Germany, with almost 50% of silver firs in the 26–60% needle/leaf loss group. In 1986, up to 25% loss was occurring right across Europe.

In the UK, the Forestry Commission was initially dismissive, attributing the damage to successive drought years. It is now accepted that the observed damage is the result of several factors but that acid deposition contributes significantly to forest decline. The Forestry Commission is monitoring 141 plots including Sitka spruce, Norway spruce and Scots pine and classifies damage in five groups (Table 5.12).

A French study (Leclercq, 1988) based on boring tree trunks and evaluating annual growth cast doubt on the theory that only natural conditions were responsible for decline. In particular, the annual growth of spruce in the Vosges and of Silver fir in Luchon did not correspond with the stress events of drought in 1921, 1948 and 1976, nor with the severe frost of 1956.

There is no short-term link with SO_2 levels. In Hesse, damage to trees occurred at daily average levels of 800–1600 µg/m³, whereas in Freiburg damage was observed at an annual mean level of only 12 µg/m³. This must be significant despite the different methods of averaging. Direct damage to

Table 5.11 Forest damage assessment in five European countries (Innes, 1987)

	\[0–10%\] 83	84	85	86	\[11–25%\] 83	84	85	86	\[20–60%\] 83	84	85	86	\[61–100%\] 83	84	85	86
	\[Needle/leaf loss\]															
United Kingdom																
Sitka spruce		65	83	45		28	12	39		6	5	15		1	0	1
Norway spruce		71	84	32		26	15	36		3	1	31		1	0	1
Scots pine		49	74	25		29	18	41		16	7	32		5	1	3
West Germany																
Norway spruce	59	49	48	46	30	31	28	32	10	19	21	20	1	2	3	2
Pine	56	41	43	46	32	38	41	40	10	20	15	13	1	1	2	1
Silver fir	25	13	13	18	27	29	21	22	41	45	50	49	8	13	16	11
Beech	74	50	46	40	22	39	40	41	4	11	13	18	0	1	1	1
Oak	85	57	45	39	13	35	39	41	2	9	16	19	0	0	1	1
Other trees	83	69	69	65	9	24	23	25	8	7	7	9	0	1	1	1
Switzerland																
Norway spruce		65	63	50		28	29	36		6	6	12		1	2	2
Pine		50	35	34		31	47	43		16	13	19		1	5	4
Silver fir		62	60	47		27	28	36		9	8	13		2	4	4
Larch		64	66	39		28	23	44		7	7	12		1	4	5
Beech		74	69	52		23	27	40		3	3	7		0	1	1
Oak		71	60	37		28	33	50		1	6	11		0	1	2
Maple		86	86	73		11	11	25		2	1	1		1	2	1
Ash		84	77	57		13	20	36		3	2	7		0	1	7
Netherlands																
Norway spruce		62	48	49		28	41	34		7	9	12		3	2	4
Scots pine		34	48	50		51	36	33		12	14	13		2	2	3
Corsican pine		57	40	19		34	42	29		8	15	40		1	3	12
Douglas fir		50	33	17		39	43	27		9	22	45		2	2	11
Beech		71	72	68		24	21	26		4	6	5		1	1	2
Oak		57	40	30		38	39	42		5	19	20		1	2	9
Luxembourg																
Norway spruce		79	84	87		17	12	10		3	3	2		2	1	1
Oak		59	77	81		34	20	16		6	3	2		2	1	0
Beech		66	70	67		28	28	27		5	5	6		1	1	1

Table 5.12 Classification of damage to trees (Innes, 1990)

Class	Percentage needle loss
0 Healthy	0
1 Slight damage	11–25
2 Medium–serious	26–60
3 Dying	61–99
4 Dead	100

trees may occur in fog and cloud water, which can have 20 times the SO_2 concentration of rain water and a pH in the range 2.8–3.1.

It is difficult to define the effect of individual factors because a considerable time usually elapses between exposure and the emergence of visible signs such as discoloration, shortening of needles and shoots, dry buds and crown die-back. Current thinking is that ozone (Section 5.1) is implicated but also that NO_x and/or SO_2 may act synergistically and that long-term fumigation tests of these agents are needed. A recent study (Spence *et al.*, 1990) which employed ^{11}C labelling in a test cabinet promises a way to test the toxicity of individual pollutants.

In spite of the undoubted damage done by acid rain to trees close to industrialized areas, the tendency to attribute all forest decline (Waldsterben) to air pollution post-1970 should be resisted (Kandler and Innes, 1995). The situation is complex and changes for the worst, and for the better, may be species specific. Actual improvements in forest quality have been recorded in Europe post-1980, while photographs of forest stands, taken as far back as the 1920s, provide evidence of foliage deficits comparable to those seen today.

A note on units. It is important when levels of air pollution are measured that the prevailing temperature and pressure of the air body are recorded. This is very necessary for SO_2 levels in mountain forests because a given mass of a gas will occupy a greater volume under the lower air pressure. It is better to quote levels in ppm (vol./million vol.) rather than in mass (mg)/volume (m^3). (NB. Unless the temperature and pressure of the air are recorded at the time of measurement subsequent conversion between these units will be inaccurate.) A table for conversion of these units is given in the appendix; it is based on the following calculation:

At 20°C (293 K) one atmosphere of a gas *which obeys the gas laws* contained 4.16×10^{-2} mol/l, hence 1 ppm will contain:

$$= 4.16 \times 10^{-2} \times 10^{-6}$$
$$= 4.16 \times 10^{-8} \text{ mol/l}$$
$$= 4.16 \times 10^{-5} \text{ mol/m}^3$$

Therefore, if the molecular weight of the gas is M,

$$1 \text{ ppm as } \mu g/m^3 = 4.16 \times 10^{-5} \times M \ \mu g/m^3$$
$$= 4.16 \times 10^{-2} \times M \text{ mg/m}^3 \tag{5.31}$$

For SO_2 at 20°C, 1 ppm $= 4.16 \times 10^{-2} \times 64$

$$= \mathbf{2.66} \text{ mg/m}^3 \tag{5.32}$$

At 0°C, in accordance with the gas law, this relation is

$$1\,ppm = 2.66 \times 293/273 = 2.85\,mg/m^3 \qquad (5.33)$$

Damage to buildings. In the UK, this has been the subject of detailed analysis by the Building Effects Review Group. The buildings worst affected are those finished with limestone, which is freely soluble (Equation 5.36). St Paul's cathedral is exposed to acid rain at pH 4.0 (Table 5.9) and the surface pH has been shown to rise as a result of solution of the building stone, which suffers surface erosion at the rate of 200 μm/year. Damage to the limestone of the colleges of Oxford University cost £7 million to repair in 1991.

Rusting of iron and steel is primarily caused by oxygen and water but it is accelerated by SO_2, nitric acid and chlorides.

Studies of these effects are ongoing through the National Materials Exposure Programme. At the time of the London smog the Beaver Committee estimated losses at £250 million. A review of stone weathering in south-east England is available (Janes and Cooke, 1987).

Monitoring and control. Air sampling work in the UK was formerly undertaken by the Warren Spring Laboratory; this has now been merged with part of AEA Technology to form the National Environmental Technology Centre (AEA, Seventh Report; Bower *et al.*, 1995). The UK monitoring network now includes over 1500 sampler measurement sites and 49 automatic monitoring stations.

Legislative requirements. Control of SO_2 and NO_x are best considered together as they depend on common procedures and set limits for emissions. In Europe for the future these must conform to the EC Framework Directive (84/360/EEC; Speakman, 1990), which defines emission limits to be achieved in three phases by the year 2003 (Tables 5.13 and 5.14).

The above limits apply only to power stations and, therefore, exclude about one-third of the emitted SO_2 (p. 179) and half of the NO_x, for which motor vehicles are largely responsible (p. 175).

Table 5.13 EC targets for major producers of SO_2 and NO_x (Speakman, 1990)

	1980 SO_2 (K tonnes)	Ceiling (K tonnes)	Reduction (%)	1980 NO_x (K tonnes)	Ceiling (K tonnes)	Reduction (%)
Germany	2225	668	−70	870	522	−40
Spain	2290	1440	−50	366	277	−40
France	1910	573	−70	400	240	−40
Italy	2450	900	−70	580	428	−40
UK	3883	1553	−60	1016	711	−30

Table 5.14 Air quality standards for NO_2 and SO_2

	Level (ppb)		
	98 percentile	23 h av.	1 h av.
NO_2			
EU directive	105	–	–
US EPA		53	600 [a]
WHO	–	78 [b]	210 [b]
SO_2			
EU directive	94 ($>$ 128) [c]		
	132 ($<$ 128) [c]		
US EPA		140	
WHO			122

[a] EPA 'alert' level.
[b] Guidelines only.
[c] The figure in parentheses takes account of synergism by the associated level of particulates (in $\mu g/m^3$).

Control measures. *Precombustion removal of sulphur* is chemically feasible for oil through the pressurized hydrogenation of the fuel and alkaline trapping of liberated hydrogen sulphide:

$$R–S–R + 2H_2 \longrightarrow 2R–H + H_2S \tag{5.34}$$

However, the costs are high and are estimated to add 20% to the price of electricity (Tolba, 1983).

There has been considerable research into the cleaning of coal; water washing dissolves part of the iron pyrites (Section 5.23) and can reduce SO_2 emissions by 10%. The chemical methods largely depend on oxidation and subsequent neutralization of sulphuric acid with lime:

$$4FeS_2 + 15O_2 + 8H_2O \longrightarrow 2Fe_2O_3 + 8H_2SO_4 \tag{5.35}$$

The aqueous slurry is pressurized with oxygen in an autoclave and heated to 150–200°C, when most of the inorganic sulphur is removed (Equation 5.35).

Organic sulphur compounds (Fig. 6.3, p. 254) include thiols (RSH), which react completely under these conditions, and some alkyl thioethers (RSR), which react in part, but of the remaining known compounds, aryl ethers (PhSPh) and sulphur heterocycles are unaffected. An added disadvantage is the loss of heating value of the order of 10–20% for only 10–20% sulphur removal. It is, therefore, policy to control emissions by flue gas desulphurization.

Flue gas desulphurization (FGD). This was pioneered in the Battersea power station, which achieved 80% reduction by scrubbing with water from the river Thames. Figure 5.10 outlines the material flows for the 4000 MW Drax power stations. The scrubbing plant injects limestone slurry, which converts SO_2 into calcium sulphite:

$$CaCO_3 + SO_2 \longrightarrow CaSO_3 + CO_2 \tag{5.36}$$

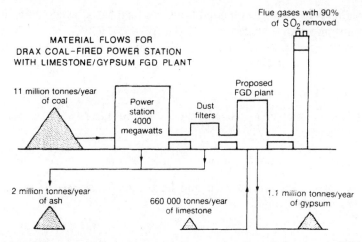

Figure 5.10 Material flows for a Drax power station (from Roberts *et al.*, 1990).

and in the same process air is also injected to produce the more tractable calcium sulphate (gypsum):

$$CaSO_3 + \tfrac{1}{2}O_2 \longrightarrow CaSO_4 \tag{5.37}$$

Major new environmental problems arise since extraction of limestone on this scale risks despoilation of the nearby Peak National Park and movement of limestone, gypsum and ash entails some 200 000 lorry movements a year.

Fluidized bed combustion (FBC).   This technique was researched by British Coal and again depends upon the capture of SO_2 by lime or limestone to form calcium sulphate. Coal in the furnace bed is mixed with limestone and fluidized by passage of air (*cf.* Equation 5.37) at 800–1000°C. A higher temperature must be avoided for two reasons:

1. calcium sulphate begins to decompose and release CO_2:

$$CaSO_4 \xrightarrow{c.1200°C} CaO + SO_2 + \tfrac{1}{2}O_2 \tag{5.38}$$

2. in the region of 1400°C, thermal fixation of nitrogen occurs to produce NO_x; this may be 40% of the total.

Control of NO_x.   For fuel oils, which have little combined nitrogen, the emission is governed by the furnace temperature and can be kept as low as $200 \, mg/m^3$ by operating in the lower region.

As seen above, FBC can prevent the fixation of nitrogen during coal burning, but combined nitrogen is in the range of 1–2% and will decompose to release ammonia and other nitrogen compounds, which are subsequently oxidized to NO_x. If fuel-rich conditions are established in the early stages, these substances are reduced to release nitrogen which escapes at the lower operating temperature, reducing the NO_x emission by up to 30%.

Coal volatiles will reduce NO:

$$2NO + 2C \longrightarrow N_2 + 2CO \tag{5.39}$$

and if further control is needed other reducing agents such as ammonia can be injected:

$$6NO + NH_3 \xrightarrow{c.\ 900°C} 5N_2 + 6H_2O \tag{5.40}$$

An FBC plant can process coal and keep NO_x emissions below the EC requirement of 650 mg/m³.

5.3
Heavy metals

5.3.1   General properties

'Heavy metals' is a general collective term applying to the group of metals and metalloids with an atomic density greater than 6 g/cm³. Although it is only a loosely defined term, it is widely recognized and usually applied to the elements such as Cd, Cr, Cu, Hg, Ni, Pb and Zn which are commonly associated with pollution and toxicity problems. An alternative (and theoretically more acceptable) name for this group of elements is 'trace metals' but it is not as widely used. Unlike most organic pollutants, such as organohalides, heavy metals occur naturally in rock-forming and ore minerals and so there is a range of normal background concentrations of these elements in soils, sediments, waters and living organisms. Pollution gives rise to anomalously high concentrations of the metals relative to the normal background levels; therefore, presence of the metal is insufficient evidence of pollution, the relative concentration is all important.

Apart from aerosols in the atmosphere and direct effluent discharges into waters, the concentrations of heavy metals available to terrestrial, aquatic and marine organisms (i.e. their bioavailability) is determined by the solubilization and release of metals from rock-forming minerals and the adsorption and precipitation reactions which occur in soils and sediments. The extent to which metals are adsorbed depends on the properties of the metal concerned (valency, radius, degree of hydration and coordination with oxygen), the physico-chemical environment (pH and redox status), the nature of the adsorbent (permanent and pH-dependent charge, complex-forming ligands), other metals present and their concentrations, and the presence of soluble ligands in the surrounding fluids.

Although heavy metals differ widely in their chemical properties, they are used widely in electronics, machines and the artefacts of everyday life as well as 'high-tech' applications. Consequently, they tend to reach the environment from a vast array of anthropogenic sources as well as natural geo-

Table 5.15 Primary production of metals and global emissions to soil (10^3 t/year) (from Nriagu, 1988)

Metal	Production in		Global emissions to soil in 1980s
	1930	1985	
Cd	1.3	19	22
Cr	560	9940	896
Cu	1611	8114	954
Hg	3.8	6.8	8.3
Ni	22	778	325
Pb	1696	3077	796
Zn	1394	6024	1372

chemical processes. Some of the oldest cases of environmental pollution in the world are caused by heavy metal use, such as Cu, Hg and Pb mining, smelting and utilization by ancient civilizations, such as the Romans and the Phoenicians.

The primary production of heavy metals in 1930 and 1985 and the global emissions of metals to soils in the 1980s are shown in Table 5.15.

The data in Table 5.15 show that production of all the metals has increased over the 55 year period. Nickel showed the greatest increase ($\times 35$), followed by Cr ($\times 17$) and Cd ($\times 14$), whereas that for Hg production had not even doubled. The greater emission to soil than annual production for Cd and Hg is probably explained by the occurrence of Cd as a contaminant of fertilizers and other metal ores and Hg being both a constituent of other metal ores and also being emitted naturally from volcanoes.

5.3.2 Biochemical properties of heavy metals

Some of the elements in this group are required by most living organisms in small but critical concentrations for normally healthy growth (referred to as 'micronutrients' or 'essential trace elements') but excess concentrations cause toxicity. Those metals which are unequivocally essential, whose deficiency causes disease under normal living conditions include Cu, Mn, Fe, and Zn for both plants and animals, Co, Cr, Se and I for animals; B and Mo for plants. Most of the micronutrients owe their essentiality to being constituents of enzymes and other important proteins involved in key metabolic pathways. Hence, a deficient supply of the micronutrient will result in a shortage of the enzyme, which leads to metabolic dysfunction causing disease.

Some other elements have been shown to have some beneficial effect under rigorous experimental conditions but are not likely to be responsible for deficiency disorder under normal conditions. Elements with no known essential biochemical function are called 'non-essential elements' but are

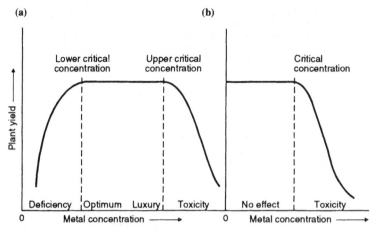

Figure 5.11   Typical dose–response curve for (a) micronutrients and (b) non-essential trace elements (adapted from Alloway, 1990).

sometimes also referred to (incorrectly) as 'toxic' elements. These elements, which include As, Cd, Hg, Pb, Pu, Sb, Tl and U, cause toxicity at concentrations which exceed the tolerance of the organism but do not cause deficiency disorders at low concentrations like micronutrients. These are clearly shown by the typical dose–response curves in Fig. 5.11.

At the biochemical level, the toxic effects caused by excess concentrations of these metals include competition for sites with essential metabolites, replacement of essential ions, reactions with –SH groups, damage to cell membranes, and reactions with the phosphate groups of ADP and ATP. Organisms have homeostatic mechanisms which enable them to tolerate small fluctuations in the supply of most elements but prolonged excesses eventually exceed the capacity of the homeostatic system to cope and toxicity occurs, which if severe can cause the death of organisms. An example of homeostasis in animals and the control of excess metals is the formation of metallothionein proteins containing –SH groups, which bind certain metals, such as Cd and Zn, and enable them to be excreted without causing biochemical dysfunction. In plants, similar compounds called phytochelatins carry out the same function, binding divalent metals, such as Cd, in physiologically inactive forms.

5.3.3   Sources of heavy metals

Geochemical sources. In geological terms heavy metals are included in the group of elements referred to as 'trace elements', which together constitute less than 1% of the rocks in the earth's crust; the macroelements (O, Si, Al,

Fe, Ca, Na, K, Mg, Ti, H, P and S) comprise 99% of the earth's crust. These trace elements occur as 'impurities' isomorphously substituted for various macroelement constituents of the crystal lattice of many primary minerals. Primary minerals are those found in igneous rocks which originally crystallized from molten magma. In sedimentary rocks, trace elements occur sorbed to the secondary minerals which are the products of the weathering (physical disintegration and chemical decomposition) of primary minerals and in their resistant fragments.

Primary and secondary minerals differ widely in their trace element content and hence the igneous and sedimentary rocks which they, respectively, form also show a wide variation in heavy metal content. Typical ranges of values for heavy metal concentrations in the earth's crust and various major rock types are given in Table 5.16.

In view of the high degree of variation in the metal contents of rocks shown in Table 5.16, there is a possibility that the soils and stream sediments in a locality suspected of being polluted may have developed from rocks with anomalously high concentrations of certain heavy metals and that pollution, in the strict sense of the definition, has not occurred. Nevertheless, the natural enrichment of metals in the soils may still give rise to harmful effects in living organisms. The natural members of an ecosystem in an area

Table 5.16 Typical concentrations of heavy metals in the earth's crust and major types of igneous and sedimentary rock (from Alloway (1990) based mainly on Krauskopf (1967) and Rose *et al.* (1979))

	Earth's crust	Igneous rocks			Sedimentary rocks		
		Ultramafic	Mafic	Granitic	Limestone	Sandstone	Shales/clays
Ag	0.07	0.06	0.01	0.04	0.12	0.25	0.07
As	1.5	1	1.5	1.5	1	1	13 (<900)
Au	0.004	0.003	0.003	0.002	0.002	0.003	0.0025
Cd	0.1	0.12	0.13	0.09	0.028	0.05	0.22(<240)
Co	20	110	35	1	0.1	0.3	19
Cr	100	2980	200	4	11	35	39
Cu	50	42	90	13	5.5	30	39
Hg	0.005	0.004	0.01	0.08	0.16	0.29	0.18
Mn	950	1040	1500	400	620	460	850
Mo	1.5	0.3	1	2	0.16	0.2	2.6
Ni	80	2000	150	0.5	7	9	68
Pb	14	14	3	24	5.7	10	23
Sb	0.2	0.1	0.2	0.2	0.3	0.05	1.5
Se	0.05	0.13	0.05	0.05	0.03	0.01	0.5
Sn	2.2	0.5	1.5	3.5	0.5	0.5	6
Tl	0.6	0.0005	0.08	1.1	0.14	0.36	1.2
U	2.4	0.03	0.43	4.4	2.2	0.45	3.7
V	160	40	250	72	45	20	130
W	1	0.1	0.36	1.5	0.56	1.6	1.9
Zn	75	58	100	52	20	30	120

Tl, thallium; V, vanadium; W, tungsten.

of geochemical enrichment will have evolved tolerance to the elevated concentrations of metals, but newly introduced plant and animal species may be adversely affected. It is, therefore, important to determine the local background concentrations of heavy metals in order to determine whether the concentrations in the soils and sediments under investigation are significantly higher than those of the area. The normal procedure is to conduct a survey of at least 15 to 20 samples from over the area (the number depending on the size of the area and the degree of heterogeneity in soil types) and to determine the arithmetic mean concentration and its standard deviation (SD) for the range of elements being investigated. Anomalously high (or low) concentrations will lie outside the value for the mean ±3 SDs, but values of 2 SDs above the mean are regarded as 'threshold' values.

Heavy metals present in the atmosphere and in the hydrosphere.   Table 5.17 gives the concentrations of metals in the atmosphere in three types of location, and the ranges reported in fresh water and sea water. The concentrations of heavy metals in the atmosphere in an area remote from anthropogenic effects (South Pole) are markedly lower than those from various locations in Europe. The maximum concentrations found in the

Table 5.17   Metals in the atmosphere and hydrosphere (from Bowen, 1979)

Metal	Atmosphere (ng/m^3)			Hydrosphere ($\mu g/l$)[a]	
	South pole	Europe	Near volcano	Fresh water	Sea water
Ag	<0.0004	0.2–7	3	0.001–3.5	0.03–2.7
As	0.007	1.5–53	<850	0.2–230	0.5–3.7
B	–	3.5	680	7–500	av. 4.44k
Be	–	0.9–4	–	0.01–1	2–63
Cd	<0.015	0.5–620	8–92	0.01–3	<0.01–4
Cr	0.005	1–140	45–67	0.1–6	0.2–50
Cu	0.036	8–4.9k	200–3k	2–30	0.05–12
F	–	1.5 (US<400)	High	50–2.7k	av 1.3k
Fe	0.84	130–5.9k	1k–10k	10–1.4k	0.03–70
Hg	–	<0.009–2.8	18–250	0.0001–2.8	0.01–0.22
Mn	0.01	9–210	55–1.3k	0.02–130	0.03–21
Mo	–	<0.2–3.2	–	0.03–10	4–10
Ni	–	4–120	330	0.02–27	0.13–43
Pb	0.63	55–340 (US<13k)	28–1200	0.06–120	0.03–13
Sb	0.0008	0.6–3.2	45	0.01–5	0.18–5.6
Se	0.0056	0.15–11	<2100	0.02–1	0.052–0.2
Sn	–	1.5–800	–	0.0004–0.09	0.002–0.81
U	–	0.2(US<0.5)	–	0.02–5	0.04–6
V	0.0015	5–92 (US<2k)	79	0.01–20	0.09–2.5
W	0.0015	0.35–1.5	–	<0.02–0.1	0.9–2.5
Zn	0.03	13–16k	10k	0.2–100	0.2–48

[a] $\mu g/l$ is equivalent to ppb. $k = 10^3$

USA are quoted for some elements to indicate the highest concentrations monitored for technologically advanced countries. In the case of the chalcophyllic elements (those normally found as sulphides in ore bodies, such as Cd, Cu, Pb and Zn) the concentrations were particularly high near volcanoes (Etna and Hawaii) as a result of the lavas containing high concentrations of these elements. The freshwater samples generally demonstrated a wider range than sea water owing to the effect of salinity in controlling solubility products.

Anthropogenic sources. Although heavy metals are ubiquitous in most natural materials, the following are significant sources of metals to the environment (see also Table 2.1, pp. 19, 20).

Metalliferous mining. The metals utilized in manufacturing are obtained from either the mining of ore bodies in the rocks of the earth's crust, or the recycling of scrap metal originally derived from geological sources. Ore bodies are naturally occurring concentrations of minerals with a sufficiently high concentration of metals to render them economically worthwhile exploiting. With increasing demand, generally rising prices and improvements in the technology of mineral extraction there is a trend for orebodies with progressively lower metal contents to be used. Generally, these lower grade orebodies are larger in extent and require a higher proportion of rock to be mined per tonne of metal extracted. This inevitably implies that the environmental impact of the mining operations is greater than for mines with smaller areas of much higher grade ore. There will be a need to dispose of greater amounts of tailings, which are the finely milled fragments of rock and some ore particles left behind after the extraction of the metal ore concentrates by various means based on density, magnetism and surface tension (froth flotation). Tailings disposal from current mining operations and continued weathering (chemical alteration) of ore minerals in historical and abandoned mining sites is an important source of heavy metals into the environment (see Fig. 8.2, p. 370). Some of the most common ore minerals of non-ferrous metals are shown in Table 5.18.

From this table, it can be seen that many of the major metalliferous ores are sulphide minerals and this has several environmental implications. Fragments of these minerals in tailings deposits oxidize on weathering to create acidic solutions which tend to decrease adsorption and hence increase the mobility of the metals in soils, sediments and waters. The oxidation of pyrite (FeS_2) is discussed on pages 51 and 52. The reactions for Zn and Cu ores can be written in a simplified form as:

$$\text{Sphalerite}: ZnS + 2O_2 \longrightarrow Zn^{2+} + SO_4^{2-} \qquad (5.41)$$

$$\text{Chalcopyrite}: CuFeS_2 + 4O_2 \longrightarrow Cu^{2+} + Fe^{2+} + SO_4^{2-} \qquad (5.42)$$

Frequently, bacteria such as *Thiobacillus ferrooxidans* help to catalyse the oxidation reactions in tailings deposits and in weathering orebodies.

Table 5.18   Common ore minerals of non-ferrous metals (Peters, 1978; Rose *et al.*, 1979)

Metal	Ore minerals	Associated heavy metals
Ag	Ag_2S, PbS	Au, Cu, Sb, Zn, Pb, Se, Te
As	FeAsS, AsS	As, Au, Ag, Sb, Hg, U, Bi,
	Cu ores	Mo, Sn, Cu
Au	Native Au, $AuTe_2$	Te, Ag, As, Sb, Hg, Se
	$(Au,Ag)Te_2$	
Ba	$BaSO_4$	Pb, Zn
Bi	Pb ores	Sb, As
Cd	ZnS	Zn, Pb, Cu
Cr	$FeCr_2O_4$	Ni, Co
Cu	$CuFeS_2$, Cu_5FeS_4	Zn, Cd, Pb, As, Se, Sb, Ni, Pt, Mo, Au, Te
	Cu_2S, Cu_3AsS_4	
	CuS, Native Cu	
Hg	HgS, Native Hg	Sb, Se, Te, Ag, Zn, Pb
	Zn ores	
Mn	MnO_2	Various (e.g. Fe, Co, Ni, Zn, Pb)
Mo	MoS_2	Cu, Re, W, Sn
Ni	$(Ni,Fe)_9S_8$, NiAs	Co, Cr, As, Pt, Se, Te
	$(Co,Ni)_3S_4$,	
Pb	PbS	Ag, Zn, Cu, Cd, Sb, Tl, Se, Te
Pt	Native Pt, $PtAs_2$	Ni, Cu, Cr
Sb	Sb_2S_3, Ag_3SbS_3	Ag, Au, Hg, As
Se	Cu ores	As, Sb, Cu, Ag, Au
Sn	SnO_2,	
	$Cu_2(Fe, Zn)SnS_4$	Nb, Ta, W, Rb
U	U_3O_8	V, As, Mo, Se, Pb, Cu, Co, Ag
V	V_2O_5, VS_4	U
W	WO_3, $CaWO_4$	Mo, Sn, Nb
Zn	ZnS	Cd, Cu, Pb, As, Se, Sb, Ag, Au, In

When the sulphide ore minerals are smelted to obtain the metals, it is necessary to roast (or sinter) the ores in air in order to convert the sulphides to oxides, which are subsequently reduced to the metal. This process gives rise to large amounts of SO_2 and metal smelters are among the major sources of this important atmospheric pollutant. Although smelter fumes can now be scrubbed to remove the SO_2, this is by no means the practice everywhere and much severe environmental pollution has occurred and still occurs from this source. In the past, the acidic fumes together with metal aerosol emissions exacerbated the environmental impact of the metals for many kilometres from the smelter source. This is clearly demonstrated at Zlatna in Romania, where a Cu–Zn smelter has given rise to severe soil erosion due to emission of acidic fumes.

Another feature shown in Table 5.18 is that most of the major ore minerals often have several other metals associated with them. Many of these metals will have contaminated the environment in the vicinity of mines and smelters, although nowadays it is more likely that the minor constituents will also be refined and marketed as well as the major metal constituent of the ores. Old mine waste deposits have been reworked in some countries, such as the UK, where the mineral separations used in Victorian times were inefficient and some minerals in the ore deposit were considered worthless at that time (gangue minerals). For example, hydrothermal Pb–Zn deposits in Carboniferous limestone rocks in Derbyshire, England, contained large amounts of fluorspar (CaF_2) and barite ($BaSO_4$), which were dumped at the mines. With the increased demand for fluorine as an additive in potable water (NaF), PTFE non-stick coatings and chloro-fluorocarbon compounds in refrigerators and aerosol propellants (p. 160) it became profitable to rework old mine waste tips. In the process, economically worthwhile amounts of barite and PbS were also extracted; the former was utilized in deep drilling as a high-density medium (e.g. 4.5 g/cm), and the latter was ultimately smelted for Pb metal production. A consequence of the reuse of mined material anywhere in the world is the production of large quantities of tailings, which need to be disposed of in an environmentally appropriate manner. Modern mineral separation methods involve the use of large volumes of water but much of this is normally recycled within the process, although smaller volumes of effluents containing metals, frothing agents and other chemicals (including cyanides in gold extraction) do need disposal eventually.

Agricultural materials. Agriculture constitutes one of the very important non-point sources (NPS) of metal pollutants. The main sources are:

- impurities in fertilizers: Cd, Cr, Mo, Pb, U, V, Zn (e.g. Cd and U in phosphatic fertilizers)
- pesticides: Cu, As, Hg, Pb, Mn, Zn (e.g. Cu, Zn and Mn-based fungicides, Hg seed dressings, historical Pb–As orchard sprays)
- desiccants: As for cotton
- wood preservatives: As, Cu, Cr
- wastes from intensive pig and poultry production: Cu, As, Zn
- composts and manures: Cd, Cu, Ni, Pb, Zn, As
- sewage sludge: especially Cd, Ni, Cu, Pb, Zn (but many other elements)
- corrosion of metal objects (e.g. galvanized metal roofs and wire fences: Zn, Cd).

Fossil fuel combustion. A wide range of heavy metals is found in fossil fuels, which are either emitted into the environment as particles during combustion (Section 6.1), or accumulate in ash, which may itself be transported in air and contaminate soils or waters or may be leached *in situ*. Some of the metals arising as pollutants from fossil fuel combustion are Pb, Cd, Zn, As, Sb, Se, Ba, Cu, Mn and V. The combustion of petrol (gasoline) containing Pb additives gives rise to large amounts of Pb particulates, mainly PbBrCl. Lead-containing particles in the exhausts of petrol vehicles are 0.01–0.1 μm in diameter, but these primary particles can cluster to form larger particles (0.3–1 μm). Diesel smoke normally contains these larger size particles. The Pb-containing particles in motor vehicle exhausts tend to be larger in rural areas and near motorways than in urban areas (Fergusson, 1990).

Coal combustion gives rise to a wide range of metals, including U in some coal, which accounts for coal-burning power stations being responsible for the emission of significant amounts of radioactive pollutants. Coal ash can contain relatively high concentrations of soluble compounds, including oxides of B, As, Se, which can be leached and cause toxicity in sensitive crops and aquatic organisms. Coal ash is an important source of Cr ($< 172\,\mu g/g$). Crude oil can contain relatively large amounts of V which is emitted during the combustion of some oil products and accumulates in the ash of oil-fired boilers.

Steinnes (1987) reported concentrations of Pb, Cd, Zn, As, Sb and Se to be around ten times higher in moss, soil humus and topsoils along the southern coast of Norway than in the centre of the country. This was ascribed to the deposition of aerosols arising largely from fossil fuel combustion in north-western Europe.

Metallurgical industries. Many heavy metals are used in specialist alloys and steels: V, Mn, Pb, W, Mo, Cr, Co, Ni, Cu, Zn, Sn, Si, Ti, Te, Ir, Ge, Tl, Sb, In, Cd, Be, Bi, Li, As, Ag, Sb, Pr, Os, Nb, Nd and Gd. Hence both the manufacture and disposal, or recycling, of these alloys in scrap metal can lead to environmental pollution with a wide range of metals. Steel manufacture usually involves a lot of recycling of scrap, and so steel works are often discrete point sources of atmospheric aerosols of metals.

Non-ferrous metal production causes marked environmental pollution not only of the metals being manufactured, but also of other minor associated metals (see Table 5.18) such as As, Cd, Cr, Cu, Co, Ni, Pb, Sb, Tl, Te, U, V, Zn and Se. Nowadays, many of these metals are extracted from the ores, refined and sold, but in the past this was not usually the case. For example, Cu smelters have a long history of causing As pollution in the surrounding countryside. Estimates of 1.5 kg As emitted per tonne of Cu produced are typical, with values of up to 16.8 kg As/t Cu from time to time at a smelter in Washington State, USA (O'Neill, 1990). In Austria, cancerous conditions have been reported in livestock grazing near to Cu smelters. Zinc ores often contain relatively high concentrations of Cd ($< 5\%$) and so Zn production (and, to a lesser extent, Pb and Cu smelting) can give rise to significant environmental pollution with Cd. Pacyna (1987) estimated that primary non-ferrous metal production gave rise to atmospheric emissions of 1630 t Cd/yr. Anomalously high Cd concentrations were found in soils and vegetation up to 40 km downwind from historic smelting activities in the Lower Swansea Valley in South Wales and 15 km from the Avonmouth Pb–Zn Smelter near Bristol, in the UK.

Electronics.   A large number of trace elements, including the heavy metals, are used in the manufacture of semi-conductors and other electrical components. These include Cu, Zn, Au, Ag, Pb, Sn, Y, W, Cr, Se, Sm, Ir, In, Ga, Ge, Re, Sn, Tb, Co, Mo, Hg, Sb, As and Gd. Environmental pollution can occur from the manufacture of the components and their disposal in waste. There is now a growing industry concerned with the recovery of valuable metals from decommissioned items of complex electrical equipment, such as computers. However, it must be remembered that old electronic equipment will include capacitors and transformers containing PCBs, which are toxic and persistent organic environmental pollutants (p. 293).

Other sources.   Other significant sources of heavy metal pollution in manufacture (sometimes in use) and disposal include:

batteries – Pb, Sb, Zn, Cd, Ni, Hg, Pm (promethium)
pigments and paints – Pb, Cr, As, Sb, Se, Mo, Cd, Ba, Zn, Co, I, Ti
catalysts – Pt, Sm (samarium), Sb, Ru, Co, Rh, Re, Pd, Os, Ni, Mo, I, Rh
polymer stabilizers – Cd, Zn, Sn, Pb (from incineration of plastics)
printing and graphics – Se (Xerox process), Pb, Cd, Zn, Cr, Ba
medical uses: dental alloy – Ag, Sn, Hg, Cu and Zn
            drugs/medicinal preparations – As, Bi, Sb, Se, Ba, Ta, Li, Pt
additives in fuels and lubricants – Se, Te, Pb, Mo, Li.

Waste disposal.   Many metals, especially Cd, Cu, Pb, Sn and Zn, are dispersed into the environment in leachates from landfills, which pollute soils and groundwaters, and in fumes from incinerators (Section 8.3.2, pp. 358–9). Sewage sludge contains many metals, including Zn, Cu, Pb, Cr, As and Mo, but the greatest cause for concern is currently considered to be Cd. Although present in many sludges in quite low concentrations ($< 10\,\mu g/g$), Cd is relatively easily taken up by food crops, especially leafy vegetables, and enters the human diet. Under the European Community Directive 86/278, the maximum permissible concentration of Cd in sludged soils used for food production is $3\,\mu g/g$ (Table 8.7, p. 363).

5.3.4   *Environmental media affected*

Heavy metals originating from the sources listed above generally have their most significant effects on the following environmental media:

mining	air (fumes, ore dusts and fine tailings) waters (effluents and tailings)
agriculture	air (fungicide droplets)
	waters (fungicide spillages and wash-off, livestock, slurry, Cu, As, Zn, spillage and seepage)

	soils (fertilizers, wastes, manures and composts)
fossil fuel combustion	air (aerosol particles from combustion)
	water (ash-pollutants leached into water courses)
	soils (deposited aerosol particles, ash disposal and leaching)
metallurgical industries	air (aerosol particles from furnaces, dusts from resuspension of deposited larger particles)
	waters (effluents, wash-off of particles)
	soil (deposited aerosols and larger particles, metal-rich sewage disposal, waste dumps)
electronics	air (aerosols from manufacturing processes)
	waters (effluents and corrosion of decommissioned electrical components)
	soils (wastes and corrosion of decommissioned electrical components)
chemical industries	air (volatilization of electrodes and catalysts, explosions)
	waters (effluents, spillages)
	soils (wastes)
pigments and paints	air (droplets of sprayed paint, particles of weathered paints)
	waters (effluents, anti-fouling paints, weathering of pigments and paints)
	soils (wastes, spillages, weathering of pigments and paints)
waste disposal	air (aerosols from incineration of metal-containing wastes)
	waters (leachates from landfill, runoff, corrosion of waste dumped in wet pits)
	soils (disposal/utilization of waste, e.g. sewage, composts from wastes, fallout of aerosols from incinerators, ash from bonfires).

5.3.5 Heavy metal behaviour in the environment

On reaching the environment after emission, or being released, the metal pollutants can behave in a number of ways.

Atmospheric aerosol particles. These remain suspended for varying lengths of time determined by the size of the particle, the windspeed, relative humidity and precipitation (Section 6.1). Aerosol particles range in diameter from 5 nm to $20\,\mu m$ but most are in the size range $0.1\text{--}10\,\mu m$. Particles

>10 μm tend to settle out under gravity relatively rapidly, but those <10 μm remain in the atmosphere for 10–30 days being removed by washout, settlement, impaction and, in the case of very small particles (<0.3 μm), by diffusive deposition. Under some circumstances, such as high humidity, smaller particles may sometimes cluster and form larger particles which are deposited more rapidly. In the 10–30 day period during which aerosol particles may remain suspended in the atmosphere they can be transported thousands of kilometres, depending on the circulation of air masses. It is for this reason that elevated concentrations of Pb, Cd, Zn, As and Se are found in the soils of southern Norway; particle transport occurring from the industrialized areas of Europe.

While suspended in the air, these metal aerosol particles may be inhaled by humans and animals and subsequently absorbed into the bloodstream through the alveoli of the lungs. Particles falling onto foliage may also enter plant tissues by absorption through the cuticle but this depends on the presence of moisture and its pH, the type of plant and other parameters. For example, children are at severe risk from the inhalation of aerosol sized particles of lead compounds in the exhaust emissions of motor cars (Section 6.1). These particles also settle onto crop foliage and may be consumed with the plant (e.g. on lettuce leaves). Most particles reach the soil eventually and may be ingested with incompletely washed vegetables, by children eating soil intentionally (pica), or accidentally from unwashed adult or children's hands after gardening or playing with contaminated soil.

On being deposited onto the soil, the metal compounds in the aerosol particles react with the soil constituents and become incorporated into the soil system. Metal particles absorbed by plants also reach the soil through the mineralization of plant litter.

Aqueous and marine environments.   Aerosol particles deposited into water, either directly or washed off surfaces into water courses, either react with the constituents of the water or settle to the bottom where they react with the sediments. The solubility of metal ions in solution will depend on the concentrations of anions and chelating ligands present in the water, its pH and redox status, and the presence of adsorbent sediments. Several metal ions are adsorbed and coprecipitated with hydrous oxides of Fe, Mn and Al, in both sediments and soils, for example Fe oxides coprecipitate V, Mn, Ni, Cu, Zn, Mo; Mn oxides coprecipitate Fe, Co, Ni, Zn, Pb.

Calcium carbonate, either originating from limestone rock fragments in soils and sediments or by precipitation from soil water in the soils of semi-arid and arid regions, also adsorbs a range of metals, including V, Mn, Fe, Co and Cd. Clay minerals in soils and sediments are responsible for the adsorption and coprecipitation of V, Ni, Co, Cr, Zn, Cu, Pb, Ti, Mn and Fe.

Metal ions in solution can be absorbed into aquatic plants and animals and can cause toxicity if the concentration is sufficiently high. This factor is

Table 5.19 The ranges of metals in sea water at 35 g/l salinity (Rainbow, 1993)

Metal	Concentration (nmol/kg)
As	15–25
Cd	0.001–1.1
Cr	2–5
Cu	0.5–6.0
Hg	$(2–10) \times 10^{-3}$
Ni	2–12
Pb	5–175
Sn	$(1–12) \times 10^{-3}$
Zn	0.05–9.0

exploited in the use of $CuSO_4.5H_2O$ to control algal blooms in lakes and reservoirs.

Accumulation of heavy metals in marine invertebrates. Trace metals are accumulated by marine invertebrates to concentrations many times higher than that in the sea water in which they live. Table 5.19 gives data for the ranges of metals in sea water. Concentration factors (whole body: water) for Cd as high as 15 000 have been found in some invertebrate species including *Anodata grandis grandis*. The metals are mainly accumulated from solution and from food.

Species vary greatly in their ability to accumulate metals: mussels and prawns accumulate lower concentrations of Zn than oysters and barnacles.

Some organisms such as barnacles (e.g. *Elminius modestus*) are net accumulators of almost all metals taken up and can contain concentrations of < 11 700 mg/kg Zn in dry matter, where it is stored in a detoxified form as phosphate granules. In contrast, the caridean decapod (prawn) *Palaemon elegans* is able to regulate the accumulation of metals in its body by controlled excretion and has Zn concentrations in the range 70–138 mg/kg. Likewise, mussels excrete relatively large amounts of Zn (< 12 675 mg/kg). Cadmium concentrations in deep-sea caridean decapods such as *Systella debilis* tend to be much higher (< 12 mg/kg dry matter) than those in coastal species (*Palaemon elegans* < c. 1 mg/kg).

Three crustaceans of different taxa (a decapod, an amphipod and a barnacle) differed in their accumulation of biologically essential Cu and Zn but all accumulated non-essential Cd without regulation. The amphipod *Echinogammarus pirloti* and the decapod *Palaemon elegans* both showed an ability to regulate Cu and Zn, but the barnacle accumulated very large concentrations of both Cu and Zn (Rainbow and White, 1989). Apart from the ecophysiological aspects, Cd-enriched sea food is of considerable importance to human health.

A Mussel Watch programme has monitored the concentrations of Cd, Cu, Pb, Ni, Ag and Zn in bivalve molluscs at sites around the coastline of the USA. For 50 sites where data were available for a 3-year period in the late 1970s, and another in the late 1980s, some interesting trends with time were noticed. Overall, mean concentrations of Pb showed a decrease, and for Cd and Pb more sites showed a decrease than an increase. Ag, Ni and Zn showed no change over the period, but Cu showed a significant increase. Unlike all the other metals monitored, Cu is the only one to show an increase in consumption in the USA over this period. The Mussel Watch programme has the central hypothesis that molluscs serve as 'sentinels' of change in environmental contamination (Lauenstein et al., 1990).

Tributyl tin (TBT) is used in modern 'antifouling' paints, which are applied to the hulls of boats and ships to inhibit the growth of algae and other marine organisms which would increase the frictional drag on the hull. TBT rapidly replaced Cu-based biocidal compounds during the 1970s. The TBT is slowly released into the sea water around the hull and is known to have marked toxicological effects on several marine organisms. In oysters, symptoms include reduced tissue growth and enhanced shell growth; the effects differ for different species and can include shell thickening and curling of outer shell margins. The Pacific Oyster *Crassostera gigas*, introduced into Europe in the 1970s, was found to be particularly vulnerable to TBT.

Oysters and other bivalves are more vulnerable to the effects of organotin and trace metals because these pollutants tend to be sorbed on sediment particles rather than remaining dissolved in sea water. In turbid estuaries, the cyclical resuspension and deposition of sediment particles containing sorbed pollutants results in benthic feeders like oysters taking in considerable quantities of the pollutants. Therefore, their vulnerability is mainly a result of their habitat and mode of feeding (Stebbing, 1985).

Another effect of TBT toxicity occurs in female marine gastropods (snails) and is the development of male sexual features, referred to as 'imposex'. This renders the female infertile and leads to the decline of species of gastropods, such as *Nucella lapillus* (Cockerham and Shane, 1994). Although TBT is lethal to shellfish at concentrations of $0.02\,\mu g$ TBT-Sn/l, shell effects and imposex occur at much lower concentrations. TBT accumulates in sediments with half-lives of a number of years (De Mora, 1992).

In humans, TBT is readily absorbed through skin and can cause a rash. Its toxic effects are thought to include interfering with mitochondrial function (Manahan, 1991).

Heavy metals in drinking water. A survey of drinking water in the UK showed that first-draw drinking water in 24 towns had Pb > $50\,\mu g/l$. The worst situation was in Scotland where 34.4% of homes had Pb concentrations in water above the limit (Fergusson, 1990). This is because of the use of

Pb pipes and solder and because the water is relatively acidic. This is clearly shown in the data from Packham (1990) who reported that in hard and soft waters from a total of 235 towns in the UK, the mean Cu in flushed water was 24 μg/l for towns with soft water and 14 μg/l for towns with hard water. Copper intake from drinking water can be around 1.4 mg/day from soft water and 0.05 mg/day from hard water (Davies and Bennett, 1983). The main differences in constituents between hard and soft waters are that hard waters contain high concentrations of Ca and Mg (< 500 mg/l as $CaCO_3$) whereas soft waters contain higher concentrations of Al, Mn and Pb. Where soft waters are also acid, concentrations of copper are likely to be much higher. The pH of domestic water from public supplies should be between 6.5 and 8.5. The metal content of water in houses is dependent upon the following factors (adapted from Mattson (1990)):

- original metal content, pH, alkalinity and oxygen contents, total hardness and temperature of the water
- duration of time the water remains in the pipe (still-stand)
- length of flushing time after still-stand
- materials, age, internal diameter and total length of the pipes in the house, and the type of solder.

Table 5.20 Guideline and maximum acceptable concentrations for metal and other inorganic pollutants in water for human consumption

	Concentration (μg/l)[a]					
	WHO Guide	EC Guide	EC Max	Canada	UK	USA
As	10	–	–	25	50	50
B	300	1000	–	5000	2000	1000
Ba	700	100	–	1000	1000	–
Cd	3	–	–	5	5	10
Cu	2000	3000[b]	–	<1000	3000	1000
Cr	50	–	–	50	50	50
Fe	–	50	200	<300	200	50
Hg	1	–	–	1	1	–
Mn	500	20	50	<500	50	50
Pb	10	–	–	10	50	5
Zn	–	5000[b]	–	<5000	5000	5000
Ca	–	100000	–	–	250000	–
Mg	–	30000	50000	–	50000	–
CN	70	–	–	200	50	–
F	1500	–	700–1500[c]	1500	1500	800–1700[c]
NO_3^-	50000	25000	50000	–	50000	–
NO_2^-	3000	–	100	–	100	–
SO_4^-	–	25000	250000	–	250000	–

From: WHO (1993); Murley (1995); Manahan (1991); Canada Council of Ministers of the Environment (1993); EC (1980).
[a] Maximum permissible concentrations except for WHO and EC guideline values.
[b] Cu and Zn in water after it has been standing for 12 hours at point of consumption (first draw from tap). A guide value of 100 μg/l for both Cu and Zn in water leaving the treatment works.
[c] F maximum concentrations depend on ambient temperature (lower values at higher temperatures).

The guideline and maximum acceptable metal levels in water for human consumption given by a number of agencies is outlined in Table 5.20.

Heavy metal ions in soils. Heavy metal pollution can affect all environments but its effects are most long lasting in soils because of the relatively strong adsorption of many metals onto the humic and clay colloids in soils. The duration of contamination may be for hundreds or thousands of years in many cases (e.g. half-lives: Cd, 15–1100 years; Cu 310–1500 years and Pb 740–5900 years depending on the soil type and their physicochemical parameters). Unlike organic pollutants, which will ultimately be decomposed, metals will remain as metal atoms, although their speciation may change with time as the organic molecules binding them decompose or soil conditions change.

The extent to which metal ions are adsorbed by cation exchange (non-specific adsorption) depends on the properties of the metal concerned (valency, radius, degree of hydration and coordination with oxygen), pH, redox conditions, the nature of the adsorbent (permanent and pH-dependent charge, complex-forming ligands), the concentrations and properties of other metals present, and the presence of soluble ligands in the surrounding fluids (Section 2.6.1).

The selectivity of clay mineral and hydrous oxide adsorbents in soils and sediments for divalent metals generally follows the order Pb > Cu > Zn > Ni > Cd, but some differences occur between minerals and with varying pH conditions. The selectivity order for peat has been shown to be Pb > Cu > Cd = Zn > Ca. However, in general, Pb and Cu tend to be adsorbed most strongly and Zn and Cd are usually held more weakly, which implies that these latter metals are likely to be more labile and bioavailable (Table 5.21).

It is usually found that the adsorption of metal ions onto soil solids is described by either the Langmuir or the Freundlich adsorption isotherm equations. The Langmuir equation is:

$$\frac{M}{x/m} = \frac{1}{Kb} + \frac{M}{b} \tag{5.43}$$

Table 5.21 Transfer coefficients of heavy metals in the soil–plant system (from Kloke *et al.* (1984))

Element	Transfer coefficient	Element	Transfer coefficient
As	0.01–0.1	Ni	0.1–1
Be	0.01–0.1	Pb	0.01–0.1
Cd	1–10	Se	0.1–10
Co	0.01–0.1	Sn	0.01–0.1
Cr	0.01–0.1	Tl	1–10
Cu	0.01–0.1	Zn	1–10
Hg	0.1–1		

where M = the activity of the ion, x/m = the amount of M adsorbed per unit of adsorbent, K = a constant related to the bonding energy and b = the maximum amount of ions that will be adsorbed by a given adsorbate.

The Freundlich equation is:

$$\log x = k + n \log c \qquad (5.44)$$

where x = the amount adsorbed per unit of adsorbent at concentration c of adsorbate and k and n are constants. These isotherms do not provide any information about the adsorption mechanisms involved and both assume a uniform distribution of adsorption sites on the adsorbent and absence of any reactions between adsorbed ions.

In general, nearly all metals (except Mo) are most soluble and bioavailable at low pHs and, therefore, toxicity problems are likely to be more severe in acid environments. In the case of pollution by particles of sulphide ore minerals, the weathering of the sulphide exacerbates the problem by increasing the acidity of the soil. In agricultural soils this situation can be mitigated to a considerable extent by liming.

Methylation of heavy metals in the environment.   Arsenic, Hg, Co, Se, Te, Pb and Tl can be methylated in the environment through the action of enzymes secreted by microorganisms (biomethylation) and also by abiotic chemical reactions. There is a possibility that Cd, In, Sb and Bi can also be methylated. The bacteria associated with methylation of these elements are found in the bottom sediments of rivers, lakes and the coastal waters of the sea, soils and the digestive tracts of animals (including humans). This methylation radically affects the behaviour of the elements in the components of the environment, their bioavailability and their toxicology. For example, monomethyl mercury (CH_3Hg^+), the most toxic form of Hg, is lipophyllic and, therefore, accumulates in body fats and is the only heavy metal to show bioaccumulation along the food chain (Fergusson, 1990).

The uptake of heavy metals by plants.   Transfer coefficients (concentration of metal in the aerial portion of the plant relative to total concentration in the soil) are a convenient way of quantifying the relative differences in bioavailability of metals to plants. Kloke *et al.* (1984) gave generalized transfer coefficients for soils and plants (Table 5.21); however, soil pH, soil organic matter content and plant genotype can have marked effects on metal uptake. The transfer coefficients are based on root uptake of metals but it should be realized that plants can accumulate relatively large amounts of metals by foliar absorption of atmospheric deposits on plant leaves.

From Table 5.21, it can be seen that Cd, Tl and Zn have the highest transfer coefficients, which is a reflection of their relatively poor sorption in the soil. In contrast, metals such as Cu, Co, Cr and Pb have low coefficients because they are usually strongly bound to the soil colloids.

5.3.6   *Toxic effects of heavy metals*

The sensitivity of organisms to metal toxicity varies widely with species of plants and animals and genotypes within species (e.g. cultivars of crops) and many factors can modify the response to the toxic dose of metals. Some individuals are genetically adapted to tolerating anomalously high concentrations of certain metals. Homeostatic mechanisms in animals frequently involve special proteins, metallothioneins, which are low-molecular-weight (*c.* 7000 Da) proteins containing a high proportion of cysteine (–SH groups) but no aromatic amino acids. They can bind a range of metals (Cd, Cu, Hg, Ag and Zn) in a non-bioavailable form and their formation can be induced by exposure to some, or all, of these elements. Metallothioneins are found in mammals, fish and invertebrates, but similar proteins occurring in fungi, algae and plants differ in structure and are referred to as phytochelatins (Wright and Welbourn, 1994). Nevertheless, there is a convention to use the term metallothionein (MT) for all compounds sharing similar properties with the mammalian form. The order of decreasing metal affinity for MT is: Hg > Cu > Cd > Zn.

It is, therefore, difficult to generalize about toxicity. However, an indication of relative toxicity of metals to mammals is provided by the LD_{50} values for a wide range of heavy metals, given in Table 5.22.

Phytotoxicity.   The most toxic metals for both higher plants and several microorganisms are Hg, Cu, Ni, Pb, Co, Cd, and possibly Ag, Be and Sn (Kabata-Pendias and Pendias, 1984). Although the occurrence of toxicity will depend on soil factors, such as pH, the plant genotype and the conditions under which it is growing (pot in greenhouse, or under field conditions), a general indication of the toxic levels of some metals is given in Table 5.23 and the biochemical effects of excess heavy metals in plants are given in Table 5.24.

The data in Table 5.22 provide an indication of the relative toxicity of different elements but this will be affected by considerable variation in tolerance of individuals and, in the case of diets, in the composition of the diets. The doses injected into rats and other experimental mammals probably provide a more accurate comparison of toxicity. From these data for injected doses, U and [239]Pu are jointly the most toxic elements, followed by Cd and Se and then Hg. From the human diet data, As, Cu and Hg are the most toxic but homeostatic mechanisms will often affect the extent of toxicity.

Examples of critical (trigger) concentrations of heavy metals used in different countries are given in Tables 5.25–5.27.

Table 5.25 shows the critical concentrations used in the Netherlands for contaminated soils and waters. Until recently, a system was used for soils which involved three indicative values: *A*, the 'normal' reference value; *B*, the test value to determine the need for further investigations; and *C*, the

Table 5.22 Relative mammalian toxicity of elements in injected doses and diets (from Bowen (1979) with permission)

Element	Acute lethal doses (LD_{50}) injected into mammals[a] (mg/kg bodyweight)	Dose in human diet (mg/day)	
		Toxic	Lethal
Ag	5–60	60	1.3k–6.2k
As	6	5–50	50–340
Au	10	–	–
Ba	13	200	3.7k
Be	4.4	–	–
Cd	1.3	3–330	1.5k–9k
Co	50	500	–
Cr	90	200	3k–8k
Cs	1200	–	–
Cu	–	–	175–250
Ga	20	–	–
Ge	500	–	–
Hg	1.5	0.4	150–300
Mn	18	–	–
Mo	140	–	–
Nd	125	–	–
Ni	110–220	–	–
Pb	70	1	10k
Pt	23	–	–
(^{239}Pu)	1	–	–
Rh	100	–	–
Sb	25	100	–
Se	1.3	5	–
Sn	35	2000	–
Te	25	–	2k
Th	18	–	–
Tl	15	600	–
U	1	–	–
V	–	18	–
Zn	–	150–600	6k

[a] Injected into the peritoneum to avoid absorption through the digestive tract. Chemical form of the element will affect its toxicity.

Table 5.23 Normal and phytotoxic metal concentrations generally found in plant leaves (Alloway (1995), based largely on Bowen (1979))

Element	Concentration in leaves (µg/g)	
	Normal range	Toxicity
As	0.01–0.8	1–4
As(III)	0.02–7	5–20
Cd	0.1–2.4	5–30
Cu	5–20	20–100
Cr	0.03–14	5–30
Hg	0.005–0.17	1–3
Ni	0.02–5	10–100
Pb	5–10	30–300
Sb	0.0001–2	1–2
V	0.001–1.5	5–10
Zn	1–400	100–400

Table 5.24 Biochemical effects of excessive concentrations of heavy metals in plants (from Kabata-Pendias and Pendias (1984) and reprinted from Fergusson, *The Heavy Elements: Chemistry, Environmental Impact and Health Effects,* © 1990, p. 40, with kind permission from Pergamon Press Ltd, Headington Hill Hall, Oxford OX3 0BW, UK)

Elements	Biochemical process affected
Ag, Au, Cd, Cu, Hg, Pb, F, I, U	Changes in the permeability of cell membranes
Hg	Inhibition of protein synthesis
Ag, Hg, Pb, Cd, Tl, As(III)	Bonding to sulphydryl groups
As, Sb, Se, Te, W, F	Competition for sites with essential metabolites
Most heavy metals, Al, Be, Y, Zr, lanthanides	Affinity for phosphate groups, and ADP, ATP groups
Cs, Li, Rb, Se, Sr	Replacement of essential atoms
Arsenate, selenate, tellurate	Occupation of sites for essential groups, e.g. PO_4^{3-}, tungstate, bromate, fluorate
Tl, Pb and Cd	Inhibition of enzymes
Cd, Pb	Respiration
Cd, Pb, Hg, Tl, Zn	Photosynthesis
Cd, Pb, Hg, Tl, As	Transpiration
Cd, Co, Cr, F, Hg, Mn, Ni, Se, Zn	Chlorosis
Al, Cu, Fe, Pb, Rb	Dark green leaves

Table 5.25 Guide values and quality standards used in the Netherlands for assessing soil and water contamination by heavy metals (Netherlands Ministry of Housing, Physical Planning and Environment, 1991; Moen *et al.*, 1986)

Metals	oils (mg/g)[a]				Surface waters (µg/l)[b]				Groundwater (mg/l)[c]
	A	B	C	STV	TotTV	TotLV	DisTV	DisLV	
As	20	30	50	29	5	10	4	8.6	10
Ba	200	400	2000	200	–	–	–	–	50
Cd	1	5	20	0.8	0.05	0.2	0.003	0.005	1.5
Co	20	50	300	10	–	–	–	–	10
Cr	100	250	800	100	5	20	0.5	2	1
Cu	50	100	500	36	3	3	1	1.3	15
Hg	0.5	2	10	0.3	0.02	0.03	0.003	0.005	0.05
Mo	10	40	200	10	–	–	–	–	5
Ni	50	100	500	35	9	10	7	7	15
Pb	50	150	600	85	4	25	0.2	1.3	15
Sn	20	50	300	20	–	–	–	–	10
Zn	200	500	3000	140	9	10	2	2	150

[a] A, reference value; B, test requirements; C, intervention value, from 1986 Scheme; STV, target value for soils. Target values for soils are based on 'standard soil' (10% organic matter and 25% clay).
[b] TotTV, total content target value; TotLV, total content limit value; DisTV, dissolved content target value; DisLV, dissolved content limit value.
[c] Groundwater dissolved content target value (from 1991 Environmental Quality Standards for Soils and Waters).

intervention value above which the soil definitely needs cleaning-up (Moen *et al.*, 1986). This system has been superseded by an effect-oriented scheme of 'Environmental Quality Standards for Soil and Water' (Ministry of Housing, Physical Planning and the Environment Directorate General for Environmental Protection (Netherlands), 1991. These standards are based

Table 5.26 UK Department of the Environment ICRCL trigger concentrations for environmental metal contaminants (total concentrations except where indicated) (Department of the Environment, 1987)

Contaminant	Proposed uses	Threshold trigger concentration $(\mu g/g)^a$
Contaminants which may pose hazards to human health		
As	Gardens, allotments	10
	Parks, playing fields, open space	40
Cd	Gardens, allotments	3
	Parks, playing fields, open space	15
Cr (hexavalent[b])	Gardens, allotments	25
	Parks, playing fields, open space	—
Cr (total)	Gardens, allotments	600
	Parks, playing fields, open space	1000
Pb	Gardens, allotments	500
	Parks, playing fields, open space	2000
Hg	Gardens, allotments	1
	Parks, playing fields, open space	20
Se	Gardens, allotments	3
	Parks, playing fields, open space	6
Phytotoxic contaminants not normally hazardous to health		
B (water soluble)	Any uses where plants grown	3
Cu (total)	Any uses where plants grown	130
(extractable[c])		50
Ni (total)	Any uses where plants grown	70
(extractable[c])		20
Zn (total)	Any uses where plants grown	300
(extractable[b])		130

[a] Total concentrations except where indicated.
[b] Hexavalent Cr extracted by 0.1 M HCl adjusted to pH at 37.5°C.
[c] Extracted in 0.05 M EDTA.

on ecological function and comprise target values (TV) for soils which represent the final environmental quality goals for the Netherlands. Both target and limit values for waters are given because it is intended that the target value is reached by progressively lowering the limit values. This is easier for waters than for soils owing to the long residence time of most pollutants in soils. The intervention values (C) of the 1986 scheme still stand. In some cases the target values based on risk assessment are lower than the background levels (A values) given in the 1986 scheme. This is in part because of the wide differences in sensitivity to the contaminants between organisms within ecosystems.

The critical values for soils given in Table 5.26 for the UK by the Department of the Environment Interdepartmental Committee for the Reclamation of Contaminated Land List of Trigger Concentrations for Contaminants (DOE, 1987) are more pragmatic and based mainly on the risk to human health.

The new Canadian National Classification System for contaminated soils shown in Table 5.27 is intended for use in the evaluation of contaminated sites. The values in the tables enable sites to be classified as high, medium or

Table 5.27 Selected values from the Canadian interim environmental quality criteria for soil (Canadian Council of Ministers of the Environment, 1993)

Metal	Soil concentration (μg/g)			
	Background	Agricultural	Residential	Industrial
As	5	20	30	50
Ba	200	750	500	2000
Be	4	4	4	8
Cd	0.5	3	5	20
Cr^{6+}	2.5	8	8	_a
Co	10	40	50	300
Cu	30	150	100	500
CN^- (free)	0.25	0.5	10	100
CN^- (total)	2.5	5	50	500
Pb	25	375	500	1000
Hg	0.1	0.8	2	10
Mo	2	5	10	40
Ni	20	150	100	500
Se	1	2	3	10
Ag	2	20	20	40
Sn	5	5	50	300
Zn	60	600	500	1500

a Criteria not recommended.

low risk according to their impact (current or potential) on human health and ecosystems. It is a screening system and is not intended to be a quantitative risk assessment for individual sites.

5.3.7 Analytical methods

The most commonly used analytical methods for determining the concentrations of heavy metals in extracts and acid digests are AA spectrophotometry or flame AA spectroscopy for most jobs (0.1–10 μg/ml) but electrothermal atomization AA spectroscopy (also called graphite furnace AA spectroscopy) is more useful for low concentrations (in some cases down to ng/ml) (p. 130). Hydride generation equipment can be linked to AA spectroscopy to allow As, Hg, Sb and certain other elements to be determined. Multi-element analysis (< 22 elements) can be carried out by inductively coupled plasma–atomic emissions spectrometry (ICP-OES) and by X-ray fluorimetry (Chapter 4).

5.3.8 Examples of specific heavy metals

Arsenic. A toxic, non-essential element that has been used as a pigment, pesticide, wood preservative and a livestock growth promoter (pigs and poultry, phenylarsonic acid, **2**) and has caused environmental pollution in these roles. It is also present in many sulphide ores of metals and is, therefore, emitted from metal smelters as an atmospheric pollutant (40% of the anthropogenic emissions) (O'Neill, 1990). Coals also contain significant amounts of As and its combustion accounts for 20% of atmospheric emissions. Coal ashes are also a significant source of As, which can be leached out into waters or the soil. The toxicological importance of As is partly because of its chemical similarity with phosphorus, which means that As can disrupt metabolic pathways involving phosphorus. Both acute and chronic toxicity are recognized and the continual inhalation of airborne forms of As is known to be carcinogenic. Respiratory cancers have occurred in occupationally exposed workers. Arsenic is also associated with skin cancers, and the WHO (1993) has estimated that populations consuming drinking water with a concentration of $0.17\,\mu g/l$ As had an excess lifetime skin cancer risk of 10^{-5}. With the WHO guideline limit of $10\,\mu g/l$, the excess lifetime skin cancer risk is 6×10^{-4}. Assuming a 20% As intake in drinking water, a provisional maximum tolerable daily intake (PMTDI) is $2\,\mu g/kg$ body weight of inorganic As.

$$Ph{-}\overset{\displaystyle\overset{O}{\|}}{\underset{\displaystyle\underset{OH}{|}}{As}}{-}OH$$

2

Cadmium.   A highly toxic non-essential metal which accumulates in the kidneys of mammals and can cause kidney dysfunction. In humans, kidney damage diagnosed by the presence of microglobulin proteins in the urine is the main toxic effect resulting from chronic emposure to the metal. High concentrations of inhaled Cd aerosols can cause emphysema and related acute lung conditions. Cadmium becomes very volatile above 400°C and hence is likely to be dispersed as an aerosol when mixtures of metals containing Cd are heated or cast. Cadmium is a 'modern' metal, having been used increasingly in corrosion prevention, polymer stabilization, electronics and pigment applications since the 1960s. It tends to be less strongly adsorbed than many other divalent metals and is, therefore, more labile in soils and sediments and more bioavailable. There is more danger from this metal moving through the human food chain from contaminated soils than most other metals. Sewage sludge-amended soils can contain sufficiently high concentrations of Cd to cause elevated concentrations of Cd in food crops and there is a European Community Directive limiting the maximum Cd content of sludged soils to $3\,\mu g/g$. Although sewage sludge applications are considered a major source of Cd in the soils receiving sludges, the most important sources overall are phosphatic fertilizers and industrial emissions.

A serious case of Cd poisoning occurred in the Jintsu Valley in the Toyama Prefecture in Japan, where Pb–Zn mining and smelting had caused widespread Zn and Cd contamination of the alluvial soils, most of which were used for paddy rice production. The farmers in the valley live mainly on rice grown in the contaminated paddies and also relied on the metal-polluted river for their drinking water. After the Second World War, it was found that more than 200 elderly women who had all had several children had developed kidney damage and skeletal deformities. The condition was known as 'itai-itai' disease, which literally means 'ouch-ouch' because of the pain caused by the deformed bones. The Cd toxicity was exacerbated by a low protein and vitamin D diet and the birth of several children. The rice which they consumed contained ten times more Cd than local controls and the contaminated water was an additional intake. It was estimated that the people in the valley had a Cd intake of around 600 µg/day which is around ten times greater than the maximum tolerable intake of 60–70 µg/day. A survey of paddy soils in the whole of Japan revealed that 9.5% of the area was significantly contaminated with Cd, with a further 3.2% of upland rice-growing soils and 7.5% of orchard soils. The source of the Cd in these soils is probably phosphatic fertilizers and industrial/mining pollution.

Copper. A micronutrient which can be deficient in some soils causing severe loss of yield in several crops, especially cereals. Toxicity problems can occur in crops in polluted soils and in livestock grazing herbage growing on polluted soils. Sheep are the agricultural livestock most sensitive to Cu toxicity, but they (and cattle) are also prone to deficiency disorders. Herbage with < 5 µg/g Cu can lead to Cu deficiency in both sheep and cattle but if the herbage contains > 10 µg/g Cu then toxicity is likely to occur in sheep. Copper pollution can arise from Cu mining and smelting, brass manufacture, electroplating and excessive use of Cu-based agrichemicals (e.g. Bordeaux Mixture). Copper sulphate is used widely as an algicide in ornamental ponds and even in water supply reservoirs which are affected by blooms of toxic blue-green algae. Copper is used widely in houses for piping water and although the concentrations of Cu in the drinking water are higher in soft water, this is not considered to be a hazard so long as the pH is within the normal limits (pH 6.5–8.5). More acid waters could create problems with excessive concentrations, but none has been reported with public water supplies. Copper contamination of soils can cause phytotoxicity in crop plants. A commonly affected crop is the vine, which is frequently sprayed with Cu-containing fungicides and high concentrations can accumulate in the topsoil. While old vines can tolerate this Cu accumulation because their deep roots reach far below the metal-rich topsoil, newly planted young vines or other crops can be killed by Cu toxicity. It has been found that liming Cu-contaminated soils to pH 7 can mitigate this toxicity by reducing the bioavailability of the Cu. Copper is also highly toxic to the soil microbial biomass and this can affect various aspects of soil fertility, including the mineralization of plant debris (and hence the cycling of nutrient elements) and nitrogen fixation in legumes. However, although some crops, such as vines and potatoes, have traditionally received numerous applications of Cu-containing fungicides, it is the accumulation of Zn from frequent applications of sewage sludge which constitutes the more ubiquitous toxicity hazard to the soil microbial biomass.

Chromium.   A micronutrient, which is essential for carbohydrate metabolism in animals. Pollution of soils occurs as a result of the dumping of chromate wastes, such as those from tanneries or electroplating and from sewage sludge disposal on land. However, unlike many other heavy metals, Cr can exist in a trivalent (chromite) and a hexavalent form (chromate) and the Cr(VI) form is more phytotoxic than the Cr(III) form. Soluble chromate concentrations of $0.5\,\mu g/ml$ have been shown to be significant. Hence the redox conditions in the environment are very important, waterlogged soils with reducing conditions will have the less toxic Cr(III). However, in many freely drained aerated soils the predominant form is also Cr(III) because soil organic matter results in the reduction of Cr(VI) to Cr(III). Soils developed on ultramafic rocks, such as serpentinites, can contain very high concentrations of Cr of geochemical origin and cannot be considered polluted but either 'naturally' or 'geochemically' enriched.

Chromite(III) appears to be more toxic to fish than Cr(VI), especially salmon, but toxic concentrations for several species of fish range from 0.2–$5\,\mu g/ml$. Municipal wastewater can contain concentrations of $< 0.7\,\mu g/Cr$ ml, mainly in the Cr(VI) form, which are toxic to many species of marine animals, algae and microorganisms; but reduction of Cr(VI) to Cr(III) usually occurs if there is organic matter present (Langard, 1980).

Chromium (VI) can be carcinogenic, causing cancer of the respiratory organs in chromate workers chronically exposed to Cr-containing dusts (Langard, 1980).

Mercury.   A non-essential element. An important source of air pollution is the chlor-alkali process for the production of Cl_2 and NaOH from brine (Section 6.4.1) where 0.1 to 0.2 kg Hg are lost into the environment with every 1000 kg Cl_2 produced. Mercury is used as a catalyst in the production of some plastics and there are cases of severe environmental pollution resulting from the discharge of Hg-containing liquid wastes. In the Minamata Bay in Japan, $HgSO_4$ in effluents reaching the bay was first of all precipitated as insoluble HgS, which later underwent biomethylation through the action of bacteria in the sediments to form CH_3Hg^+. This methylated form is very volatile and lipophylic, and accumulated in the food chain of fish resulting in the fish catches in the bay having high concentrations of Hg which were harmful to humans eating them. Mercury is a neurotoxin and has teratogenic effects; at least 78 people were severely affected by the CH_3Hg^+ poisoning, seven of them fatally. Another coastal area nearby, Niigata, was similarly affected. Other heavy metals can be methylated in the environment, but Hg appears to cause the most dangerous problems for human health (Fergusson, 1990). Other sources of Hg pollution in lakes and rivers is the use of Hg-containing slimicides in pulp paper mills. Several cases of pollution from this source have been reported around the world, including Canada and Sweden.

Until recently, alkyl-Hg was used as an agricultural seed dressing to prevent fungal disease in germinating seeds. This resulted in significant amounts of the metal being added to highly productive, intensively farmed agricultural soils in technically advanced countries. Its use has been discontinued because of its toxicity to humans and wildlife. A serious case of mass poisoning (probably the most serious case of chemical poisoning through the diet) occurred in 1971–2 when a consignment of Hg-treated seed exported to Iraq for growing was mistakenly milled into flour and used for making bread. By the end of 1972, 6530 people had been admitted to hospital suffering from Hg poisoning and 459 of these patients died (Kazantis, 1980). Other modern sources of Hg are small batteries for use in cameras, hearing aids, etc. and these could easily be swallowed by small children. Steinnes (1995) reviews the behaviour of mercury in soils.

Lead.   A non-essential element; it is a neurotoxin and a good example of a multimedia pollutant. The main sources of Pb pollution in the environment are petrol (air pollutant, but can also be water or soil pollutant from spillages), particulates in exhaust fumes from petrol combustion (air pollutants, inhaled by humans), particulates from petrol, fossil fuel combustion in soil (soil pollutant – taken up by plants and also ingested with plant food crops), paint flakes from old paint containing a high percentage of Pb, some traditional ethnic cosmetics (e.g. surma – skin absorption), constituent of solders and varnishes used on interiors of food cans (food contaminant), Pb pipes for potable water (water pollutant), pesticide (historic use of Pb/As-containing pesticide sprays in orchards), lead shot used in guns for game and clay pigeon shooting (soil pollutant, but also a food contaminant if inadvertently consumed with the game flesh) and, finally, Pb pollution from mining and smelting of the ore (usually PbS) – this includes acid mine drainage with soluble Pb (water pollutant), tailings from ore dressing (particulate water pollutant and soil pollutant), which oxidize to release soluble Pb, and smelter fumes – Pb aerosols (air and soil pollutants). Heavily used clay pigeon shooting club sites can contain very high concentrations of Pb (and also antimony used to harden the Pb). Where millions of cartridges are fired each year, soils can accumulate several percent of Pb. This and the Sb may be recoverable, at least to a certain extent if suitable machines are developed, but the weathering of the shot pellets in the soil does result in relatively large amounts of Pb being dispersed within the soil–plant system and, therefore, potentially bioavailable.

On a comparative basis, Pb is neither as toxic as many other heavy metals (Tables 5.22 and 5.23) nor as bioavailable (Table 5.21); however, it is generally more ubiquitous in the environment and is a cumulative toxin in the mammalian body. Toxic concentrations can accumulate in the bone marrow, where red blood cell formation (haematopoiesis) occurs. At least five stages in the formation of the haem part of haemoglobin are affected by Pb but the two enzymes most affected are δ-amino laevulinic dehydratase (ALAD) and ferrochelatase (Waldron, 1980). This inhibition of haem synthesis results in anaemia. Kidney damage also occurs as a result of exposure to Pb. Lead, like Hg, is a powerful neurotoxin and a range of pathological conditions is associated with acute Pb poisoning, most characteristic of which is cerebral oedema. However, the absorption of Pb in amounts which are not high enough to cause acute poisoning may induce behavioural abnormalities, including learning difficulties.

The critical concentrations for Pb in blood are the EC recommended level of 35 µg/dl and the UK threshold level for follow-up investigations of 25 µg/dl in at least half the population and less than 30 µg/dl in 90% of the population. The critical level in blood for occupational exposure is higher with values of 60, 80 and even 100 µg/dl being used in different countries.

Zinc.   A micronutrient which results in the most serious deficiency problem in crops in the world as a whole, especially in tropical regions and where the soils have developed on sandstones or sandy drift. Humans can also be affected by deficiencies and in extreme cases short stature and delayed sexual maturity can be caused by it. In the context of pollution, Zn is mainly a cause of phytotoxicity and has a relatively low toxicity to animals and humans. Zinc pollution is often associated with mining and smelting and Cd is always present as a guest element in ZnS and other ores. Mining causes pollution of air, water and soil with fine tailings particles which ultimately undergo oxidation to release Zn^{2+}. Zinc sulphide (sphalerite) ore often occurs together with PbS (galena), the main ore of Pb, and so Zn pollution is often associated with Pb and also Cu, in some cases, as well as traces of Cd. The geochemical association with Cd implies that impure Zn compounds may contain Cd, but also that Zn–Cd antagonism may mitigate some of the effects of Cd contamination.

A relatively ubiquitous source of Zn in the environment is galvanized steel, which in the form of wire fencing gradually dissolves in rain and drops to the ground below, but in the case of roofing, the Zn^{2+} usually drains away via gutters and drainpipes. Water collected from these roofs may not be suitable for drinking in areas of acid rain because of the high Zn and relatively high Cd concentrations which could arise.

Sewage sludges and some animal manures are very significant sources of Zn in soils. Soils which have received several applications of sewage sludge can accumulate relatively high concentrations of the metal. In most cases, this is not high enough to cause phytotoxicity, but several species of soil microorganisms are susceptible to Zn toxicity. It has been found that the nitrogen-fixing bacteria in root nodules of legumes, such as clover, can be adversely affected by Zn toxicity at concentrations of Zn within the maximum permissible value of 300 mg/kg used by the EU. The micro-organism species found to be most susceptible to heavy metal toxicity soils (which have received heavy applications of sewage sludge) is *Rhizobium leguminosum* bv *trifolii*, which is the nitrogen-fixing symbiotic bac-terium found in nodules on the roots of white clover. However, although the order of decreasing toxicity in this microorganism was found by Chaudri *et al.* (1992) to be Cu > Cd > Ni > Zn, the last metal is usually present in the highest concentrations and constitutes the greatest toxicity hazard to the soil microbial biomass. As a result of the findings of field experiments on sludged soils in the UK and Germany, the UK is planning to reduce the maximum allowable total Zn concentration in sludged soils to 200 mg/kg. This implies that much less sewage sludge can be applied to any particular site, unless it has a significantly lower Zn content. Although significant reductions have been made in the concentrations of most heavy metals in sewage sludges in the UK over the period 1982–3 to 1990–1 (26% in the median Zn concentration: 1205 mg/kg down to 889 mg/kg Zn), it is relatively difficult to make major further reductions in the Zn contents of sewage sludges (Ministry of Agriculture, Fisheries and Food, 1993). This is because Zn is widely used in both domestic products (e.g. baby creams, shampoos) and industrial products and processes. Consequently, a greater number of farm fields will need to be used for regular applications of sludge.

**5.4
Other metals and
inorganic pollutants**

5.4.1   Aluminium

Aluminium (atomic weight 27, comprises 8.2% of earth's crust) is widely used as an alternative to steel for cladding structures, and as an alternative to copper in electrical and thermal conduction. Its oxide form (alumina) is widely used in industry as an abrasive. Aluminium hydroxide is also used as an antacid by people with digestive disorders. Aluminium is used extensively in water treatment and normally levels in the drinking water should be below 100 µg/l, although the WHO guideline value for Al in drinking water is a compromise value of 200 µg/l (WHO, 1993). An accidental introduction of 20 tonnes of aluminium sulphate into the water supply of the Camelford District of Cornwall in July 1988 caused many reported health effects, although allegations of impaired mental function have not been substan-tiated. Not only did this accident give rise to a very high Al concentration in the drinking water, but it also increased the acidity of the water, which caused the solution of Cu from pipes and Pb from solders. In addition to the effects on people, thousands of fish were also killed when the erroneously contaminated water was discharged into a river.

Acid soil conditions exacerbated by acid precipitation (SO_x and NO_x) (Section 5.2.2) lead to increased concentrations of Al in stream and river waters (range 8–3500 µg/l), and these may be used for drinking water supplies.

Toxicity.   Aluminium dust can cause lung disease (pulmonary fibrosis), which can lead eventually to emphysema. However, the majority of the population is exposed to Al mainly from food and drink, rather than from particulates in air, as a result of the use of Al in cooking pans, drink cans and as a constituent of foods and beverages. It is recognized that patients undergoing kidney dialysis could develop dementia as a result of the accumulation of Al from the water. Some investigations of the composition of the brains of people who had died from Alzheimer's disease (senile dementia) revealed high concentrations of aluminium. However, others have failed to find these accumulations of Al and so there is much uncertainty about a causal link between this metal and onset of this disease.

Exposure to aluminium could be large in some cases where people consume large doses of antacids (< 5000 mg Al/day) which is 42–250 times greater than the average daily intake from other sources. Drinking water obtained from rivers or lakes in acidified catchments may have considerably elevated Al concentrations. The tea plant is a natural accumulator of Al and so regular tea drinkers probably have an increased intake of Al (and fluorine, which the plant also accumulates).

5.4.2   Beryllium

Beryllium (atomic number 4, atomic weight 9, concentration in the crust 2 µg/g) is the lightest of all the stable metals, it is resistant to corrosion, very hard and is used in various industrial applications, such as in alloys with Cu and Ni. Until 1949, it was used in phosphors in fluorescent lamps but this was discontinued owing to its hazardous nature. Coal contains small concentrations of Be (average 2.5 µg/g) and its combustion probably acts as the largest source of the metal in the environment. However, industries extracting or using the metal are the major sources of potentially hazardous concentrations.

Toxicity.   The main hazard is in the inhalation of particles of the metal, its oxides or other compounds into the lungs. A chronic form of pneumoconiosis called berylliosis appears many years after exposure (< 17 or even 25 years). Once the condition has commenced it is progressive, even in the absence of further exposure. An acute condition can develop within 72 hours of exposure to high concentrations of Be fumes, but cases are relatively rare. Although studies on the carcinogenicity of Be have been inconclusive, it is classified as a probable human carcinogen.

The main danger of exposure is in workers involved in the extraction, smelting or other uses of the metal, but residents living near to Be-using industries could also be at risk. The emission and transport of dusts and fumes from industrial plants using the metal, therefore, needs to be monitored and kept to an absolute minimum. However, the imposition of strict occupational hygiene standards has already reduced the risk of disease from the metal, but it remains a potential problem wherever it is manufactured or used (Harte *et al.*, 1991). Tepper (1980) has written a detailed review of the toxicology and environmental behaviour of Be.

5.4.3 Fluorine

Fluorine (atomic number 9, atomic weight 18.9984, crustal abundance 0.0460%) is a pale yellow gas which is the most reactive of all elements and is used for the manufacture of various fluoride compounds such as UF_6, SF_6 and chlorofluorocarbons.

Various fluoride compounds cause environmental pollution problems, these include HF, which can be generated during the manufacture of phosphatic fertilizers from the rock phosphate fluorapatite $Ca_5F(PO_4)_3$:

$$Ca_5F(PO_4)_3 + H_2SO_4 + 2H_2O \longrightarrow CaSO_4.2H_2O + HF + 3H_3PO_4 \quad (5.45)$$

HF is very toxic and corrosive, it severely irritates the respiratory tract, eyes and other tissues and brief exposure to $1000 \, \mu g/ml$ may be fatal.

Silicon tetrafluoride, SiF_4, is a pollutant from steelmaking plants using fluorspar (CaF_2) as a flux:

$$2CaF_2 + 3SiO_2 \longrightarrow 2CaSiO_3 + SiF_4 \quad (5.46)$$

Electrolytic Al smelting by the Hall–Heroult process uses cryolite (Na_3AlF_6) as a non-aqueous solvent for molten Al_2O_3 and consequently almost all Al smelters are significant sources of atmospheric F pollution. The method uses $70 \, kg$ (Na_3AlF_6) per tonne of Al metal produced (Fergusson, 1982).

Brick kilns are a common source of F compounds and they emit mixtures of HF, SiF_4 and H_2SiF_6 fumes and particulates (together with other pollutants including SO_2 and alkane thiols). In the early 1980s, two large brickworks situated in close proximity on the Oxford Clay in Bedfordshire, in the English Midlands, were emitting a combined total of 430 t F/year (and 43 000 t SO_2/yr). Average concentrations of F in the air around the works were in the range $1–8 \, \mu g/m^3$, with maximum short period concentrations of $< 35 \, \mu g/m^3$ (Owen, 1981 and S. Cray, personal communication). Fluoride-containing particulates settle out onto pastures surrounding the brickworks and are ingested by grazing livestock causing a toxic condition called fluorosis, which affects the bones and teeth. Plants, including cereals and trees, can also be affected by F toxicity and a synergistic effect of combined SO_2 and F pollution is recognized in crops in areas affected by atmospheric pollution from brickworks. The Oxford clay used for brickmaking in Bedfordshire contains a relatively high concentration of organic matter which is beneficial in brickmaking because it increases the calorific value of the clay, but it is also a source of oxides of S and Mo in the fumes from the kilns. The Mo can give rise to molybdenosis in livestock, which is a pathological condition caused by excess Mo.

Fluorine is added to many public water supplies to improve dental health, especially to reduce the incidence of dental caries (which require fillings). In the USA, 60% of potable waters have around the optimal concentration of F (1 μg/ml) either naturally, or by addition (Harte *et al.*, 1991). Tea leaves are a relatively rich source of F (and Al). In areas where local water supplies have naturally high concentrations of F(<1.5 mg/l) people tend to have mottled teeth. Excessive intakes from water with $>3-6$ mg F/l can cause skeletal fluorosis (deformed limbs), which becomes crippling with water concentrations >10 mg F/l. However, in some places where this occurs, contamination of food (drying cobs of maize) by the smoke from cooking stoves burning high F-containing coal is the main source of F, rather than the drinking water (I. Thornton, personal communication).

5.5 Radionuclides

5.5.1 History and nomenclature

The first observation of radioactivity was made in 1896 by Becquerel, who noted that salts of uranium were phosphorescent and emitted radiation which penetrated black paper, opaque to light, and which reduced a photographic plate. For pure salts, this activity is independent of the mode of chemical combination and of changes in temperature and magnetic flux. The activity always relates to the amount of the source element present and hence radioactivity is described as an infra-atomic property.

A principal ore of uranium is pitchblende, which includes its black oxide U_3O_8. Marie and Joliot Curie with Bemont noticed that the intensity of emission varied with the source. Pitchblende from the Joachimstahl mine in the Czech Republic was four times stronger than that from Cornwall and both were more active than pure uranium nitrate. They concluded that these ores contained a further more active source of radioactivity; this led them to the discovery of polonium (Po) and then to radium (Ra), which was isolated as its sulphate and separated from the closely related barium salt by a patient fractionation of the chlorides. One ton of pitchblende afforded only 0.3 g of radium but its activity is over 10^6 times that of U.

5.5.2 Types of radioactive emission

The α-particles emitted by radium and uranium have the mass of He atoms but carry two positive charges; 1 g Ra emits 3.7×10^{10} α-particles and

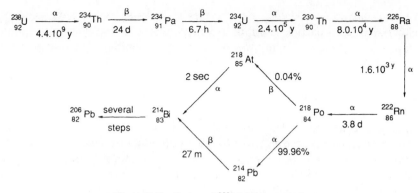

Figure 5.12   Decay of ^{238}U (Mellor, 1932).

2×10^9 cal, which is 2×10^5 times the calorific value of coal. The other principal source of radioactivity is the β-particle, which is an electron carrying a
single negative charge. Radioactive decay may give rise to other entities including the positron, but these are not of concern as pollutants. The chemical
consequences can best be shown by considering a decay series (Fig. 5.12).

The predominant naturally occurring form of uranium is ^{238}U; it loses
four mass units and two nuclear charges on emission of an α-particle to
give an isotope of thorium, written as ^{234}Th. This product decays with the
ejection of a β-particle, which entails no mass change but a gain of one
nuclear charge to form protactinium (Pa); a further β-emission leads to
^{234}U. This isotope contributes only 0.7% of natural uranium and this can be
accounted for by considering the half-lives ($t_{1/2}$) of its predecessors. The two
β-emitters have very short lives relative to that of ^{238}U, while they sustain
the sequence through their activity their rate of formation is less than their
rate of decay and so only small amounts survive in the equilibrium. Because
$^{234}_{92}$U decays much more slowly than $^{234}_{91}$Pa, it will accumulate to a relatively
higher level than this precursor, nevertheless $^{234}_{92}$U (with $t_{1/2} = 2.5 \times 10^5$
years) is shorter-lived than $^{234}_{92}$U by over 10^4 times and so survives at a lower
concentration than the parent uranium.

The series is continued by the successive emission of two α-particles to
produce ^{230}Th and then ^{226}Ra with eight fewer mass units than ^{234}Th and
atomic number 88. Radium is in group II of the periodic table with
chemistry similar to barium, loss of an α-particle gives $^{222}_{86}$Rn in the inert gas
group zero. ^{218}Po in group VI is formed by further α-loss but this isotope
has a dual decay path and yields the more electronegative ^{218}At by β-loss
and also $^{214}_{82}$Pb by α-loss. Radioactive decay ends after several further steps
with the stable isotope ^{206}Pb. There are two other natural decay series, one
begins with ^{235}U and the other with ^{232}Th.

An important member of the ^{238}U series is the short-lived inert gas radon
($t = 3.8$ days) and owing to its release from granitic rocks we have to consider
its toxicity within buildings erected on these foundations (Section 7.9, p. 346).

Gamma-radiation. When a new species is produced by the emission of either an α- or β-particle it will be at an energy level above its stable ground state and to achieve stability the excess energy is emitted as γ-radiation.

5.5.3   Units of energy and measurement of toxicity

Energy is measured in electron-volts: 1 eV is the energy acquired when unit electronic charge is accelerated through a potential difference of 1 V and is equivalent to 1.60×10^{-12} ergs. The emission energies of radioactive particles are greatly in excess of this and are conveniently expressed in MeV or 10^6 eV. Table 5.28 gives some representative values including those for a number of nuclides in the ^{238}U series. Beta-particles have a range which depends on the emission energy; for those of 0.5 MeV it is about 1 m and for those of 3.0 MeV it is about 10 m in air. When taken up by obstacles they produce a more penetrative secondary radiation known as bremsstrahlung. The best protective barriers are solids of low atomic number such as aluminium, perspex and rubber. A 2.5 cm piece of perspex will protect against β-emission of energy as high as 4 MeV.

Exposure to radiation is measured in grays (Gy) (the rad is the non-SI unit and is equivalent to 0.01 Gy). Absorption of 1 Gy leads to the release of 10 000 erg/g of body tissue. This is also equivalent to exposure to a level of one Roentgen (R).

The current usage is to refer to exposure in rems. The rem takes account of a quality factor, which for the more dangerous α-particles is 10, so that exposure to them at the level of 1 rad is rated as 10 rem.

Gamma-rays are more difficult to stop because they are not ionic, but their intensity falls away exponentially and a shield must be thick enough to reduce the intensity to acceptable levels.

Table 5.28   Representative emission energies (Wilson, 1986)

Nuclide	Energy (MeV)	Half-life
Beta-emission		
^{234}Th	0.19	24.1 days
^{234}Pa	2.31	1.18 min
^{210}Pb	0.063	22 years
^{35}S	0.167	87 days
^{36}Cl	0.71	3×10^5 years
^{40}K	1.32	1.3×10^9 years
Alpha-emission		
^{238}U	4.2	4.5×10^9 years
^{234}U	4.77	2.5×10^5 years
^{226}Ra	4.78	1.6×10^3 years
^{222}Rn	5.48	3.8 days

Materials of high atomic number may exhibit photoelectric absorption whereby the energy of γ-rays is transferred to an electron, which is then ejected from the atomic shield. The isotope ^{60}Co is a dangerous product of nuclear fission (see below) and emits γ-rays of mean energy 1.26 MeV; to reduce their dose tenfold 4.6 cm of lead is needed and for each subsequent tenfold reduction a further 4.6 cm of shielding is necessary.

5.5.4   Radioactive potassium

Although not a member of a series, ^{40}K presents a minor environmental risk:

$$^{40}\text{K} \xrightarrow[t_{1/2} = 1.3 \times 10^9 \text{ years}]{\beta} {}^{40}\text{Ca} \qquad (5.47)$$

In Europe and the USA ^{226}Ra occurs in concrete at an average level of 0.9–2.0 pCi/g (*cf.* Section 7.9), with ^{232}Th at 0.8–2.3 and ^{40}K significantly higher at 9–19 pCi/g (Ci, curie). A 70 kg human contains 140 g of K, mostly in muscle, and the ^{40}K component emits 0.1 µCi and delivers 20 mrem to the gonads and 15 mrem to bone. Fortunately, although K salts are important and widely used, the proportion of the radioactive isotope is only 0.0118%.

5.5.5   Production of radionuclides by artificial means

Partly because of their industrial value as tracers, many radionuclides have been synthesized; they can be obtained by neutron bombardment. If there is a fruitful collision with a nucleus, loss of a proton or an α-particle will lead to chemical change while loss of γ-radiation will produce an isotope. An example of each is given below:

$$\text{n,p reaction } {}^{35}_{17}\text{Cl} + {}^{1}_{0}\text{n} \longrightarrow {}^{35}_{16}\text{S} + {}^{1}_{1}\text{p} \qquad (5.48)$$

$$\text{n,α reaction } {}^{27}_{13}\text{Al} + {}^{1}_{0}\text{n} \longrightarrow {}^{24}_{11}\text{Na} + {}^{4}_{2}\text{He} \qquad (5.49)$$

$$\text{n,γ reaction } {}^{23}_{11}\text{Na} + {}^{1}_{0}\text{n} \longrightarrow {}^{24}_{11}\text{Na} + \gamma\text{-rays} \qquad (5.50)$$

$^{35}_{16}$S produced as in Equation 5.48 is oxidized to SO_2 and is then injected into plumes to trace the extent of transport (p. 29). Its half-life of 87 days and strong β-emission make it well suited for this purpose and since it is chemically indistinguishable from unlabelled SO_2 its response to the environment is identical.

5.5.6   Nuclear fission

Discovery.   Marie and Frederic Joliot-Curie came close in 1934 when they noted that the product from neutron bombardment of $^{238}_{92}$U was not the expected $_{89}$Ac but was more like $_{57}$La. Hahn and Strassman, while attempting the synthesis of transuranium elements in 1938, were led to the first firm description of a fission reaction:

$$_{92}U + {}_0^1n \longrightarrow {}_{56}Ba + {}_{36}Kr \tag{5.51}$$

In 1940, McMillan and Abelson observed that some deuterons recoiled with high energy resulting from fission when they were used to bombard uranium, but a new element was formed (Equation 5.52) which lacked the energy to escape from the body of uranium. This was named neptunium (Np) after the first planet beyond Uranus:

$$_{92}^{238}U + {}_1^2H \longrightarrow {}_{93}^{238}Np + 2\,{}_0^1n \tag{5.52}$$

Subsequently Seaborg showed that:

$$_{93}^{238}Np \xrightarrow[\text{2.1 days}]{\beta} {}_{94}^{238}Pu \tag{5.53}$$

This isotope of plutonium (Pu) named after the furthest planet, is an α-emitter with $t_{1/2} = 86$ years; that required for nuclear energy generation is $_{94}^{239}Pu$.

Criteria for fission.   Slow fission is possible but such events are rare and of no use for power generation; thus $_{92}^{238}U$ undergoes spontaneous fission to give $_{54}Xe$, which accumulates in natural sources and which can be used for dating.

E_{critical} is that energy required to split a nucleus in two. When this energy is less than the binding energy the atom only requires to capture a neutron of zero energy to undergo fission; it is said to be *fissile*. The isotope $_{92}^{235}U$ in the only natural material with this property, capturing a neutron and undergoing fission as ^{236}U. The synthetic nuclide $_{94}^{239}Pu$ is also fissile.

A nucleus is said to be *fissionable* if it reacts only after capturing an energetic neutron. In order to sustain an energy-generating chain reaction in fissionable nuclei, one energetic neutron must survive to initiate the next generation. In practice, more than one such neutron must be produced on fission since some escape from the system and others are captured by accumulating non-fissionable products. Figure 5.13 shows the rapid process which follows the release of two neutrons leading to four reactions in the second generation.

The availability of energetic neutrons is expressed as the *multiplication factor (k)* = number of fissions in one generation/number in the preceding generation. Provided that $k > 1$, the conditions for maintaining a chain reaction will persist and be enhanced with each generation. This *super-critical* state was necessary in the $_{92}^{235}U$ bomb which was detonated over Hiroshima on 7 August 1945.

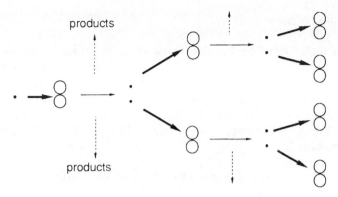

Figure 5.13   Ideal fission without neutron loss.

When $k = 1$, the system is said to be *critical* and this is the state required for the controlled release of energy.

When $k < 1$ the system is *subcritical* and unproductive.

Quantitative yield.   To appreciate the potential of fission for energy production it is helpful to analyse a typical reaction:

$$\underbrace{^{235}_{92}U + {}^{1}_{0}n}_{236.0526} \longrightarrow \underbrace{^{95}_{42}Mo + {}^{139}_{57}La + 2\,{}^{1}_{0}n}_{235.8292} + \beta\text{-particles} \quad (5.54)$$

235.0439 1.0087   94.9057 138.9061 2.0174

By subtraction, a mass loss Δm of 0.2234 units is obtained and this may be converted into its energy equivalent using the Einstein formula $E = mc^2$, where $c =$ the velocity of light $(3 \times 10^{10}\,\text{cm/s})$ and m must be divided by Avogadro's number (6×10^{23}) to put it on an atomic scale. Hence the energy yield $= 0.2234 \times 1/(6 \times 10^{23}) \times 9 \times 10^{20}$

$$= 0.335 \times 10^{-3}\,\text{ergs} = 0.335 \times 10^{-10}\,\text{J}$$

Fission of 1 g ^{235}U will produce:

$$0.335 \times 10^{-10} \times \frac{6 \times 10^{23}}{235} \times 10^{-6}\,\text{MJ}$$

A typical coal yields 32 MJ/kg and its equivalent mass is:

$$\frac{0.335 \times 6 \times 10^7}{235 \times 32} = \mathbf{2680}\,\text{kg or 2.68 tonnes}$$

5.5.7   Power generation in nuclear reactors

In this book only a brief summary of power generation in nuclear reactors is given; for a more detailed treatment see Nero (1970) and Lamarsh (1983).

Preparation of the fuel

Gaseous diffusion. The formidable problem of enriching the low level (0.7%) of $^{235}_{92}$U in natural uranium was initially dependent on conversion of the oxides to UF_6. Then the corrosive mixed gases were passed through Ni columns with pores of 0.01 μm when the lighter $^{235}UF_6$ escaped faster than $^{238}UF_6$; since the fractional separation was only 1.002 a multi-stage process was needed. After enrichment the hexafluoride was converted to the dioxide by hydrolysis:

$$UF_6 + \text{steam} \longrightarrow UO_2.F_2 \xrightarrow{H_2} UO_2 + 2HF \qquad (5.55)$$

Alternatively the tetrafluoride may be reduced to uranium metal (Greenwood and Earnshaw, 1984):

$$UO_2 + 4HF \xrightarrow[-H_2O]{550°C} UF_4 \xrightarrow[700°C]{Mg} U + 2MgF_2 \qquad (5.56)$$

Centrifugation. A more economical process is to place the UF_6 in a series of gas centrifuges where now the heavier isotope migrates more readily to the boundary and $^{235}UF_6$ becomes concentrated near the axis of rotation.

Laser separation. This depends on the raising of uranium metal to an excited state rather in the manner of AA spectroscopy (p. 124). A laser emitting green light can be tuned precisely to a wavelength of 502.73 nm so as to excite only ^{235}U. It is then ionized by a second source and electromagnetically separated from residual uncharged ^{238}U. In practice, much of the original uranium vapour is also collected and so the net change is one of enrichment and not complete separation.

5.5.8   Nuclear reactor types

The pressurized water reactor (PWR). This is widely used today around the world. Ordinary water is used as a moderator for the production of low energy or *thermal* neutrons, that is those with energy comparable to the system as a whole. Water is favoured because it is cheap and being composed of lighter atoms it is efficient. In Fig. 5.14, the oxide fuel containing 3% $^{235}UO_2$ is supplied as 1 cm pellets packed within 3.5 m long hollow zircalloy tubes or pins, which are grouped into bundles or elements of about 20 cm^2. The reactor (A) contains about 200 of these elements with spaces for control rods, coolant flow and neutron detecting instruments. Moderating water enters at 290°C and leaves at 320°C and hence the reactor chamber is at a high pressure of about 2000 psi or 15 MPa (see appendix). The containing steel vessel is some 12 m high and 5 m in diameter.

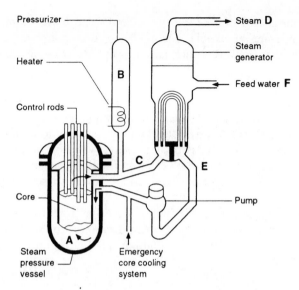

Figure 5.14   Simplified diagram of a PWR (Collier and Hewitt, 1987).

The pressurizer (B) is fitted to ensure that there is no loss of coolant by evaporation; this would be especially serious as neutron energies would also rise wherever steam replaced water.

The issuing hot water (C) is passed through a heat exchanger to produce steam which leaves (D) to drive the turbine. After exchanging its heat, the cooling water is returned at (E), while fresh cooling water and turbine condensate enter the exchanger at (F). The primary coolant in contact with the reactor is isolated from the secondary heat exchanger circuit.

The multiplication factor (see above) may be rapidly varied by movement of the control rods, which are withdrawn to raise the power level and inserted to reduce it. If need be they can shut the reactor down completely. The rods are neutron-absorbing materials such as boron steel or silver alloys and their action is supplemented by injection of boric acid solution into the moderator.

The boiling water reactor (BWR).   At pressures of about 7 MPa, the reactor can be stabilized even when the moderator boils and the emitted steam can be used directly to drive the turbines. However, these and all the linking condensers and pipework must also be shielded as the circulating water becomes radioactive.

The advanced gas-cooled reactor (AGR).   This was developed originally in England and the reactor at Hinckley Point B is of this type with enriched UO_2 fuel in steel jackets. The moderator is constructed from graphite bricks and CO_2 is circulated to cool the system and to generate steam. Although this reactor operates at an efficiency of 40%, it does not produce electricity more cheaply than a PWR, which has been the preferred design in the UK since 1979.

The heavy water reactor. The Canadian design is the CANDU, an acronym for CANada and Deuterium Uranium. It was developed there after World War II to take advantage of available natural uranium without the need for enrichment; this is because heavy water has a low absorption for thermal neutrons. It also follows that to reach thermal energies energetic neutrons must undergo more collisions and so travel further in D_2O than in light water.

A simplified diagram (Fig. 5.15) shows the fuel rods passing through a tank, or calandria, containing the moderator at a low pressure while a high coolant pressure is maintained only within the fuel channels. This has the

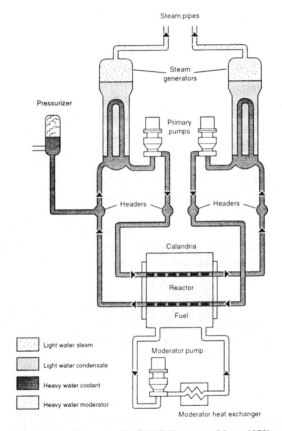

Figure 5.15   Diagram of a CANDU reactor (Nero, 1979).

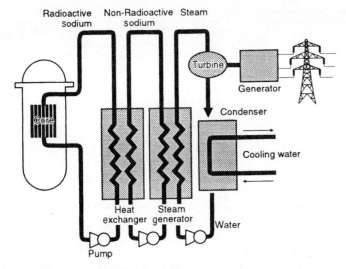

Figure 5.16   Diagram of a fast-breeder reactor (Nero, 1979).

advantage of limiting the volume of the vessel but it also means that the steam pressure and temperature are less than in the PWR and the efficiency is only 30%.

Breeder reactors – the liquid metal fast-breeder reactor.   This one type (LMFBR) will be considered to illustrate the principle. These reactors are attractive to nations with no indigenous fuel supply because they produce fissile material in the course of their operation and can use depleted fuel from thermal reactors.

Figure 5.16 broadly outlines a typical arrangement. Here the driving force is obtained from fast neutrons and there is, therefore, no moderator; the cooling and heat-transfer circuits are charged with metallic sodium (m.p. 98°C). From a purely chemical point of view this is a hazardous choice because of its high reactivity to water and oxygen and the container must be flooded with nitrogen gas to limit these risks. However, it has engineering advantages in that sodium has only a slight moderating effect and at the operating temperature of 500°C a high-pressure vessel is not required. In the breeder reactor, sodium within the inner cooling circuit captures some neutrons to form the higher nuclide:

$$^{23}_{11}\text{Na} + ^{1}_{0}\text{n} \longrightarrow ^{24}_{11}\text{Na} \tag{5.57}$$

Hence a double-looped system is required to prevent circulation of active material outside the screen.

If a breeder reaction is to be sustained, the ratio of fissile atoms produced to the average number of fuel atoms consumed must be > 2. This is because

Table 5.29   Some fission products

Element	Symbol	$t_{1/2}$	Activity $\times 10^3$ (Ci/ton uranium)	Risk level
Strontium	^{89}Sr	50 days	93	2
	^{90}Sr	28 years	92	2
Ruthenium	^{106}Ru	1 year	760	2
Caesium	^{134}Cs	2.1 years	203	2
	^{137}Cs	30 years	130	2
Iodine	^{129}I	1.7×10^7 years	Low	
	^{131}I	8 days	72	
Plutonium	^{238}Pu	86 years	5	
	^{239}Pu	2.4×10^4 years	Low	
	^{240}Pu	6.5×10^3 years	Low	1
	^{241}Pu	13 years	175	
	^{242}Pu	3.8×10^5 years		

Ci = curies.

of the demands: one neutron is required to sustain the critical state, one neutron is required to produce fissile material and some excess is necessary to make up losses.

Under typical conditions the reactor will be charged with ^{238}U mixed with about 20% by weight of ^{239}Pu. When ^{239}Pu captures a neutron of essentially zero energy, the product ^{240}Pu has binding energy of 6.4 MeV, which is in excess of the energy required for fission (4.0 MeV) and so the self-sustaining reaction begins. In a year of operation, about 50 times the initial weight of ^{239}Pu will be produced (Lamarsh, 1983) together with other products (Table 5.29).

Fission products

Strontium 90.   This presents a risk inherent in its relatively long half-life and its chemical affinity with calcium, which leads to its concentration and persistence in bone; particulate emission also presents a hazard to lungs. After the Urals disaster in 1958, ^{90}Sr was detected in soil at levels up to 43 Ci/ha and it was distributed up the food chain in plants and small mammals (Medvedev, 1979).

Caesium 137.   This is similar to potassium in its chemistry and, therefore, it penetrates neural tissue and presents a hazard to muscle (Adelman, 1987).

Release by leakage from the Windscale (now Sellafield) plant of British Nuclear Fuels peaked at 1.2×10^4 Ci/month in 1975 (Crouch, 1986). Fish caught in this part of the Irish Sea were marketed with ^{137}Cs levels of 10 pCi/g compared with a typical level in North Sea catches of 0.1 pCi/g.

In the aftermath of the Chernobyl incident, concern arose from the spread of ^{137}Cs and the levels at some distant sites are given in Table 5.30.

Table 5.30 Spread of ^{137}Cs from Chernobyl (Nuclear Energy Agency, 1989)

Country affected	Distance from Chernobyl (Median km)	Activity in soil (kBq/m²)	
		^{137}Cs^a	^{131}I
Austria	1250	23	120
Norway	2000	11	77
UK	2250	1.4	5.0
France	2000	1.9	7.0
Ireland	2750	5.0	7.0

^{a} With ^{134}Cs.

Iodine 131. Iodine is combined and concentrated in the thyroid gland following the ingestion of milk from cows grazing on polluted pasture, as was the case following the Windscale release of 1957.

Ruthenium 106. This isotope has been found in seaweed at levels of 200 pCi/g – a concentration factor of 2000, with consequences for those who consume laverbread.

Plutonium nuclides.   These do not follow a natural path in a food chain as they are artificial; nevertheless, ingestion of their water-soluble forms presents a hazard to bone and to liver. Insoluble forms inhaled as dust induce lung cancer.

5.5.9   The future of nuclear power

The first successful generator of power for the public supply came on stream in 1956 at Calder Hall, Cumberland. This stimulated international enthusiasm and massive capital investment. One motivating factor in the climate of the 'cold war' was the production of plutonium for nuclear weapons. Today a more sober attitude is taken largely as a result of the public perception of risk of exposure to radiation and also the properly critical assessment of cost of nuclear-generated electricity. Although other resources of fossil fuels have been identified, shortages of power and serious social consequences will be experienced in the long term if no other alternative is available. While energy-demanding research into fusion energy is continued, there remains a pressing need for energy conservation. In its report of 1981, the House of Commons Select Committee on Energy observed that 'The Department of the Environment has no clear idea as to whether spending £1300 million on a new nuclear plant is as cost effective as spending the money on energy conservation'.

Table 5.31   Some nuclear events as placed on the IAEA scale of risk

Level	
7	Chernobyl, USSR, 1986
6	Ural mountains waste explosion, 1958
5	Fire at Windscale (Sellafield), 1957
	Three Mile Island, USA, 1979
4	Fatal accidents at Los Alamos, Wood River and Idaho Falls in period 1945–64
	and Saint Laurent, France, 1980
3	Unauthorized release at Vandellos, Spain, 1989; Tomsk-7, Russia, 1993
2	Incidents necessitating safety reassessment at UK Magnox stations, 1968, 1983, 1989
1	Management deficiencies in waste reprocessing, Windscale, 1986

Risk of exposure to radiation.   The International Atomic Energy Agency based in Vienna has proposed a system of classification for nuclear accidents (Table 5.31) on a scale from 4 to 7, with less severe incidents on a scale from 1 to 3.

Owing to the perceived risk, nuclear stations are built as far as possible from major conurbations at sites where large quantities of cooling water are available. The distribution of stations in the USA (Fig. 5.17) and the UK (Fig. 5.18, at five times the US scale) shows that distance affords more protection to the former community than it does to the latter.

The nuclear industry recognizes the concept of 'maximum acceptable risk' based on an assessment of probabilities used originally in the chemical industry (Jackson, 1992). This, for example, put the risk of death from exposure to radiation in the UK at 1 in 10^7, the same as that from a lightning strike. These evaluations are necessarily based on notional data

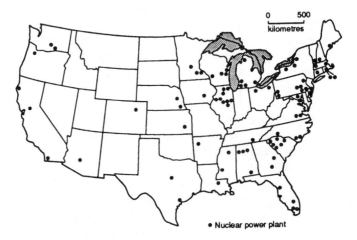

Figure 5.17   Distribution of nuclear stations in the USA (*Nuclear News*, 1992).

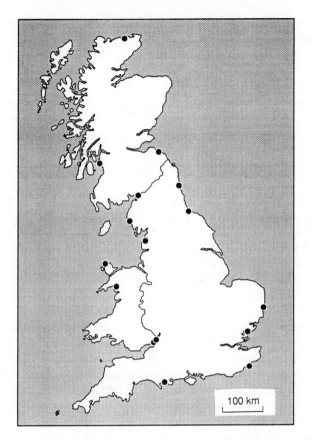

Figure 5.18 Distribution of nuclear stations in the UK (Oppenshaw, 1986).

and not on experimental facts. They do not take account of the uniquely widespread effects and long-term pollution arising from nuclear accidents. The uneasy public perception will inevitably be based on the evident consequences of a few major accidents rather than the purely statistical approach.

5.5.10 Observations on major accidents

The fire at Windscale (8 October 1957). This occurred in the number one plutonium production reactor and was the result of human error coupled with inadequate operating instructions (Patterson, 1976). The reactor required purging by first raising and then lowering the power, but it appeared when the temperature was falling that the operation was incomplete. The crucial error was that power was boosted again to raise the temperature; however, the indicated temperature was not maximal and hotter spots elsewhere in the core took fire. No sign of malfunction was evident until 10 October when radioactivity was detected at the top of the massive discharge stack.

In the interim a major fire had taken hold and was consuming the uranium metal fuel and the graphite moderator. The operators attempted to maintain secrecy for 24 hours after the fire was discovered but as the supply of CO_2 available to them only fuelled the fire they had to involve off-site fire brigades who quenched it with water injected through fuel entry ports.

It was estimated that the most hazardous release was that of 20 000 Ci ^{131}I (see Table 5.29) and the Government ordered the disposal of all milk from dairy herds within a radius of about 25 miles. The full report of the subsequent enquiry was never published but a book detailing the causes and consequences has appeared (Arnold, 1992). A recent review (Chamberlain, 1996) shows that activity in topsoil from fallout of ^{239}Pu declined from 700 Bq/m^2 near the plant to 70 Bq/m^2 at a distance of 15 km, where it is equivalent to that arising in the general area from nuclear weapons testing. This should be compared with radioactive levels indoors (p. 347).

The Urals disaster (1958).   This affected an area of some 1000 square miles between Cheliabinsk and Sverdlovsk, causing death and destruction of property. An explosion occurred in waste from a plant which was most probably producing weapons-grade plutonium. Many years elapsed before any information reached the West, but a full account by Medvedev was published in 1979; his early advice was rejected as 'rubbish' by the then Chairman of the UKAEA, Sir John Hill.

Three Mile Island (28 March 1979).   The reactor was operated by Metropolitan Edison on the Susquehanna river close to the small community of Middletown and 7 miles from Harrisburg (population 60 000) and about 75 miles from Baltimore and Philadelphia (see Fig. 5.19).

It is worth noting that the Atomic Energy Commission experts who published the accident probability assessment WASH-740 (*Theoretical Possibilities and Consequences of Major Accidents in Large Nuclear Power Plants*, 1957) were themselves of the opinion that a medium-sized accident would occur in their own lifetimes. However, no measures were in place to protect the local community; in the event, the company informed the Middletown authorities that 'Even the worst possible accident postulated by the AEC would not require evacuation of the town'.

The official report into the accident (*The Report of the President's Commission on 3 Mile Island*, Pergamon, 1979) concluded that operator error was responsible but that inadequate training, confusing procedures and failure to appreciate earlier accidents were factors. A review of the incident has been published (Stephens, 1980) and the following abridged sequence of events is taken from it:

Figure 5.19   Three Mile Island nuclear power station (the damaged reactor is in the building on the left adjacent to the cooling towers) (photo: B. J. Alloway).

1. 31 December 1978: plant operations begin just in time to earn a $40 million tax credit for the year.
2. 28 March 1979: during a routine operation to renew clogged ion-exchange resin in one of the large columns used to remove minerals from the feedwater, a small volume of water was forced into the compressed air control system for the facility. After a time-lapse this led to failure of one of the feedwater pumps and stopped the circulation of cooling water. Up to 1,000 gallons/min could be supplied to the whole system.
3. Within seconds the temperature and pressure in the reactor rose rapidly. A release valve ejected a blast of excess steam to the open air.
4. The boron and silver control rods functioned and dropped amongst the 37 000 zirconium-clad UO_2 fuel pellets, so closing down the reactor. Pressure fell to a safe level but continuing radioactive decay still generated 6% of the operating thermal capacity.
5. Owing to the fault in the control system, a further valve failure led to the release of a quarter of a million gallons of radioactive coolant onto the floors of the containment and subsidiary buildings.
6. Still only 1 minute into the incident, the 1 m bore coolant inlet pipes hammered as pumps tried to force water past input values closed for a previous test and *not reopened*. Indication of the closure was given on the control panel but was overlooked. Owing to the fault in the control system very many visual and oral alarms were activated at the same time, so confusing the engineers.
7. The engineers were unaware of the loss of coolant and concluded that the problem was the result of flooding of the pressurizer (this keeps the pressure close to the optimum either by spraying water to condense

excess steam or by evaporation of excess water by electrical heating). Hence more water was drained and the second flow pump was cut. The probability of failure of both feedwater valves had been put at 1 million to 1 against.

8. Steam bubbles forced out the remaining water. The overall temperature was recorded as 400°C but some thermocouples registered 1100°C and were disbelieved – here meltdown was occurring.

9. The intense radiation converted steam into hydrogen, which accumulated dangerously, and oxygen, which reacted with the zirconium cladding so loosing the UO_2 pellets inside the reactor. Under these conditions, the silver and boron control rods were all destroyed. Fortuitously the release of the iodine nuclides was limited by combination with this silver to give involatile silver iodide.

10. At 2 hours into the incident, radiation at a level of 1000 millirem was detected in an auxiliary building.

11. On the Friday following the incident, a helicopter at 600 feet above the station recorded activity at 1200 millirem/hour – 4 hours at this level would be equivalent to the EPA approved dose for a worker for a year. Owing to a failure of communication, this activity was held to be that at ground level off-site. In the ensuing panic, it is estimated that 40 000 people fled the area, although no official order was given to evacuate.

In conclusion, it should be noted that although the core came close to meltdown no one was physically injured; the major health effect was one of mental stress. There was, however, the loss of the reactor and the cost of clean-up, put at $2 to $3 billion levied on the taxpayer. As a result, 2000 truckloads of waste were transported across Washington State and 4000 gallons of polluted water were diluted and discharged into the river. Bruce Babbit, in a minority view to the report of the President's Commission, queried the running of nuclear plant by the private sector.

Chernobyl (26 April 1986) (Nuclear Energy Agency, 1987; Marples, 1986). The town dates from the twelfth century and lies in a sparsely populated area 135 km north of Kiev. The generating complex was active in October 1977 and by 1986 four 1000 MW units were producing 10% of the total USSR capacity.

The accident was the result of operator error while conducting an experiment in obtaining back-up power from the spinning turbines after shut-down. This was necessary because back-up diesel power for the pumps took 50 seconds to maximize. The release of information by the authorities was said to have been made with the principal aim of preserving a nuclear capacity in the country.

During the experiment the operator failed to set the power level control, which led to steam condensation and loss of productive neutrons. While purging the system of xenon, which acts as a moderator, more loss of neutrons occurred and *contrary to the safety procedure* the control rods were withdrawn and the safety system was closed off in order to complete the experiment against the clock.

In 4 s, the power output rose from 200 MW to over 300 000 MW. Similar disruption of the fuel took place as at Three Mile Island and under the pressure of hydrogen, steam and burning graphite the 1000 ton top plate was blown off. Radioactive steam was distributed throughout the whole plant via the ventilation system and 300 on-site workers were treated for radiation poisoning, while measuring devices for firemen went off-scale.

The public demand for information could not be resisted in the light of reported radioactivity 800 miles away over Sweden. Nevertheless, the responses tended to vary with the national interest – France with 65% of nuclear-generating capacity said nothing for 14 days. In the UK, Secretary of State Kenneth Baker told the House of Commons 'If there is no further discharge the effects will be over within a week', but 2 months later on June 20th grassland in the Lake District, North Wales and Scotland was so polluted that one million sheep and lambs were too radioactive to be slaughtered for food (see Table 5.30).

Six years after the accident, children in the region of Kiev are being born with birth defects including adrenal and thyroid cancers. The Ukrainian authorities put the number of radiation-related deaths at several thousand, with 15 000 affected by disease to a lesser extent.

Other sources of radionuclide pollution. Pollution has arisen from the scuttling of nuclear submarines in the Sea of Japan by the former USSR, whilst others have been left without maintenance in the port of Murmansk.

5.5.11   Social aspects of nuclear power generation

The principal factors which weigh aganist an expansion of nuclear power are:

1. the recorded major accidents
2. doubts about the safe ultimate disposal and reprocessing of radioactive waste, as evidenced by the THORP facility at Sellafield
3. lack of confidence in Government agencies as a result of concealment and denial of information
4. inaccurate accounting and the recent admission that nuclear electricity is not cost effective.

Nevertheless, the developed nations depend to a significant extent on nuclear stations for their electricity supplies (Table 5.32). Italy and Switzerland have halted nuclear development. Should other nations with intense urban and industrial development follow suit, considerable construction of new fossil fuel units would be required. Their building

Table 5.32   Proportion (%) of power generated by nuclear fission in some developed nations

Belgium	60	West Germany	31
Sweden	43	Spain	22
UK	21	France	65
USA	15		

costs would rise because of the need to control CO_2 emissions. Meanwhile China, France and Spain continue to build; Japan is aiming for 40 nuclear units by 2010 and to be 40% nuclear by 2050. Czechoslovakia has eight stations and plans a further eight, although there are doubts about the safety of their plant at Bohunice, which is close to the Austrian border. Similarly North Korea plans four reactors at Sinpo, 300 miles from Vladivostok, and has rejected supervision by the Atomic Energy Authority. This still leaves the risk of pollution from other suspect sources; for example there are 15 PWR stations of the type existing at Chernobyl that remain in use in Eastern Europe.

In the UK, increasing doubts have been expressed about the commercial viability of nuclear power. Critics have seized upon the difficulties of safe disposal of radioactive waste and the high cost of building nuclear units, as compared with gas-fired stations. Independent assessments put the cost of nuclear-generated electricity at 4–8 p/kW hour in excess of fossil-fuelled alternatives. The UK regulator capped their price at 2.4 p/kW hour in 1994.

In December 1995, the private sector operator/British Energy abandoned plans to build the proposed nuclear power stations Sizewell C and Hinckley Point C. At that time, a gas-powered station could have been built at one-third the cost of its nuclear alternative. Other factors in the decision were the decommissioning costs of a nuclear station and the lower price of gas-generated electricity (BBC 1, 1995)

It is not possible to enlarge on the engrossing history of nuclear power in this book but the reader is referred for details to Hall (1986), Pocock (1977) and Patterson (1985).

5.5.12   Power from thermal fusion

The great density and high temperature of the sun's plasma leads to a continous release of energy through the combination of protons. This occurs in a controlled way rather than as a massive explosion as the probability of the reaction is low. This depends on one of the colliding protons shedding its positive charge as a positive electron or positron, energy release then occurs through reaction of the residual neutron with its partner proton:

$$_0^1 n + _1^1 H \longrightarrow \quad _1^2 H \quad + \quad energy \tag{5.58}$$
$$\text{a deuteron (d)}$$

The new deuteron will react rapidly with one of the many protons available in the dense material:

$$\,^2_1\text{H} + \,^1_1\text{H} \longrightarrow \,^3_2\text{He} + \gamma\text{-rays} \qquad (5.59)$$

In the sun, energy release occurs through the sequence:

$$\,^3_2\text{He} + \,^3_2\text{He} \longrightarrow \,^4_2\text{He} + 2\,^1_1\text{H} \qquad (5.60)$$

The two protons formed re-enter the cycle, with ^4_2He providing more energy by combinations such as:

$$\,^4_2\text{He} + \,^3_2\text{He} \longrightarrow \,^6_3\text{Li} + \,^1_1\text{H} + \text{energy} \qquad (5.61)$$

Terrestrial experiments on thermal fusion are of a similar type but differ in detail. Two atoms of deuterium may combine with similar probabilities to produce helium or tritium:

$$\,^2_1\text{H} + \,^2_1\text{H} \longrightarrow \,^3_2\text{He} + \,^1_0\text{n} + 2.45\,\text{MeV} \qquad (5.62)$$

$$\,^2_1\text{H} + \,^2_1\text{H} \longrightarrow \,^3_1\text{H} + \,^1_1\text{H} + 3.0\,\text{MeV} \qquad (5.63)$$

These energy outputs are about one million times that obtained from a typical chemical change and require an input of only a few thousandths of that produced.

The reaction between deuterium and tritium is even more productive:

$$\,^2_1\text{H} + \,^3_1\text{H} \longrightarrow \,^4_2\text{He} + \,^1_0\text{n} + 18\,\text{MeV} \qquad (5.64)$$

but tritium has disadvantages as a source in that its production is expensive in terms of capital and of the energy input required.

Apart from the low level of radioactivity of tritium, the fusion processes avoid the hazards inherent in nuclear fission. This gives it a degree of priority which is heightened by public fears and political problems over the extension of existing nuclear fission programmes. Thermal fusion faces formidable experimental difficulties if a continous gain is to be won.

Although only a few kiloelectronvolts are needed to bring two nuclei together, for continuous production the fuel must be of a density high enough to generate sufficient collisions and exist at temperatures in excess of that in the sun's plasma. The destructive body of gas is contained *in vacuo* out of contact with the reaction vessel within a torus of magnets. The critical level requires that a temperature of the order of 10^8 degrees and density of 10^{20} ions/m^3 are maintained for 1 second. These conditions have been met singly in the Joint European Torus (JET) but not in a concerted manner.

5.5.13   Cold fusion

The formidable difficulties encountered with hot fusion encouraged the search for an alternative based on an idea of Frank in 1947, which depends on the effect of muons on atomic properties. A muon bears a single electronic charge but has 207 times the mass of the electron. On interaction with deuterium or tritium, the muon enters an orbit much closer than that of the electron, which is ejected since it is no longer attracted by the balanced charges. The resulting atoms of deuterium or tritium are reduced in size by the factor of 207 and resemble neutrons. This change greatly enhances thermal fusion but unfortunately has yet to produce more energy than that required to generate a muon in a particle accelerator. The lifetime of a muonic atom is only a few microseconds, which limits the number of collisions that it can catalyse.

Close (1990) gives a fuller account of fusion processes and of the history of events, including the recent mistaken claim that fusion occurred when electrolytically generated deuterium becomes concentrated in a Pd electrode.

5.6 Mineral fibres and particles

5.6.1   General aspects

Mineral pollutants, as classified here, enter the environment as fibres or particles of crystalline inorganic compounds (frequently as silicates) of relatively fixed chemical composition. In most cases, they will remain as fibres or particles and exert their potentially harmful effect in this form, although all will undergo chemical weathering at rates determined by the mineral and the environment. For example, quartz particles will be very resistant to weathering, but calcite particles will rapidly dissolve in moist acid environments. Other mineral pollutants, such as metallic ore minerals, whose major toxic effect is caused by the release of metal ions on weathering in either water or soil, are dealt with elsewhere in the book (Sections 3.3 and 5.3).

Mineral fibres are commonly used for the thermal insulation of buildings, pipes and storage tanks, fireproofing, and reinforcement of construction materials. Examples of uses include glass fibre-reinforced plastics (GRP or 'fibreglass') and asbestos-reinforced cement products, including roofing panels and pipes for drinking water. Asbestos withstands the high temperatures caused by friction and is used in the manufacture of brake linings for motor vehicles; it is also woven into cloth for use in fire protection suits. Rock wool is used as cavity wall thermal insulation (blown into the cavity) and as a rooting medium for hydroponically grown plants.

Unlike most other pollutants reviewed in this book, the health hazard from mineral fibres and particles does not result from a biochemical toxic reaction but from an irritational effect related to the size, shape and surface of the particles, which causes an inflammatory reaction in body tissues (especially the lung) and the formation of scar tissue or even a carcinoma. Inhalation of separate dry fibres in a confined air space is the major hazard. When the fibres are bonded they constitute a relatively low hazard because there will be few loose fibres. However, damaged thermal insulation (e.g. during alterations or demolition), and accumulated fibres from the manufacture or wear of asbestos or other mineral fibre materials (e.g. brake shoe/pad dust) constitute a major hazard. Appropriate safety masks and skin protection should always be worn when handling mineral fibres.

Dust particles can range in size up to 150 μm but those with diameters (or lengths) of < 10 μm are more hazardous to health because they tend to remain suspended in air for long periods and those smaller than 5 μm can penetrate the respiratory system reaching the bronchioles and alveoli of the lungs. An aerosol particle 10 μm in diameter has a mass 1000 times that of a 1 μm particle and a deposition velocity 100 times greater. Eighty percent of aerosols comprise particles > 1 μm, of which 45% are in the range 10–50 μm (Jaenicke, 1982). Mineral fibres have an aerodynamic configuration which results in them being very easily suspended and resuspended in air and, therefore, exacerbates their hazard to health.

The trachea and larger bronchi are lined with cilia (fine hairs) which help to filter inhaled air and transport particles removed from this air in mucus carried up and out of the respiratory tract by coordinated beating of the cilia. However, although no longer a potential hazard to the lungs, these particles can still affect health when they are swallowed into the stomach in mucus.

Many inhaled dusts accumulate in the lungs without stimulating any local reaction and, therefore, do not cause any recognizable disease. Urban dwellers commonly have accumulations of dark material in their lungs. Benign (non-cancerous) pneumoconioses are recognized as being caused by prolonged exposure to $CaCO_3$, $CaSO_4$, Fe, Sn and Sb dusts. For further discussion of particulate pollution, see Chapter 7.

5.6.2   Examples of mineral pollutants

Silica mineral particles

Source.   Quarrying, rock crushing, ceramics.

Toxicity.   Fine particles of quartz and other forms of silica (SiO_2) can accumulate in the lungs and cause the formation of localized nodules of scar tissue (fibrosis), which enlarge, merge and reduce the respiratory function of the lungs. This disease is called silicosis and it has a long latency period so it could be well advanced before the symptoms are recognized. Silicosis is frequently associated with infections such as bronchitis and even tuberculosis.

Coal dust

Coal is a complex organic polymer containing bound trace metals (< 12% inorganics) sometimes with sulphide mineral inclusions. It has a high calorific value (*c.* 28 MJ/kg) and, therefore, is a combustion or explosion hazard (Section 6.2).

Source.   Mining, transport and handling of coal.

Toxicity.   Prolonged exposure to coal dust produces a fibrosis condition of the lung called pneumoconiosis, which is distinguishable from silicosis because the lung lesions are black. Lung tissue is obliterated and becomes prone to recurrent infection, causing severe respiratory disability. This disease is a major occupational health hazard of coal mining. The specific underlying cause of death from lung disease in males over 24 years old in the USA in 1986 was attributed as follows (Wagener, 1990):

pneumoconiosis (coal workers)	882
asbestosis	501
silicosis	135
asthma	1409
malignant neoplasms	564

References   Adelman, G. (ed) (1987) *Encylopedia of Neuroscience*, Vol. 2. Birkhauser, Boston.
Advances in Chemistry Series (1959) 21: Ozone chemistry and technology. American Chemical Society, Washington, DC.
Alloway, B. J. (ed.) (1995) In *Heavy Metals in Soils* (2nd edn), Ch. 3. Blackie and Son, Glasgow.
American Conference of Government Industrial Hygienists (1990) *Documentation of the TLV and Exposure Indices* (5th ed), Cincinnati.
Armstrong-Brown, S., Rounsevell, M. D. A. and Bullock, P. (1995) Soil and greenhouse gases: management for mitigation, *Chem. Ind.*, 647–650.
Arnold, L. M. (1992) *Windscale 1957. Anatomy of a Nuclear Accident*. Macmillan, London.
Battarbee, R. W., Flower, R. J., Stevenson, A. C. and Rippey, B. (1985) Lake acidification in Galloway: a paleoecological test of competing hypotheses. *Nature*, **314**, 350–352.
BBC 1 (1995) *News Broadcast*. 11 December.
Bowen, H. J. M. (1979) *The Environmental Chemistry of the Elements*. Academic Press, London.
Bower, J. S., Broughton, G. F. J., Willis, P. G. and Clark, H. (1995) *Air Pollution in the United Kingdom: 1993/4*, 7th Report. AEA Technology, Culham, Oxford.
Callender, G. S. (1958) On the amount of CO_2 in the atmosphere. *Tellus*, **10**, 243–248.
Canada Council of Ministers of the Environment (1992). Canadian Environmental Quality Criteria for Contaminated Sites. Report CCME EPC-C534, Winnipeg, Manitoba.
Chamberlain, A. C. (1996) Emissions from Sellafield and activities in soil. *Sci. Total Environ.*, **177**, 259–280.
Chaudri, A. M., McGrath, S. P. and Giller, K. E. (1992) Metal tolerance of isolates of *Rhizobium leguminosarum* bv *trifolii* from soil contaminated by past applications of sewage sludge. *Soil Biol. Biochem.*, **24**, 625–632.
Close, F. (1990) *Too Hot to Handle*. Allen, London.
Collier, J. G. and Hewitt, G. F. (1987) *Introduction to Nuclear Power*. Hemisphere, Washington DC.
Crouch, D. (1986) Science and trans–science in radiation risk assessment: child cancer around the fuel reprocessing plant at Sellafield, U.K. *Sci. Total Environ.*, **53**, 201–216.
Davies, D. J. A. and Bennett, B. G. (1983) *Exposure Committment Assessments of Environmental Pollutants*, Vol. 3. Monitoring and Assessment Research Centre, London.

De Mora, S. J. (1992) In *Understanding our Environment*, Ch. 4. (ed. R. M. Harrison). Royal Society of Chemistry, London.

Department of the Environment (1987) *Interdepartmental Committee for the Reclamation of Contaminated Land List of Trigger Concentrations for Contaminants*. DOE, ondon.

Department of the Environment (1993) DOE, London.

Dickinson, R. E. and Cicerone, R. J. (1986) Future global warming from atmospheric trace gases. *Nature*, **319**, 109–115.

Dunne, J. M. (1990) Vehicle emission control technology. In *Energy and the Environment* (ed. J. Dunderdale), p. 208. Royal Society of Chemistry, Cambridge.

EC (1980) *Quality of Water for Human Consumption*, 80/778/EEC. Commission of the European Communities, Brussels.

Emsley, J. (1995) *Chem. in Britain*, 871.

Epstein, J. M. and Gupta, Raj (1990) *Controlling the Greenhouse Effect*. Brookings Institute, Washington DC.

Fergusson, J. E. (1982) *Inorganic Chemistry and the Earth*. Pergamon Press, Oxford.

Fergusson, J. E. (1990) *The Heavy Elements: Chemistry, Environmental Impact and Health Effects*. Pergamon Press, Oxford.

Greenwood, N. N. and Earnshaw, A. (1984) *Chemistry of the Elements*. Pergamon, Oxford.

Grubb, M. (1991) *Energy Policies and the Greenhouse Effect: Future Trends, Population and Development*, Vol. 1, Dartmouth Publications, Aldershot, UK.

Hall, A. (1986) *Nuclear Politics: The History of Nuclear Power in Britain*. Penguin, Harmondsworth.

Harte, J., Holden, C., Schnieder, R. and Shirley, C. (1991) *Toxics A to Z*. University of California Press, Berkeley, CA.

Houghton, J. T., Jenkins, G. J. and Ephraims, J. J. (1990) *Climate Change: the IPCC Assessment*. Cambridge University Press, Cambridge.

Innes, J. L. (1987) *Air Pollution and Forestry*. HMSO, London.

Innes, J. L. (1990) *Assessment of Tree Condition*. HMSO, London.

Jackson, D. (1992) In *The Chemical Industry – Friend to the Environment* (ed. J. A. G. Drake). Royal Society of Chemistry, Cambridge.

Jaenicke, R. (1982) In *Chemistry of the Unpolluted and Polluted Troposphere* (eds W. Georgii and W. Jaeschke), pp. 341–374. NATO Reidal, London.

Janes, S. M. and Cooke, R. U. (1987) Stone weathering in S.E. England. *Atmos. Environ.*, **21**, 1601–1622.

Kabata-Pendias, A. and Pendias H. (1984) *Trace Elements in Soils and Plants*. CRC Press, Boca Raton, FL.

Kandler, O. and Innes, J. L. (1995) Air pollution and forest decline in Central Europe. *Environ. Pollution*, 90, 171–180.

Kazantis, G. (1980) In *Metals in the Environment*, Ch. 8. (ed. H. A. Waldron). Academic Press, London.

Keeling, C. D., Bacastow, R. B., Carter, A. F., Piper, S. C., Whork, T. P., Heimann, M., Mook, W. G. and Roeloffzen, H. (1989) A three-dimensional model of atmospheric CO_2 transport based on observed winds. *Geophysical Mono.*, **55**, 165–236.

Kloke, A., Sauerbeck, D. R. and Vetter, H. (1984) In *Changing Metal Cycles and Human Health*, (ed. J. O. Nriagu). Springer-Verlag, Berlin.

Krauskopf, K. B. (1967) *Introduction to Geochemistry*. McGraw-Hill, New York.

Lamarsh, J. R. (1983) *Introduction to Nuclear Engineering* (2nd edn). Addison-Wesley, Reading, MA.

Langard, S. (1980) In *Metals in the Environment*, Ch. 4. (ed. H. A. Waldron). Academic Press, London.

Lauenstein, G. G., Robertson, A. W. and O'Connor, T. P. (1990) Comparison of trace metal data in mussels and oysters from a Mussel Watch Programme of the 1970s with those from a 1980s, programme. *Marine Pollution Bulletin*, **21**, 440–447.

Leclercq, I. (1988) *On the Track of the Scourge of the Forest*. DEFORPA, Paris.

Manahan, S. E. (1991) *Environmental Chemistry* (5th edn). Lewis Publishers, Chelsea, MI.

Marples, D. R. (1986) *Chernobyl and Nuclear Power in the USSR*. Macmillan, Basingstoke.

Mattson, E. (1990) In *Cu '90: Refining, Fabrication, Markets*. Proceedings of Conference in Vasteras, Sweden, October, Paper 16. Institute of Metals, London.

Medvedev, Z. A. (1979) *Nuclear Disaster in the Urals*. Angus & Robertson, London.

Mellor, J. W. (1923) *Comprehensive Treatise on Inorganic and Theoretical Chemistry*. Vol. 4, pp. 53–154.

Mellor, J. W. (1932) History of Uranium. *Comprehensive Treatise on Inorganic and Theoretical Chemistry*, Vol. 12, pp. 1–14.

Ministry of Agriculture, Fisheries and Food (1993). *Review of the Rules for Sewage Sludge Application to Agricultural Land*. Soil Fertility Aspects of Potentially Toxic Elements. Report No. P1 31561, MAFF, London.

Ministry of Housing, Physical Planning and Environment, Directorate General for Environmental Protection (Netherlands) (1991) *Environmental Standards for Soil and Water, 1991*. Leidschendam.

Moen, J. E. T., Cornet, J. P. and Evers, C. W. A. (1986) In *Contaminated Soil* (eds J. W. Assink and W. J. van den Brink). Martinus Nijhoff, Dordrecht.

Molina, M. J. and Rowland, F. S. (1974) Stratospheric sink for chlorofluoromethanes, chlorine atom catalysed destruction of ozone. *Nature*, **249**, 810.

Murley, L. (ed) (1995) *Pollution Handbook*. National Society for Clean Air and Environmental Protection, Brighton.

Nebel, C. (1981) *Encyclopedia of Chemical Technology*, (3rd edn). Wiley, New York.

Nero, A. V. (1979) *A Guidebook to Nuclear Reactors*. University of California, Berkeley, CA.

Newman, P. and Kenworthy, J. (1989) *Cities and Automobile Dependence*. Gower, Aldershot, UK.

Nriagu, J. O. (1988) A silent epidemic of environmental metal poisoning. *Environ. Pollut.*, **50**, 139–161.

Nuclear Energy Agency (1986) *Projected Costs of Generating Electricity from Nuclear and Coal Fired Stations Commissioning 1995*. OECD, Paris.

Nuclear Energy Agency (1987) *Chernobyl and the Safety of Nuclear Reactions in OECD Countries*. OECD, Paris.

Nuclear Energy Agency (1989) *Nuclear Accidents: Intervention Levels for Protection of the Public*. OECD, Paris.

Nuclear News (1992) *World List of Nuclear Power Plants*, **35**, 55–74.

O'Neill, P. (1995) In *Heavy Metals in Soils* (2nd edn), Ch. 5. (ed. B. J. Alloway). Blackie and Son, Glasgow.

Openshaw, S. (1986) *Nuclear Power Siting and Safety*. Routledge & Kegan Paul, London.

Owen, K. (1981) Technology: How the 'Bedford smell' foxed the experts. *The Times* (London), January 23.

Packham, R. F. (1990) *Pollution: Causes, Effects and Control* (2nd edn), Ch. 5. (ed. R. M. Harrison). Royal Society of Chemistry, London.

Pacyna, J. M. (1987) in Hutchinson, T. C. and Meema, K. M. (eds) *Lead, Mercury, Cadmium and Arsenic in the Environment*, SCOPE 31, John Wiley and Sons, Chichester.

Pahl, D. A., Zimmerman, D. and Ryan, R. (1990) An overview of combustion emissions in the USA. In *Emissions from Combustion Processes*, (eds R. Clement and R. Kagel). Lewis, Boca Raton, FL.

Park, C. (1987) *Acid Rain*. Routledge, London.

Patterson, W. C. (1976) *Nuclear Power*. Pelican, Harmondsworth.

Patterson, W. C. (1985) *Going Critical*. Paladin, London.

Peters, W. C. (1978) *Exploration and Mining Geology*. John Wiley, New York.

Pocock, R. F. (1977) *Nuclear Power: Its Development in the UK*. Unwin, Old Woking.

Roberts, L. E. J., Liss, P. S. and Saunders, P. A. H. (1990) *Power Generation and the Environment*. Oxford University Press, Oxford.

Rainbow, P. S. (1993) In *Ecotoxicology of Metals in Invertebrates* (eds R. Dallinger and W. Rainbow), pp. 3–23. CRC Press, Boca Raton, FL.

Rainbow, P. S. and White, S. L. (1989) Comparative strategies of heavy metal accumulation by crustaceans: zinc, copper and cadmium in a decapod, an amphipod and a barnacle. *Hydrobiologia*, **174**, 245–262.

Rose, A. W., Hawkes, H. E. and Webb, J. S. (1979) *Geochemistry in Mineral Exploration* (2nd edn). Academic Press, London.

Rotty, R. M. (1983) Annual emissions of CO_2 by fuel type. *J. Geophys. Res.*, **88**, 1301.

Royal Commission on Environmental Pollution (1991) *15th Report, Emissions from Heavy Duty Diesel Vehicles*, p. 79. HMSO, London.

Speakman, K. (1990) Emission control – statutory requirements. In *Energy and the Environment* (ed. J. Dunderdale). Royal Society of Chemistry, Cambridge.

Spence, R. D., Rykiel, E. J. and Sharpe, P. J. H. (1990) Ozone alters carbon allocation in loblolly pine: assessment with carbon-11 labelling. *Environ. Pollut.*, **64**, 93–106.

Sprague, A. (1990) *Statistical Review of World Energy*. British Petroleum, London.

Stebbing, A. R. D. (1985) Organotins and water quality – some lessons to be learned. *Marine Pollution*, **10**, 383–390.

Stephens, N. (1980) *Three Mile Island*. Junction.

Steinnes, E. (1987) In *Lead, Mercury, Cadmium and Arsenic in the Environment* (eds T. C. Hutchinson and K. M. Meema) SCOPE 31, John Wiley and Sons, Chichester.

Steinnes, E. (1995) *Heavy Metals in Soils* (2nd edn), Ch. 11. (ed. B. J. Alloway). Blackie, Glasgow.

Tepper, L. B. (1980) In *Metals in the Environment* (ed. H. A. Waldron). Academic Press, London.

Tolba, M. K. (1983) *Water Quality Bull.*, **8**, 115–120, 167.

Ulrich, B. (1980) *Soil Science*, **130**, 193–199.

Ulrich, B. (1990) Waldsterben: forest decline in West Germany. *Environ. Sci. Technol.*, **24**, 439.

United Nations (1994) *Statistical Yearbook*, 39th Issue.

Wagener, D. K. (1990) Number of deaths due to lung disease: how large is the problem? *Environ. Res.*, **56**, 68–77.

Waldron, H. A. (ed.) (1980) In *Metals in the Environment*, Ch. 6. Academic Press, London.

Warren Spring Laboratory (1994) *Air Pollution in the UK, 1992/3*.

Wayne, R. (1991) *Chemistry of Atmospheres*. Oxford University Press, Oxford.

WHO (1993) *Guidelines for Drinking Water Quality*, Vol. 1, *Recommendations*. WHO, Geneva.

Wilson, B. J. (ed.) (1986) *The Radiochemical Manual*. Radiochemical Centre, Amersham.

Wright, D. A. and Welbourn, P. H. (1994) Cadmium in the aquatic environment. A review of ecological, physiological and toxicological effects on biota. *Environ. Rev.*, **2**, 87–214.

Wright, R. F. (1976) *Ambio*, **5**, 219.

Ozone

Banks, R. E. (ed.) (1979) *Organochlorine Chemicals and Their Industrial Applications*. Ellis Horwood, Chichester.

Knunyants, I. L. and Yakobson, G. G. (eds) *Organofluorine Chemicals and Their Industrial Application*. Ellis Horwood, Chichester.

Nebel, C. (1981) Ozone. In *Encyclopedia of Chemical Technology* (3rd edn), Wiley, New York.

Oxides of carbon, nitrogen and sulphur

Tyler Miller, G. (1990) Living in the Environment (6th edn). Wadsworth, London.

Flavin, C. (1989) *Slowing Global Warming, A Worldwide Strategy, Worldwatch Paper*, **91**, Washington, DC.

Grainger, A. (1990) *The Threatening Desert*. Earthscan.

Meetham, A. R. (1981) *Atmospheric Pollution* (4th edn). Pergamon, Oxford (appendix on units).

Warrick, R. A., Barrow, E. M. and Wigley, T. M. L. (1990) *The Greenhouse Effect and its Implications for the European Community*. EC, Luxembourg.

Brimblecombe, P. (1987) *The Big Smoke*. Methuen, London.

Heavy Metals

Alloway, B. J. (ed.) (1995) *Heavy Metals in Soils* (2nd edn). Blackie and Son, Glasgow.

Bowen, H. J. M. (1979) *The Environmental Chemistry of the Elements*. Academic Press, London.

Fergusson, J. E. (1990) *The Heavy Elements: Chemistry, Environmental Impact and Health Effects*. Pergamon Press, Oxford.

Thayer, J. S. (1995) *Environmental Chemistry of the Heavy Elements*. VCH.
Waldron, H. A. (ed.) (1980) *Metals in the Environment*. Academic Press, London.

Radionuclides

Allaby, M. (ed.) (1988) *Dictionary of the Environment*. Macmillan, Basingstoke.
Burn, D. (1978) *Nuclear Power and the Energy Crisis*. Macmillan, Basingstoke.
Eisenbud, M. (1987) *Environmental Radioactivity from Natural, Industrial and Military Sources* (3rd edn). Academic Press, Orlando.
Hines, L. G. (1988) *The Market, Energy and Environment*. Allyn & Bacon, Boston.
Hoyle, F. (1977) *Energy or Extinction*. Heinemann.
Judd, A. M. (1991) *Fast Breeder Reactors, an Engineering Introduction*. Pergamon, Oxford.
Lilienthal, D. (1980) *Atomic Energy: a New Start*. Harper and Row.
Marshall, W. (ed.) (1983) *Nuclear Power Technology*, Vol. 3, *Nuclear Radiation*. Clarendon Press, Oxford.
Nuclear Installations Inspectorate (1992) *Safety Assessment Principles for Nuclear Plants*. NII, London.
OECD (1987) Chernobyl and the Safety of Nuclear Reaction in OECD Countries, OECD.
Parliamentary Select Committee on Energy (1990) 4th Report, *The Cost of Nuclear Power*. HMSO, London.
UK Royal Commission on Environmental Pollution (1976) *6th Report. Nuclear Power and the Environment*. HMSO, London.
United Nations (1994) *Statistics Yearbook*. 39th issue.
US Department of Energy (1990) *Radon Research Programme*. March.
Weart, S. R. (1988) *Nuclear Fear, A History of Images*. Harvard University Press, Cambridge, MA.
Yearbook of Science and Technology (1992). McGraw-Hill, p. 297.

Organic pollutants 6

Smoke is a term applied to particulate carbonaceous particles of $< 10\,\mu m$ in diameter produced by the partial combustion of organic substances. Smoke rises upwards into the air above the fire as a result of its thermal buoyancy; its colour depends on the material being burnt. The amount of smoke is related to the combustion conditions, including the physical form of the fuel (droplets of oil, coal pieces or powder), the degree of mixing of volatile materials released on heating with air, and the temperature. The products of the complete combustion of an organic substance are CO_2, water vapour and ash (non-combustible inorganic residue). The formation of smoke and soot particles only occurs when the combustion is incomplete.

In the flame, hydrocarbon molecules are progressively broken down to smaller fragments and combine with oxygen to form CO. In the presence of adequate O_2 the CO becomes oxidized to CO_2. However, the formation of CO is relatively fast, but its oxidation to CO_2 is slower, and so if there is an inadequate supply of O_2, CO will accumulate. Soot formation generally accompanies the formation of CO as a result of an inadequate supply of air. Partially degraded fuel molecules can polymerize to produce carbon nuclei which then accumulate further material by surface adsorption and coagulation. The soot particles formed are usually $< 1\,\mu m$ in diameter and are similar in size to the wavelength of light so they can be effective in absorbing light and scattering it (Clarke, 1992).

Spray-atomized light fuel oils normally evaporate completely and produce little soot or smoke. With heavy fuel oils only the light fractions evaporate at first, leaving a tarry residue which is gradually cracked (releases lower molecular weight volatile compounds) by the heat of the flame. However, if the temperature is not high enough, some particles $< 50–100\,\mu m$ in diameter (called smuts) are likely to be emitted from the chimney. If the oil has a high sulphur content the carbon smuts can contain sulphuric acid, which makes them very corrosive to fabrics, paintwork and many other surfaces (Clarke, 1992).

Coal combustion proceeds in a similar way to oil, the volatile compounds burn as gases and the remaining material burns slowly leaving an ash residue. If the coal is pulverized first (as in coal-fired electricity-generating stations), the combustion is more rapid and controllable and does not normally produce smoke. In contrast, an open fire in a grate with lumps of coal, a log fire, or a primitive cooking stove tend to form soot and smoke because of the inefficient mixing of air and the volatile compounds. Following the introduction of the Clean Air Act of 1956, smoke control was achieved by using smoke-free fuels in place of bituminous coal or wood (Clarke, 1992) (Sections 5.2 and 6.2).

In general, low temperature fires, such as coal or wood fires in a grate, or uncontrolled fires, such as bonfires, forest fires and burning cereal straw stubbles, tend to produce more smoke than controlled hot fires in stoves or boilers. The importance of the material itself is shown by comparing smoke from different sources: diesel engine smoke is three times darker than bituminous coal smoke and seven times blacker than petrol engine exhaust smoke (Elsom, 1987). Smoke is generally less of a problem in spark ignition petrol engines because the fuel and air are more thoroughly mixed than in diesel engines, especially when the latter are under a high torque load where there is a rich supply of fuel. The diesel smoke odours include partly oxygenated hydrocarbons such as aldehydes and with excess air available most of the carbon consumed is emitted as CO_2. However, particulate emissions at $4-14$ g/l greatly exceed those of petrol engines at 0.65 g/l. Most of the latter are from lead additives, which are declining, whilst about half of diesel particulates are of PM_{10}. These are of current concern (Section 5.6.1) because they can penetrate deeply into the lung. In the UK, there has been a growth in the usage of diesel cars from 5% of passenger vehicles in 1986 to 19% in 1993 (DOE, 1993).

Direct injection diesel engines produce less particulate pollution than indirect injection engines but at the expense of an increased release of unburnt fuel. From January 1996 emission limits (g/km travelled) for new vehicles in the UK were as follows:

	CO	HC and NO_x	Particles
Petrol	2.2	0.5	–
Diesel, indirect injection	1.0	0.7	0.08
Diesel, direct injection	1.0	0.9	0.10

Since even a small passenger diesel engine emits as much as 4 g of solid per litre, any trap in the exhaust system must be cleared during a journey. One design for this is illustrated in Fig. 6.2. This simplified drawing shows concentric porous ceramic pipe sections which trap the solids but permit gases and water vapour to penetrate the walls and so escape to air. To ensure continuing efficiency, it is necessary to burn off deposited carbon by either an electrical or diesel-fuelled heater. Table 6.1 shows a typical level of solids in urban air.

Weighted annual mean values for PM_{10} at 799 sites in the USA have been recorded (USEPA, 1994a) and show a steady fall from 57 $\mu g/m^3$ in 1987 to 46 $\mu g/m^3$ – this is, of course, despite a rise in traffic.

Figure 6.1 Burning cereal straw and stubble (photo B. J. Alloway)

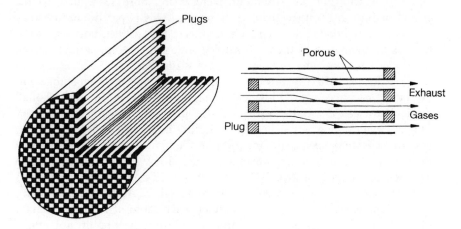

Figure 6.2 Schematic diagram of a particulate trap.

As a result of the controls brought in after the Clean Air Act of 1956 in the UK, smoke emissions decreased by 85% and SO_2 emission by 40% between 1956 and the late 1980s (Timberlake & Thomas, 1990).

It was estimated that the disposal of straw by burning in open fields in the UK in 1984 produced 18 000 t of black smoke over a short period of two or three weeks in the main cereal growing areas (Fig. 6.1) (RCEP, 1985). This significant source of smoke pollution continued annually until it was banned from 1993 for nuisance and environmental reasons.

Smog is a mixture of smoke and fog where the smog is generated from coal combustion, SO_2 is frequently present also. The notorious London smog of December 1952 caused the death of 4700 people and many thousands of others became very ill with respiratory problems. The peak

Table 6.1 Diurnal variation of the hourly mean for PM_{10} in London June 1992

Time	PM_{10} ($\mu g/m^3$)	Time	PM_{10} ($\mu g/m^3$)
0.0	38	15.0	45
5.0	33	20.0	39
10.0	42	24.0	35

Urban Air Review Group, 1993a.

daily concentrations of atmospheric pollutants reached $6000\,\mu g/m^3$ smoke and $4000\,\mu g/m^3$ SO_2 (Section 5.2). In addition to these two characteristic pollutants, other toxic substances such as HF from industrial sources may also be present in high concentrations in urban and industrial areas and they can exacerbate the health effects of smoke and SO_2 (Elsom, 1987).

People suffering from respiratory allergies, such as asthma and hay fever, and people carrying out strenuous exercise are more susceptible to the effects of SO_2 and smoke than other people. This is partly because they usually need to breathe through their mouths instead of their nostrils, and so there is more chance of the acid droplets and smoke going straight down the trachea without any filtering in the nostrils (Elsom, 1987). Studies in London during 1958–72 showed that there was a link between smoke and mortality; more recently in Philadelphia a rise of $100\,\mu g/m^3$ in the level of all particulates was associated with a 4–7% increase in deaths per day. A similar conclusion was reached in St Louis. Pope and colleagues (1992) undertook a study specifically for PM_{10} in the Utah Valley. The conditions there were favourable to the research because of low levels of other possible irritants: the daily mean for NO_2 was below 0.07 ppm (WHO: 0.08 ppm) and for SO_2 the figure was below 0.02 ppm (USEPA, 1994a: 0.14 ppm), while ozone levels were minimal in winter when those of PM_{10} were high. Most of the 260 000 people in the area were Mormons who do not smoke cigarettes. The finding of an increase in mortality of 12% for an increase of $100\,\mu g/m^3$ in PM_{10} was comparable to the result for St Louis.

In the UK during 1979–87, deaths from asthma/million in men over 64 years rose from 95 to 110 and in women of this age group they rose from 120 to 145. It must be noted that this increase could be caused partly by better diagnosis and an adverse reaction to medication. Furthermore, although the trend is worldwide, there is no direct causal link between pollutants and asthma. The geographical record is not always clear either: for UK males aged 35–64 years, the lowest mortality is in Greater London and the highest is in rural Wales; conversely for the 5–34 years group mortality is highest in Greater London and lower in rural Wales (Higgins and Britton, 1995).

The WHO short-term maximum concentrations of both SO_2 and smoke are $250\,\mu g/m^3$ (at which the condition of patients with respiratory illness would be expected to deteriorate) and $100\,\mu g/m^3$ for long-term exposure. At $500\,\mu g/m^3$ of smoke and SO_2, excess mortality can be expected among elderly and chronically sick people (Elsom, 1987). This correlates with the USEPA Pollutant Standards Index (1993), which treats a level of PM_{10} of $500\,\mu g/m^3$ as an emergency.

Soot and smoke particles can have numerous organic pollutants adsorbed onto their surfaces and most important among these are PAHs (Section 6.2.5). These also include benzo-[a]-anthracene and benzo-[a]-pyrene, which have been shown to induce cancer in humans (IARC group 2A), together with 2-nitrofluorene and several nitropyrenes (pp. 267–270). The latter are classed as group 2B substances because they have been shown to induce cancer only in animals. The group A substances in soot are probably responsible for the higher incidence of scrotal cancer in chimney sweeps which was observed over 200 years ago (in 1775) by Percival Potts, a London surgeon (Rodricks, 1994).

Tobacco smoke (Section 7.6) also contains numerous toxic substances including nicotine (a supertoxin), tar (with PAHs), formaldehyde, NO_x and CO (Murley, 1995). There is an overwhelming body of evidence linking smoking with the incidence of cancer of the lung and of certain other tissues. It has also been suggested that the higher concentrations of PAHs in urban air compared with those in rural areas are responsible for a higher incidence of lung cancer in urban residents. Fortunately, however, the introduction of clean air policies in both the UK and the USA have resulted in marked decreases in airborne PAH, with mean annual benzo-[a]-pyrene concentrations decreasing by a factor of ten between 1935 and 1965 in London, and by a factor of 3 between 1966 and 1975 in the USA. Table 6.2 shows the principal contributors to organic aerosols. Air quality standards for particulates are given in Table 6.3.

Although the WHO guideline value for particulates (smoke) in indoor air is a daily mean of $120\,\mu g/m^3$ for total solid particles and $70\,\mu g/m^3$ for PM_{10}, rural dwellings in Third World countries frequently exceed these values and concentrations of up to $14\,000\,\mu g/m^3$ have been recorded. These high levels of smoke result from burning wood, animal dung or coal in inefficient stoves and poor ventilation restricting smoke escape. Respiratory diseases in both children and adult are linked to indoor air pollution of this type.

Table 6.2 Principal contributors to organic carbon aerosols for the greater Los Angeles area (Cass, 1993)

Source type[a]	Contribution (%)	Amount (kg/day)
Meat cooking and charbroiling	16.6	4938
Paved road dust	15.9	4728
Pinewood burning	11.2	3332
Cars without catalytic cleaning	7.0	2088
Heavy-duty diesel vehicles	4.2	1242

[a] With significant inputs from forest fires (2.9%), cigarettes (2.7%) and catalytically cleaned vehicles (2.6%).

Table 6.3 Air quality standards for particulates

Organization	Daily mean ($\mu g/m^3$)	Annual average[b] ($\mu g/m^3$)
EU directive	$38\text{–}56^a$	30 (>34)
89/427/EEC		45 (<34)
WHO[a]	120 TSP	–
	70 PM_{10}	
USEPA	150 PM_{10}	50 PM_{10}

[a] Guide only.
[b] Figures in parentheses are interactive SO_2 levels.

In developed countries, indoor pollution can include tobacco smoke and many of the pollutants listed in Table 2.8 (p. 26). With increased awareness of the need for energy conversion, houses are generally less well ventilated nowadays through the exclusion of draughts and so indoor pollutants are more likely to accumulate, although some do undergo precipitation or transformation within the house (Section 2.4.3 and Chapter 7).

6.2
Methane and other hydrocarbons – coal and oil as sources

6.2.1   The formation of coal

Coal is formed by the compaction and metamorphosis of the residues of woody plants ultimately under high temperatures and the pressure of overlying sediments accumulated over 50 million years or longer. The initial components are cellulose and lignin. *Cellulose* is a linear polymer (**1**) of glucose with a molecular weight in the range 40×10^3 to 150×10^3 and typical composition $(C_6H_{10}O_5)_n$. *Lignin* is a three-dimensional polymer interwoven with cellulose and constructed from units (**2**) derived from cinnamic acid. The growth of the polymer is achieved largely through oxygen in ether linkages. One typical section of lignin is shown as the dimer unit in structure (**2**).

Table 6.4 Changes in the composition of coal substances as they mature

| | Typical composition (%) | | | | | Calorific value (kJ/g)[a] |
	C	H	O	N	S	
Wood	56	6.5	37.5	–	–	20
Peat (humus)	55	6	30	1	1.3	21
Lignites	73	4	21	1.5	0.5	29
Bituminous coal	83	5	12	1.5	0.5	33
Anthracite	93	2.5	2.5	1	0.5	36

[a] 430 kJ/g = 1 Btu/lb.

A representative composition of lignin is $(C_6H_{10}O_4)_n$. Anaerobic bacteria depend for their oxygen on that combined in these plant residues and hence there is a progressive reduction in combined oxygen as coalification proceeds (Table 6.4).

Lignite is the lowest ranking coal and anthracite the highest. The majority lie between these two extremes and contain 85–90% carbon.

6.2.2 Petroleum

Historical note. Petroleum has been known in years BC to occur in surface seepages and was first obtained in pre-Christian times by the Chinese. The modern industry had its beginnings in Romania and in wells sunk by Colonel E. A. Drake in Pennsylvania in 1859. The principal early use was to replace expensive whale oil for lighting, but today its consumption as a fuel and its dominance of the market for chemicals has led since 1965 to a doubling of proven reserves to 140 Gt. In the UK, about 70 Mt is used for fuel and 20 Mt for export and as chemical feedstock. The daily worldwide consumption is now about 65 million barrels or 3 Gt a year. Table 6.5 gives figures for recent production and of proved reserves.

Formation. Petroleum is largely formed biogenetically at temperatures below 200°C from matter deposited in shallow seas and subsequently compressed by the overburden of deposited clays and shales. An intermediate coal-like material formed by bacterial action on the deposits is known as kerogen. This may be one of three types formed from (i) algae, (ii) marine plankton or (iii) higher plants. Hydrogen sulphide, a typical product of organic decay, may also be formed.

Properties. The major compounds are saturated alkanes; alkenes are absent but aromatic hydrocarbons may occur to a significant extent, for example in Borneo deposits. The environmentally significant polycyclic aromatic hydrocarbons are discussed at the end of this chapter. Other

Table 6.5 Current reserves and production of petroleum [a]

	Proved reserves 1994 (Gt) [b]	Production (10^6 t)
North America	12	652
South and Central America	11	267
Western Europe	2.2	287
Eastern Europe	8.1	375
Middle East	89.4	957
Africa	8.3	330
Asia	6.1	341
World total	137.3	3209

[a] BP, 1995.
[b] 137.3 Gt (10^9 t) = 1.009×10^9 barrels.

organic components of petroleum, including some which contain nitrogen and sulphur, are shown in Fig. 6.3. Table 6.6 lists the boiling points of some typical substances.

Petroleum deposits are complex mixtures which may include as many as 40 individual substances which are not economically separable by

Aliphatic compounds include the volatiles:

methane CH_4 through the range of straight and branched chains
ethane C_2H_6 up to $C_{76}H_{154}$
propane C_3H_8

Alicyclic and aromatic components include:

cyclopropane   cyclopentane   benzene, toluene, phenol, stearic acid

Nitrogen compounds are typified by:

pyrrole   pyridine

Sulphur compounds include:
thiols R-SH; thioethers R-S-R, Ph-S-Ph; and aromatics including:

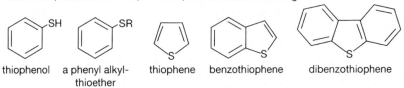

thiophenol   a phenyl alkyl-   thiophene   benzothiophene   dibenzothiophene
thioether

Figure 6.3 Some organic components of petroleum.

Table 6.6 Boiling points of typical components of petroleum (°C)

Methane	−161	Benzene	80
Propane	−42	Toluene	111
n-Butane	−0.5	p-Xylene	138
n-Hexane	69	Naphthalene	218 (mp 80)
n-Octane	126	Anthracene	340 (mp 216)

distillation. Hence commercial practice is to subdivide within a given boiling range (°C):

natural gas: methane, propane, butanes
light naphtha: 20–100
heavy naphtha: 100–150
kerosene: 150–235
light gas oil: 235–345
heavy gas oil: 345–565 (steam assisted)

Release to the environment

Clearly manipulation of these volatile fractions on the present massive scale inevitably leads to leakage. Losses in the oil industry are estimated at 6.5% of the total of about 10 Mt/year, while major subsequent losses are attributed to cars and HGVs (41%) and to the industrial use of solvents (40%). Natural gas and minor sources such as domestic, incineration and power stations account for 12% of the remainder. The rapid rise in the bulk of hydrocarbon processing is shown by the development of the petrochemicals industry in the UK from its beginnings in the 1920s to its eight-fold growth in 1940–55, when 42 Mt were processed (*cf.* Fig. 6.12, p. 283). In 1986, the North Sea field produced 177 Mt oil and 96×10^9 m^3 of natural gas and met 20% of western Europe's energy needs, equivalent to the 400 Mt of coal mined at that time (Martin, 1990). These figures may be compared with those given for the USA (BP, 1991), which produced 417 Mt oil and 444 Mt coal-equivalent of natural gas in 1990 (Table 6.5). Table 6.7 gives some recent comparative figures (Eurostat, 1995) for total energy demand.

Table 6.7 Energy needs of some major users

	Energy needs (10^6 t oil-equivalents)			
	1986	1988	1990	1992
USA	1800	1920	1920	2000
12 EC countries	1040	1080	1110	1200
Japan	370	400	420	440

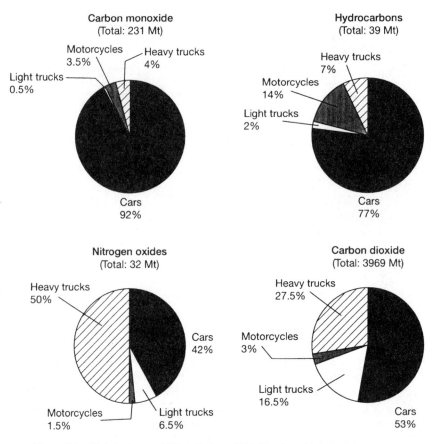

Figure 6.4 Global motor vehicle emissions, 1990 (Ferris and Wiederkehr, 1995).

Control of vehicular release to the environment

Figure 6.4 shows how the principal releases from motor vehicles are related to type (OECD, 1995). The measures for their control include:

- use of hybrid vehicles (see below)
- use of natural gas as fuel
- other alternative fuels
- cleaner engines, emission control.

Hybrid vehicles.   The petrol-fuelled internal combustion engine is only 30% efficient and, when allowance is made for inefficiency of transmission and idling losses, this is reduced to 12% in the EC urban cycle. The diesel engine is slightly better at 15% efficiency. The battery-powered vehicle is ideal in that it runs with zero emissions but is disadvantaged by its limited range – about 80 miles under ideal climatic conditions. The principle of hybrid vehicles is that a conventional petrol engine and a battery power source may be used in combination or independently. At low load, the excess capacity of a petrol engine is used to charge the battery, for short journeys in cities the battery alone can be used for power (Jones, 1995).

Use of natural gas as fuel (Carslaw and Fricker, 1995). One million of these units are now in use worldwide with Argentina in the lead with 330 000, followed by the former Soviet Union with 315 000. The fuel pump and carburettor/injector are replaced by a gas regulator and mixer/injector. The fuel can be dispensed quickly from compressed gas tanks or over several hours, including 'home-fill'. The typical composition is:

methane 93%
nitrogen 3%
ethane 3%
higher alkanes 1%
CO_2 0.3%.

Research in the USA showed that ozone formation from leakage of this fuel was 10% of that arising from the use of petrol and diesel. At the same time, benzene, butadiene and higher alkane release would be negligible, but there was the disadvantage of methane loss at a rate of 0.2 g/km travelled. Although methane has a greenhouse warming potential of about 27 times that of CO_2, it still shows to advantage when the balance is struck (Table 6.8).

Hence for each kilometre travelled, the natural gas vehicle releases the equivalent of 159 g CO_2 compared with 200 g from its petrol-powered equivalent.

Other alternative fuels. Methanol and ethanol permit higher compression ratios (p. 177) and lean combustion so reducing VOC emission and ozone formation. Particulate release is also reduced and SO_2 emission is eliminated. For methanol produced from wood, the greenhouse burden is reduced by 75% compared with that from petrol usage.

Emission control, cleaner engines. These measures include the more efficient electronic control of ignition timing and fuel injection. Fuel release can be restricted by attachment of activated carbon traps to vehicles and to the pumps at filling stations.

Cleaner operation is achieved through the use of catalytic conversion of exhaust gases (McCabe and Kisenyi, 1995). Originally platinum or palladium suspended on alumina was used to catalyse the oxidation of CO and hydrocarbons:

$$CO(RH) \xrightarrow[O_2]{} CO_2 + (H_2O) \tag{6.1}$$

Table 6.8 Comparison of vehicles powered by petrol and natural gas

	Emission (g/km)		CO_2 equivalents
	CO_2	CH_4	
Natural gas vehicle	154	0.2	154 + 5.4
Petrol powered	200	Nil	200

More recently, three-way systems have been introduced in which control of nitric oxide is effected using a rhodium (Rh) catalyst in reducing conditions produced by feedback of fuel.

$$NO + CO(H_2) \longrightarrow N_2 + CO_2(H_2O) \qquad (6.2)$$

A compromise has to be accepted between ideal oxidizing conditions for control of CO with RH and reducing conditions for control of NO. The best balance is obtained by placing an oxygen sensor upstream of the catalyst.

In the EC it is proposed that the emission limit for RH + NO_x be reduced from the present 1.0 g/km to 0.5 g/km by 1996 and subsequently to 0.2 g/km.

6.2.3   Methane

Owing to its importance as a greenhouse gas (Table 5.4, p. 168), methane is best considered separately. It occurs at 1.7 ppm in air and in addition to losses from processing natural gas it arises from the following biological sources:

- cultivation of rice
- anaerobic fermentation in swamps, rain forests and landfills
- in the intestine of animals, chiefly cattle.

Its effect on surface warming is enhanced by a factor of seven compared with CO_2 because it absorbs IR radiation near 3000 cm^{-1}, which coincides with a window in the spectrum of CO_2 (Fig. 5.5, p. 169).

Reactions in air.   The hydroxyl radical features critically in the chemistry of air pollution; it is formed when ozone is photolysed (Equation 5.5, p. 158) to afford excited oxygen, which combines with water vapour (Wayne, 1991):

$$O^{\bullet} + H_2O \longrightarrow HO^{\bullet} + .OH \qquad (6.3)$$

The $^{\bullet}$OH radical reacts with CO in clean air to form CO_2 (Equation 6.4) and also abstracts H$^{\bullet}$ atom from methane in polluted air (Equation 6.5) so converting it into the reactive methyl radical:

$$HO^{\bullet} + CO \longrightarrow H^{\bullet} + CO_2 \qquad (6.4)$$
$$HO^{\bullet} + CH_4 \longrightarrow {}^{\bullet}CH_3 + H_2O \qquad (6.5)$$

This in turn captures oxygen to give the methylperoxy radical. The natural outcome is the interaction between peroxy radicals to regenerate oxygen, but NO in polluted air diverts the system according to the Equation 6.6:

$$CH_3O-O^{\bullet} + NO \longrightarrow CH_3O^{\bullet} + NO_2 \qquad (6.6)$$

An important general point now follows: any alkoxy radical of this type may be converted into the corresponding carbonyl compound by donating an H atom to any acceptor molecule (M). The mechanism of the electron pairing is shown for the methoxy radical in Equation 6.7.

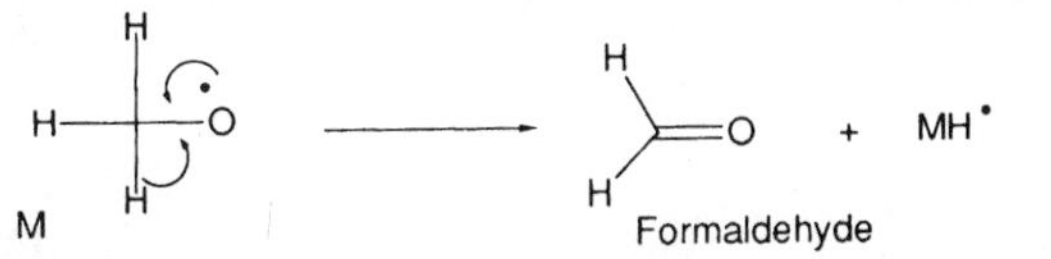

$$\text{(6.7)}$$

Formaldehyde

6.2.4 Higher alkanes

These reactions are of the same type, for example ethane is converted into acetaldehyde which, like formaldehyde, is a volatile lachrymatory substance which also contributes to smog. The further reactions of acetaldehyde are particularly significant since they lead to the formation of peroxyacetylnitrate (PAN), one of the principal irritants in photochemical smog. In the first step (Equation 6.8), a relatively stable acetyl radical is formed from acetaldehyde by abstraction of an H atom:

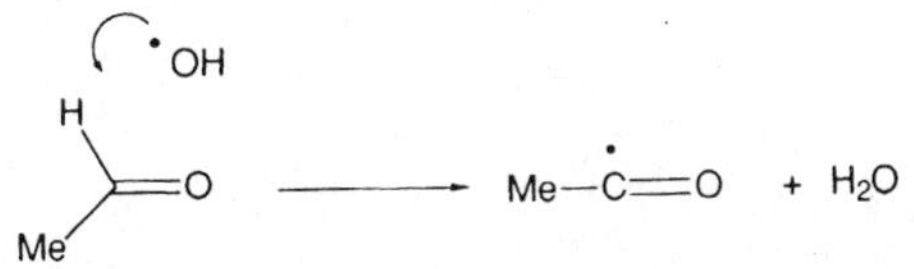

$$Me-\overset{\bullet}{C}{=}O \ + \ H_2O \qquad \text{(6.8)}$$

After capture of oxygen, which is available in high concentration, the acetylperoxy radical is obtained:

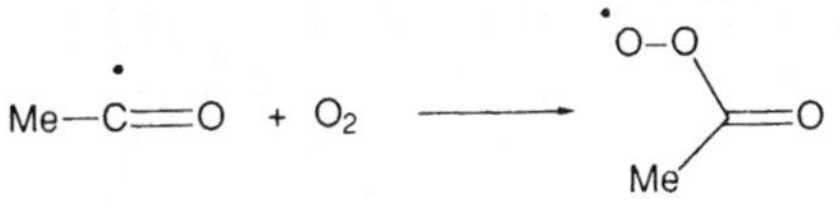

$$Me-\overset{\bullet}{C}{=}O \ + \ O_2 \ \longrightarrow \qquad \text{(6.9)}$$

Notice that NO_2 is itself an odd electron species and pairs to give PAN:

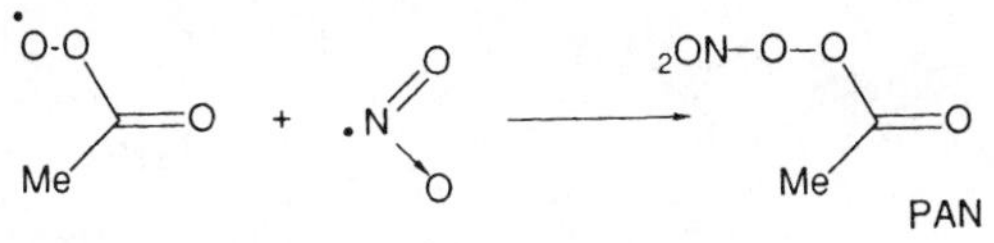

$$\text{(6.10)}$$

PAN

PAN is formed naturally by photooxidation of natural plant terpenes but at levels well below 1 ppb. In European cities, the levels reach 10 ppb and in extremely polluted areas these may reach 50 ppb (Harrison, 1992). PAN is highly phytotoxic and contributes to leaf damage (p. 184).

The formation of similar alkoxyperacetylnitrates will also take place with other volatile hydrocarbon starting materials. PAN may itself be generated following fragmentation of higher alkyl radicals obtained by the action of the same OH initiator:

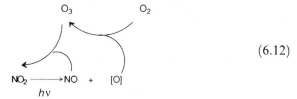

$$(6.11)$$

Urban pollution

A diurnal urban pollution cycle. Figure 6.5 shows a typical 24 hour experience in changes of pollutant levels.

In unpolluted air, O_2 and NO_x achieve a steady tropospheric state including a little ozone (Section 5.1.9, p. 163) but the injection of CO and unburnt hydrocarbons into the atmosphere diverts this equilibrium:

$$(6.12)$$

This can be explained by reference to Equation 6.6 since in this sequence NO_2 is produced via $R-H$ *without participation of ozone.* Entry of NO_2

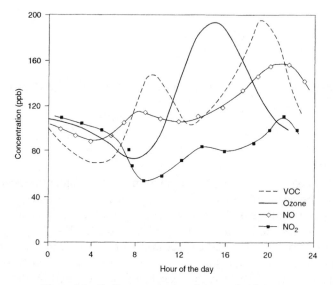

Figure 6.5 Cyclic concentration changes of pollutants.

from this source gives additional [O], which combines with oxygen to generate abnormally high levels of ozone together with aldehydes (*cf.* Equation 6.7).

A typical sequence of events based on studies made in the past in Los Angeles (Bridgman, 1990) is shown in Fig. 6.5. It is presented in this form in order to illustrate the interactions which take place throughout a day in an urban environment. The levels shown are higher than those currently experienced in cities of the developed nations, although there are exceptions, and excesses may occur during pollution episodes. The timebase will vary with the season of the year and with individual locations.

Air pollution of this kind is most severe under conditions of temperature inversion (p. 28) and is also worse in cities backed by high ground, which limits dispersion by wind. A number of episodes of historical importance are discussed by Boubel *et al.* (1994).

The concentrations given in Fig. 6.5 are maxima averaged over 1 hour, which is logical because this timescale relates to the likely exposure of motorists and pedestrians in the area. Long-term trends relating to residents will be quoted as levels averaged over a month or a year.

Levels of air pollution may be reported as percentiles. In an array of 365 observations, the 90%ile will lie 36th from the highest. On occasion, excesses are quoted, namely the number of times a given level of a pollutant is exceeded in a given period.

Residual NO_x is enhanced initially by NO emitted by early morning traffic. As the sun exerts its effect, radicals such as $R-O-O^{\bullet}$, derived from unburnt fuel, increase the proportion of NO_2 and the figure shows that this is highest in early afternoon: ozone levels reach a maximum shortly afterwards. In strong sunlight, oxidation of hydrocarbons gives rise to carbonyl compounds (Equation 6.7) and with NO_2 available PAN is also produced (Equation 6.10) through the afternoon. This dependence of pollution by NO_2 and ozone upon the release of unburnt fuel makes it appropriate to discuss the broader scene including particulates and SO_2 at this point.

Urban pollution in the USA

The USEPA sets National Air Quality Standards (NAAQS, Table 6.9, USEPA 1994a) for CO, Pb, NO_2, O_3, PM_{10} and SO_2 (See also Table 5.13, p. 187).

For CO, the annual average was satisfactory at 5 ppm in 1993 as were the maximum quarterly levels for Pb $(0.6\,\mu g/m^3)$ and the annual mean for NO_2 (20 ppb). However the position for ozone was unsatisfactory with 51 million citizens living in areas outside the standard. In the Los Angeles area the second highest 1-hour maximum was $0.20-0.40$ ppm (*cf.* Fig. 6.5), while Table 6.10 shows an episodic value of 585 ppm. For PM_{10} and SO_2 the NAAQS were largely met.

Table 6.9 National Air Quality Standards for the USA and others

Pollutant	Primary standard		Others
	Type of average	Concentration of standard	
CO	8 hour	9 ppm ($10\,mg/m^3$)	10 ppm ($10\,mg/m^3$)[a]
	1 hour	35 ppm ($40\,mg/m^3$)	25 ppm ($30\,mg/m^3$)[a]
Pb	Maximum quarterly	$1.5\,\mu g/m^3$	Annual average[b] $2.0\,\mu g/m^3$
NO_2	Annual arith. mean	0.053 ppm ($100\,\mu g/m^3$)	0.016 ppm ($30\,\mu g/m^3$)
O_3	Maximum daily	0.12 ppm ($235\,\mu g/m^3$)	Hourly mean
	1 hour		0.18 ppm ($360\,\mu g/m^3$)
PM_{10}	Annual arith. mean	$50\,\mu g/m^3$	
	24 hour	$150\,\mu g/m^3$	
SO_2	Annual arith. mean	$80\,\mu g/m^3$ (0.03 ppm)	40–$60\,\mu g/m^3$ (0.015 ppm)[b]
	24 hour	$365\,\mu g/m^3$ (0.14 ppm)	100–$150\,\mu g/m^3$ (0.038–0.056 ppm)[b]

[a] WHO guideline.
[b] EC limit.

Urban pollution in the UK

In 1991 there were 162 occasions when the NO_2 level was 'poor' and in the range of 100–299 ppb. In December of that year, an hourly maximum of 423 ppb was reached. No serious episodes occurred in 1993 but two were recorded in London in 1994 (Beevers *et al.*, 1994) (Table 6.11).

Ozone. The UK has 49 automatic recording stations and rural levels of ozone are generally about twice those in nearby urban areas; this is the result of scavenging by NO released in city streets and the time lag between release of its precursors and ozone formation. An hourly average of > 90 ppb was reached at 15 sites with a maximum of 133 ppb; this is in excess of the WHO guideline of 76 ppb (as was the EC 8-hour mean of 55 ppb) but levels were below the population warning level of 180 ppb.

Nitrogen oxides. Peaks of NO_2 above 200 ppb were recorded at three sites – Sheffield, Birmingham and Newcastle – in 1993, with other winter-time episodes below this limit. However, in 1994, five sites in London reached levels over 200 ppb. A survey based on 1200 sites gave national averages of 23 ppb at kerbsides, 17 ppb intermediate and 14 ppb at urban back-ground sites.

Volatile organic compounds. Levels of benzene in air have been reported for the first time (AEA, 1995) and peak hourly concentrations have ranged between 7 and 40 ppb. A standard for benzene in air has been set by an expert panel at an annual mean of 5 ppb. On a working day in central London total hydrocarbons range from a maximum of 70 ppb at 09.00 hours to a second peak of 97 ppb at 19.00 hours (DOE, 1993; Fig. 6.5).

Table 6.10 Comparative pollution data for some major cities of the world

City	Population (millions)	Area (km²)	Particles Annual emissions[a]	Particles Annual mean[b]	Particles Peak hourly max.[c]	SO₂ Annual emissions[a]	SO₂ Annual mean[b]	SO₂ Peak hourly max.[c]	NOₓ Annual emissions[a]	NOₓ Annual mean[b]	NOₓ Peak hourly max.[c]	Ozone Annual emissions[a]	Ozone Annual mean[b]	Ozone Peak hourly max.[c]
ATHENS (1989)	3.5	330	3.4	–	580	9.26	–	42	22.6	–	127 (NO_2) / 490 (NO)	NA	–	180
BOMBAY (1992)	11.1	603	50	235	670	157	9	40	58	15 (NO_2)	58 (NO_2)	NA	–	–
CALCUTTA (1992)	11.8	1295	200	360	742	25	60	–	40	80 (NO_2)	100 (NO_2)	NA	–	–
DELHI (1992)	8.6	591	116	370	1070	46	12	30	73	27 (NO_2)	140 (NO_2)	NA	–	–
LONDON (1993)[d]	10.6	1579	11	29[e]	140[e]	49	11	283	79	72	186 (NO_2) / 784 (NO)	NA	9	93
LOS ANGELES (1993)	10.5	16600	400	47[e]	150[e]	50	8	30[f]	440	50 (NO_2)	–	NA	–	585
MADRID (1990)	4.0	110	10	–	190	38	–	113	20	–	530 (NO_2) / 960 (NO)	NA	–	–
MEXICO CITY (1987)	19.4	2500	451	500	–	206	74	–	177	–	–	NA	300	480
NEW YORK (1993)	15.6	3585	56	100 / 47[e]	86[e]	349	14	60[f]	513	43 (NO_2)	–	NA	–	235[g]
SHANGHAI (1987)	13.3	6300	324	280	–	267	40	–	127	–	–	NA	–	–

NA, not applicable.
[a] Annual emissions in 10³ tonnes.
[b] Annual mean: particles (µg/m³); gases (ppb).
[c] Peak hourly maximum: particles (µg/m³); gases (ppb).
[d] Bloomsbury district.
[e] PM_{10} values.
[f] Over 24 hours.
[g] In 1992.

Table 6.11 Episodes of high pollution in London in 1994

	July 9–13	Dec. 23–25
CO (ppm)	10	4.8
PM_{10} ($\mu g/m^3$)	90	120
NO_2 (ppb)	200	130
Ozone (ppb)	30	120

Other air pollutants. There have been no recent excesses of the WHO 1-hour guideline of 25 ppm for CO and none of the UK stations recorded excesses of EC limits for smoke and SO_2, although an annual mean of 14 ppb for SO_2 was found alongside a diesel lorry route (p. 178) and 'poor' hourly maxima were observed at a number of locations.

Urban pollution in the EC

Within the EC, detailed data on daily pollution levels in major cities have been obtained using mobile monitoring vehicles (p. 108). This provides a continuous record in real time from which hourly and daily averages can be obtained. Data obtained in this way for Athens and Madrid are included in Table 6.10. With their more limited resources, Third World countries are inevitably exposed to higher pollution levels (page 6 and Fig. 1.1). Figures for SO_2 and suspended particles in their major cities are generally available but the record is often incomplete. When data are unavailable for one city, a rough estimate can be obtained by scaling the levels given for another city in proportion to their annual emissions/unit area.

Other members of the European Community. The Joint Research Centre of the EC countries has employed a mobile monitor (pp. 108–113) to analyse air quality in a number of major cities, including Athens (Cerutti *et al.*, 1989). Although the estimated ground-level emissions are not excessive for a city of this size, those in air frequently exceed safe limits (compare the values in Table 6.10 with those in Tables 6.3 and 6.9). This is illustrated by the maxima found for particulates ($580 \mu g/m^3$), NO_x ($240 + 610 \mu g/m^3$) and ozone (177 ppb). As a control measure, Athens city centre has recently been closed to vehicular traffic.

Madrid has also been the subject of mobile monitoring (Cerutti *et al.*, 1990). It is a city with no large industrial plants, where air pollution arises largely from vehicles, which account for 77% of CO and 53% of NO_x, and heating systems which account for 77% of SO_2 and 50% of particles (Table 6.12).

During the monitoring circuits of the city, particulates were in the range $100–150 \mu g/m^3$, with the highest at $190 \mu g/m^3$. In residential areas, SO_2 was

Table 6.12 Winter emissions from major sectors in Madrid (Cerutti *et al.*, 1990)

Sector	Emissions per year (10^3 tonnes)				
	SO_2	TSP	CO	NO_x	HC
Transport	2.96	3.6	171	11.6	23.1
Heating	50.4	6.6	50.3	3.65	2.2
Industrial	11.8	3.1	0.36	6.7	0.73
Total	65.2	13.3	221.7	22.0	26.0

at acceptable levels in the range 25–50 ppb; elsewhere levels frequently exceeded 65 ppb, with hourly maxima over 110 ppb and classified as 'poor'.

Pollution by NO_x in Madrid gave cause for concern with NO_2 levels in streets persisting in excess of 100 ppb with peaks over 300 ppb. Nitric oxide was often in the range 800–1000 ppb with short-term peaks up to 1600 ppb.

Values for ozone levels are not available for either Athens or Madrid. However, one expects that they will be raised when NO_x levels are high, since ozone synthesis is initiated by NO_2. A comparison of data for 10 of the UK sites which monitor both compounds shows NO_2 hourly maxima in the range 43–186 ppb with ozone in the range 36–133 ppb.

Developing countries

Reference to figures for the three Indian cities and for Shanghai and Mexico City (Table 6.10) illustrates the problems which arise when resources are unavoidably limited and when vehicular use is growing rapidly.

All three Indian cities suffer from heavy particulate pollution linked to the extensive use of solid fuel (68% of all sources) (Central Pollution Control Board, 1992; 1995). In 1987 there were 39 million vehicles in the 12 major urban centres and usage was rising at 1 million/year; with a total population of 800 million there is likely to be much more growth towards the levels of vehicles per head of population seen in the developed nations.

Of the 12 major Indian centres, Delhi has the most vehicles in use – 28% of the total – and in addition to heavy particulate levels those of NO_x are high and ozone must be presumed to be high also. At some of the major traffic intersections, NO is over 150 ppb for 8 hours of the day and 200 ppb was exceeded once. At these sites, NO_2 was about 80 ppb. Elsewhere in the city NO also twice exceeded 200 ppb and in seven other cities it exceeded 150 ppb, while NO_2 ranged between 70 and 90 ppb.

Shanghai suffers from very high particulate pollution, up to an annual average of 280 µg/m^3, consequent on 80% coal usage as an energy source and growth in consumption of 208% in 1970–90.

The annual hourly maximum for ozone in Mexico City was 400 ppb and in 1992 the Mexican limit of 110 ppb was exceeded on 357 days: 480 ppb was

Table 6.13 Hydrocarbons in Mexico City air

	Site 1 (ppbv)	Site 2 (ppbv)
Methane	2980	2300
Ethane	10	18
Propane	220	95
n-Butane	96	44
i-Butane	45	21
Ethylene	15	28
Propene	5.5	4.4
Butenes	10.6	2.7

reached on 16 March (*cf.* Table 5.1). General and ozone pollution is exacerbated by leakage of liquefied petroleum gas (LPG), which is widely used around the city leading to the significant levels found in inner city samples collected in February 1993 (Blake and Rowland, 1995) (Table 6.13). Of these polluting alkanes, the branched chain compounds react most rapidly with ˙OH radicals (*cf.* Equations 6.3, 6.5) to give the more stable branched chain alkyl radicals:

$$\tag{6.13}$$

and so forward the ozone-generating sequence more rapidly. The alkenes in the mix (Table 6.13) arise through the alternative elimination of a hydrogen atom from the alkyl radical and they are even more effective as ozone generators. The point was made (Blake and Rowland, 1995) that better handling procedures and the use of LPG based largely on propane would reduce ozone levels in cities around the world, including Athens.

Reactions of aromatic hydrocarbons in the air

Benzene (Section 6.3) is itself a significant air pollutant (WHO, 1987), especially as it is now used to formulate lead-free petrol. Owing to the stability of the aromatic system, it reacts with free radicals such as OH with hydrogen atom abstraction (Equation 6.5) with ring retention and the formation of phenols:

$$\tag{6.14}$$

The entry of phenols in this way into the environment has serious consequences as they are acidic and often toxic.

The side chains of benzene homologues may react in a similar fashion to methane, especially as abstraction of hydrogen gives the stabilized benzyl radical. This will capture oxygen (Equation 6.15) and then undergo analogous changes to those shown in Equations 6.6 and 6.7.

benzyl radical   benzaldehyde

(6.15)

6.2.5   Polycyclic aromatic hydrocarbons (PAHs)

The initial fusion of two benzene rings can give rise to only one product – naphthalene – but attachment of a third ring can give the linear homologue anthracene, or the branched isomer phenanthrene (Fig. 6.6). These substances occur in considerable amounts in natural fuel deposits and are formed during coking of coal, on pyrolysis of fossil fuels and incomplete combustion of organic compounds. They are not included in the definition of a PAH, which is reserved for higher homologues containing four or more benzene rings.

The number of individual PAHs is considerable, bearing in mind the permutations of the modes of ring fusion and also the various positions available for insertion of side chains. As a consequence, Fig. 6.6 shows only a limited selection, which includes those of commonest occurence and/or highest toxicity.

Historical note.   Industrial disease caused by contact with PAH was first identified in chimney sweeps by Percival Pott in 1775. These children normally worked from age eight into adulthood, although 'apprenticeships' could start at age four. A typical consequence of this cruel practice was the development of skin cancers, probably induced by benzo-[a]-pyrene (Fig. 6.6) which occurs at 0.2% in soot. In 1875, von Volkman showed that there was a link to skin cancer of workers in the German coal tar industry.

Later notable contributions were the links to lung cancer in workers in the coal tar and gas industries, which was made by the Kennaways in 1947. This was extended in 1972 by Sir Richard Doll who had made a study of mortality in a large group of workers in the same industry over a 12-year period (Table 6.14). The history of the evaluation of PAH carcinogens has been reviewed by Sir Ernest Kennaway (1995).

Toxicity.   The level of activity is given below the structures in Fig. 6.6. Benzo-[a]-pyrene (BaP) is regarded as the most dangerous of the group as it

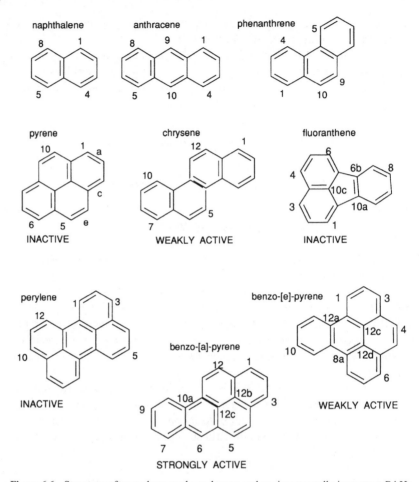

Figure 6.6 Structure of some benzene homologues and environmentally important PAHs.

Table 6.14 Mortality from cancer in gas workers in Great Britain

	Increase in deaths/100 000	
	Gas workers	National rate
Cancer of the lung	3.82	2.13
Cancer of the bladder	0.40	0.17
Cancer of the skin and scrotum	0.12	0.02

is widely distributed and strongly carcinogenic. In a test on the skin of mice, activity was noted at a dose level of 5.6×10^{-5} mmol; liver damage and teratogenicity were also observed in mice. In rats, 10 monthly intratracheal injections of 0.05 mg produced lung tumours in 28% of test animals (Osborne and Crosby, 1987).

Four fused benzene rings are a necessary but not sufficient condition for activity and a subtle structural factor is revealed since benzo-[e]-pyrene is much less active than its [a] isomer. It is known that initiation of carcinogenicity requires the binding of the PAH to cellular macromolecules DNA, RNA or protein. A key step in the activation of benzo-[a]-pyrene is its conversion into the epoxide (structure (3), *cf.* p. 61) and the ease of oxidation of PAHs by free radicals is indicative of their level of activity. Oxidized derivatives of PAHs also occur in the environment.

3

A full discussion of testing procedures and structure–activity factors is given by Harvey (1985) and by the International Agency for Research on Cancer (IARC) (1983a).

Sources and distribution.   The production of PAH in 1976 is shown in Table 6.15 (Sues, 1976). In the early 1980s, BaP emissions in the USA were of the order of 10^3 tons/year from these sources and individuals were likely to inhale up to $1.5\,\mu g$/day. The recommended acceptable level is $0.2\,\mu g/m^3$, which was always exceeded for coke oven workers. After 5 years of exposure their risk of infection was increased ten-fold.

PAHs are found at low levels (ng/g) in mineral oils and in paraffin wax. They are formed by pyrolysis in petrol and light diesel engines owing to the limited air supply. If a catalytic converter (p. 257) is not fitted, BaP is emitted at levels up to $50\,\mu g$/km travelled; the converter reduces this to a range $0.05-0.3\,\mu g$/km. Aircraft engines typically emit 10 mg BaP during each minute of operation. The principal components of particles are given in Table 6.16 (IARC, 1983b). The detection of the nitropyrene is of interest and a number of other nitroderivatives have been isolated. They clearly originate from contact with nitrogen fixation products within the engine (p. 175).

PAHs enter the water body from fallout from the above sources and also from runoff from bitumen-treated roadways. The worldwide use of this

Table 6.15   Worldwide production of PAH

Source	Quantity (10^3 tons/year)	Proportion of total %
Heating and power production	260	51
Industrial producers	105	20
Incineration and open burning	135	28
Vehicular transport	4.5	0.9

Table 6.16   Principal PAHs in emitted particles (ng/mg)

Pyrene	18–105	Benzo-[a]-pyrene	1–7
Fluoranthene	45–203	1-Nitropyrene	6–10

Table 6.17   PAH levels in water (ng/l)

	Thames (Kew)	Factory effluent	Industrial sewage	Domestic sewage
Fluoranthene	140	2200	3400	–
Pyrene	–	1960	3120	250
Benzo-[a]-pyrene	130	100	100	1

Table 6.18   Benzo-[a]-pyrene in foodstuffs (µg/kg)

Charcoal broiled steak	8	Dried flour	4
Margarine	1–36	Toast bread	0.5
Sausages	4–50	Lettuce	3–12
Roasted coffee	1–13		

material is estimated at 60×10^6 t/year with 2×10^6 t applied annually in the UK and ten times this amount in the USA. Some levels in water bodies are given in Table 6.17 (IARC, 1983b). The levels in domestic sewage vary but rise markedly during periods of heavy rain, for example pyrene then rises to 16×10^3 and BaP to 1800 ng/l.

PAHs are found in food either by deposition on leaves (lettuce) or from heated fats (steak, sausage) (Table 6.18) (IARC, 1983).

Within the house, occupants are also exposed to risk from emissions from heating, for example gas central heating in a non-smoking household generates BaP in the range 0.1–0.6 ng/m^3. For smokers these limits lie between 0.4 and 1.8 ng/m^3. Despite the recognition of their damaging effects (*cf.* Chapter 7) the production of cigarettes has increased overall over the last decade (Table 6.19).

Table 6.19   Cigarette production (millions; UN, 1994)

Country	1982	1991
UK	100 600	127 00
USA	694 200	694 500
Spain	65 270	85 000
France	62 500	53 000
Germany	146 700	177 900
Italy	80 550	57 630
China	942 000	1 599 700
India	97 390	65 270
Comparative totals	2 189 210	2 860 000 (an increase of 31%)

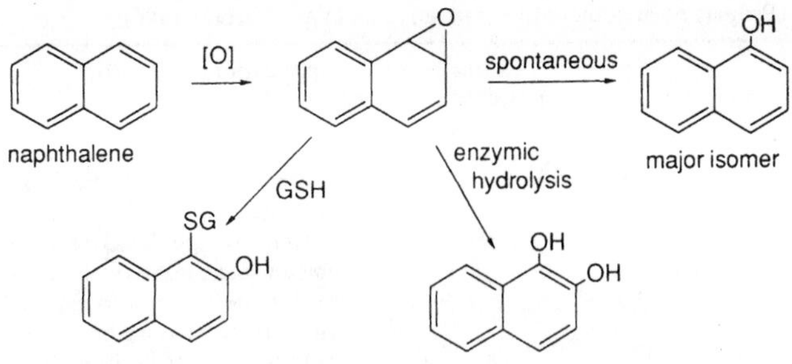

Figure 6.7   Oxidative metabolism of aromatic compounds.

Metabolism of aromatic compounds.   A principal route in the metabolism of aromatic hydrocarbons is the formation of arene oxides, which then isomerize to phenols (Fig. 6.7) (Jerina and Daly, 1974). In mammals, including humans, ingestion of these substances induces changes in the P-450 cytochrome enzymes within the endoplasmic reticulum. This is principally sited in the liver, but also in the kidneys, lungs and intestines. Oxides and phenols formed in this way become conjugated with water-solubilizing molecules and are then excreted. One such molecule is the tripeptide glutathione–glutamyl cysteinyl glycine (**4**). Its combination (as GSH) is shown in Fig. 6.7.

An alternative and competing hydroxylation path is illustrated for PCBs in Section 6.4.

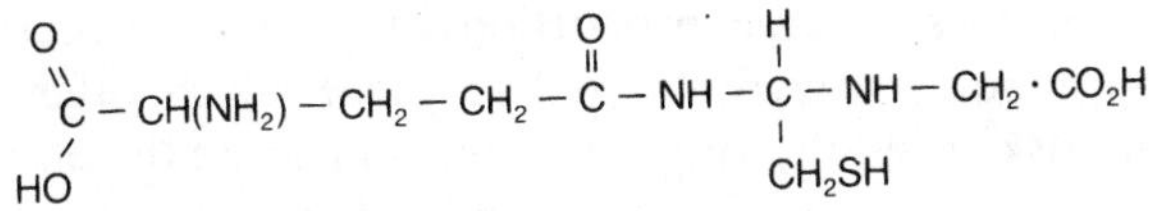

4

6.3 Organic solvents

Organic solvents are today produced almost entirely from petroleum (Fig 6.12, p. 283) and units for their production and plant for their development in chemical synthesis are commonly sited in close proximity. The relative importance of individual compounds is indicated by production levels in the USA (1984) shown in Table 6.20, which includes those of the primary sources ethylene, propylene and 1,3-butadiene.

Table 6.20   Production of organic chemicals in the USA (1000 tons)

Ethylene	13	Ethylene glycol	2.0
Propylene	6.3	*p*-Xylene	1.8
Toluene	6.1	Acetic acid	1.3
Benzene	5.5	Phenol	1.2
Styrene	3.1	Butadiene	1.0
Vinyl chloride	3.1	Acetone	0.8
Methanol	3.0	Cyclohexane	0.7

Table 6.21 Properties of some important solvents (Verschueren, 1996)

Compound	b.p.	Odour perception (mg/m^3)	Applications
Acetic acid	118	0.1–1.5	Synthesis, pharmaceutical, photographic, polymers
Acetone	56	5–50	Solvent for cellulosic resins and vulcanisates, adhesives, paints
Benzene	80	3	Motor fuels, synthesis especially maleic anhydride, flavours, perfumes, paints
Carbon tetrachloride	77	500	Manufacture of fluorocarbons
Chloroform	62	1000	Refrigerants, pharmaceuticals
Cyclohexane	81	4–10	Nylon synthesis
1,4-Dioxane	101	4–25	Lacquers, paints, cosmetics, cleaning and detergence
Ethanol	78	5–50	Synthesis, brewing, soaps, cosmetics
Ethyl ether	35	1–10	Manufacture of ethylene, perfumery, extraction
Methanol	65	1–100	Synthesis of proteins, esters and HCHO for plastics. dehydration of natural gas
Tetrahydrofuran	66	10	Polyether synthesis, PVC solvent
Toluene	111	1–100	Fuels, paints and coatings, isocyanates, benzene substitute
Trichloroethane	74	540	Metal cleaning, protection of upholstery, adhesives
p-Xylene	138	1–10	Synthesis of terephthalic acid, motor fuels

Table 6.21 lists those solvents in most common use with their principal properties and applications. As will be appreciated, a detailed study of solvents encompasses a large proportion of organic chemistry and the discussion must necessarily be limited to those which offer the greatest environmental risk. The following brief case studies are therefore concerned with those which are especially toxic and those which are most readily released to air.

6.3.1 Adhesives

These are formulated with volatile solvents so that once applied they rapidly become tacky and ready to bond. The solvents are, therefore, liable to escape when used domestically (Chapter 7) and in small factories with inadequate control measures.

In 1983, the demand in the USA was 1.53 Mt and in Europe 0.69 Mt with usage in Japan about the same as in Europe. This consumption grew at almost 4% until these outlets consumed about 3.36 Mt by 1988 (Table 6.22) (Montreal Protocol, 1991). 1,1,1-Trichloroethane (TCE) is one of the most used solvents as it is non-flammable and dries rapidly. Between 40 and 50 thousand tonnes are used annually for adhesive production. It is dangerous when inhaled by 'glue sniffers' and can cause death through congestion of the lungs or from heart failure (Glowa, 1990). Other solvents in common

Table 6.22 Demand of adhesives by sector in USA and Europe

Application	Weight (10^6 tonnes)	%
Packaging	2500	42
Non-rigid bonding	1100	19
Construction	1000	17
Tapes	500	8.5
Rigid bonding	400	7
Transportation	300	5
Domestic market	100	1.5

use are dichloromethane, toluene, *n*-hexane, methyl ethyl ketone and ethyl acetate. Major particular uses are in bonding decorative laminates to particle board or plywood, also in bonding polyurethane foam to furniture frames.

The use of TCE expanded in the USA as it was exempted from the volatile organic compounds (VOC) regulations; it is no longer a constituent of 'Tipp-Ex' correcting fluid. Apart from the uses just mentioned it replaced ethyl acetate in adhesives for packaging; it was also applied in latex adhesives based on styrene–butadiene formulations. At present in Western Europe, 75% of TCE production is used for metal cleaning, 10% in adhesives and 15% in aerosols, electronics and other applications. With concern about damage to the ozone layer, there has been a change of policy and TCE is no longer favoured; it has an ODP of 0.1 (pp. 161–2) and consequently there is a move back towards the VOC listed in Table 6.21. This substitution may protect the ozone layer, but release of these solvents enhances capture of NO (Equation 5.17, p. 163) so further increasing tropospheric ozone levels.

Other protective measures.   These include greater use of water-based latexes and emulsions. These are ineffective for the bonding of rubbers and some plastics; when sprayed they also produce fine mists which are carried around the workplace. Solid adhesives consisting of isocyanates and polyurethanes which cure when exposed to moisture are also available.

Solvent release may be avoided by the hot melting of plastics such as polyethylene and polyesters which bond to metals, plastics and paper on cooling. This technique is used in bookbinding, glazing and for the installation of car carpets and panels. Curing of acrylic, polyester and urethane resins by irradiation with UV, IR light or an electron beam is being used increasingly. These alternatives require expensive new equipment but are generally not more costly overall.

6.3.2 Coatings and inks

In 1986, almost half of coatings were solvent based. In 1989 this represented 1.7×10^6 t but only 1.2% or 21 800 t was TCE. Its use was restricted to thinning for spraying inks on to wallpapers, and for labelling bags, bottles and cartons. Here also the use of water-based materials and radiation curing controls solvent release.

6.3.3   Aerosol sprays

Figure 6.8 illustrates the much publicized concern about the damage to the ozone layer arising from the use of CFCs in spray cans (p. 160). These formulations include three components: (i) the active ingredient such as insecticide, (ii) a solvent and (iii) a propellant.

Figure 6.8   David Doniger, an attorney for the National Resources Defense Council, displays some of the 141 common household and office product aerosol cans that his organization says contain ozone-destroying chemicals. The products include hairsprays, spot removers and solvents. The cans contain 1,1,1-trichloroethane (methyl chloroform) (photo.: Associated Press).

Table 6.23 Annual waste from cleaning car engine control boards

Waste type	Conventional soldering with CFC cleaning (kg/year)	Controlled atmos. soldering (kg/year)
Volatile organics	5400	900–800
Formic acid	Nil	45–90
Lead	4–7	0.5–5
CFCs	9000–23 000	Nil
Work rate of moving belt	6 m/min	0.8–2.0 m/min

It has been estimated that in 1988 6.8 billion individual items were produced and many included TCE as solvent. At this time, production in the USA was 18 600 t and in Europe 12 400. In Japan, production was 10 800 t in 1986 and had fallen to 5000 by 1990. Alternative solvents are dimethoxyethane and low boiling alcohols, ketones and petrol fractions.

6.3.4   Metal cleaning

Trichloroethylene was phased out in 1960. At present CFCs are used for cleaning and drying a wide range of parts as also are perchloroethylene and methylene chloride. Table 6.23 shows how they can be eliminated from the waste from cleaning soldered car engine parts by the substitution of other organic solvents. A nitrogen atmosphere reduces the risk of fire but the throughput is also reduced. Other parts which require cleaning in solvent baths include adhesive spreaders, silk screen stencils, polymer formers, oil rig equipment, photocopiers and printing machines.

At present the SAAB company uses ethanol in place of CFC-113 for cleaning and Nissan (Japan) was expected to replace CFC-11 (CCl_3F), CFC-12 (CCl_2F_2) and CFC-113 ($CCl_2F-CClF_2$) with HCFC-134 (CF_3-CH_2F) by the end of 1994. The nomenclature and ozone depletion potential (ODP) of these hydrofluoro-chlorocompounds is discussed in Section 5.1.6 (p. 161).

6.3.5   Dry cleaning of clothes

Organic solvents are used as, unlike water, they do not distort the fibres. The machines have long been totally enclosed on account of cost control. The solvents available are listed in Table 6.24. According to the provisions of the Montreal Protocol (Section 5.1.7) for protection of the ozone layer, CFC-113 (ODP 0.8) and TCE (ODP 0.1) will not be acceptable and should be phased out by 1997. This presents practical problems since dry cleaning equipment has a lifetime of about 15 years and is only suitable for the solvent originally specified. Even HCFC-225, the recommended substitute, will be unacceptable by the year 2020.

Table 6.24 Some solvents of use in dry cleaning

Solvent	Boiling point (°C)	Energy to boil 1 litre (kcal)
CFC-113	48	64
Perchlorethylene	121	116
Petrol based	150–210	–
1,1,1-TCE	74	90
HCFC-225 ($CF_3CF_2CHCl_2$)	53	64

Carbon tetrachloride is still used for dry cleaning in Eastern Europe, the former USSR and in South East Asia. Otherwise it is produced by the high temperature chlorination of methane, propylene or carbon disulphide for use as a feedstock in the synthesis of CFC-11 (CCl_3F) and CFC-12 (CCl_2F_2). In 1988, the production in the USA of 283×10^3 t was almost entirely used for this purpose. As these solvents are phased out it will only be used in a limited number of special syntheses, for example that of picloram (Fig. 6.22, p. 300).

6.3.6　Solvent toxicology

The occurrence and toxicology of solvents in water has been extensively studied in the UK by the Water Research Centre based at Medmenham. Apart from protecting potable water supplies, the analysis reveals those substances released by industry whether by evaporation to air or by direct discharge to rivers as waste.

Aliphatic hydrocarbons. Some 60 compounds in this group have been detected in air and water consistent with their use in fuels. They do not in general present much risk to mammals but butane and *n*-hexane have attracted attention.

Butane accounts for 30% of all solvent abuse by glue sniffers and is obtained by them from lighter refills (Pottier *et al.*, 1992).

n-Hexane is one of the isomers present in motor fuel and in alkane mixtures used for paint thinning and for metal cleaning. It forms 1% of diesel emissions and 1.2% of the exhaust from petrol engines while losses from fuel tanks and carburettors can reach 10%. *n*-Hexane has a lifetime of about 6 hours in smog and has an odour threshold of 200–800 mg/m^3.

On exposure for 10 minutes to levels of 2000 µg/g there was no acute response in humans. The principal chronic effect in mammals (including humans) from exposure is one of neurotoxicity.

Abnormalities in the nervous system have been noted in workers occupationally exposed and also in glue sniffers. *n*-Hexane is rapidly metabolized in mammals being excreted as the alcohol (**5**); its neurotoxicity is attributed to further oxidative metabolism to form 2,5-hexanedione (**6**).

Aromatic hydrocarbons. Attention is here focused on benzene. On inhalation, acute effects include giddiness, headaches and nausea; chronic poisoning leads to depression of bone marrow function. There is evidence of its being a tumour promoter as it is linked to myeloid leukaemia and lymphomas. The strength of economic factors can be judged from the fact that benzene has been banned as a laboratory solvent but it still forms up to 5% (vol.) of lead-free petrol. Over 15×10^6 tons are produced annually in England and it contributes to exhaust emissions (2.4%, vol. of total hydrocarbons). Near the M1 at Luton, benzene reached a level of 155 ppb. The maximum allowable levels for an 8 hour day and 40 hour week range between 5 and 10 ppm ($16–32\,mg/m^3$).

Toluene is now preferred to benzene as a solvent. Death resulting from its inhalation has occurred as a result of solvent abuse; at high levels of occupational exposure (30 ppm) it can induce mild abnormalities of the CNS. Levels in petrol and air are similar to those of benzene; its lifetime in smog is about 6 hours.

6.3.7   Organochlorine compounds

Vinyl chloride.   This compound presents the most serious risk although it is still used in massive amounts for the production of PVC (Table 6.20). Its b.p. of $-14°C$ makes for a problem of containment. Vinyl chloride (Section 3.2.2, p. 69) causes liver degeneration in mammals and is carcinogenic in humans, inducing tumours of the liver and blood in those occupationally exposed; it is suspected as a cause of human mutations (Lewis, 1992). Accepted levels in the workplace are in the range 3–5 ppm but may be made more stringent. Vinyl chloride is no longer available as a laboratory reagent.

Chloroform (trichloromethane).   This occurs in air at levels up to $5\,\mu g/m^3$ and has been detected in foodstuffs up to 30 mg/kg. It is common in water samples, being produced by the haloform reaction during chlorination of water and sewage; in air it may be formed by the photochemical degradation of trichloroethylene.

Chloroform causes kidney damage and is classified as a group 2B carcinogen in mice and rats. It is teratogenic in these animals and its oral

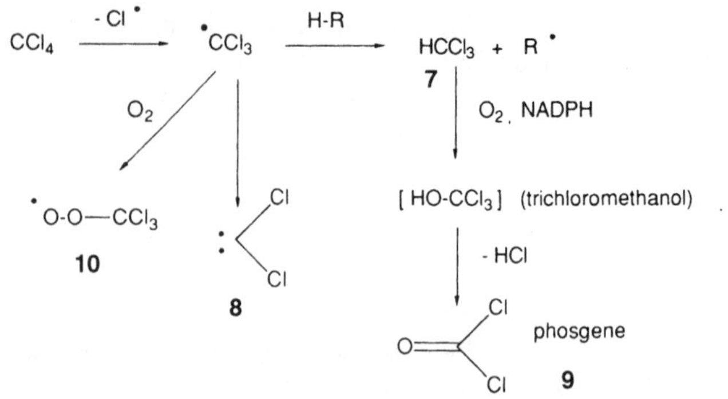

Figure 6.9　Metabolism of carbon tetrachloride.

LD_{50} in rats is 1000 mg/kg (DOSE, 1994; p. 362). Studies showed that chlorination of tap water did not lead to enhanced chloroform levels in human blood plasma; the WHO has set a guideline of 30 μg/l for chloroform in drinking water.

Carbon tetrachloride.　As seen previously (Section 6.3.5), this is on the decline as an industrial chemical. It occurs in air and food; in drinking water it is at a level of 2–3 μg/l. Acute exposure in animals produces death by depression of the CNS or through liver damage. In the liver it is metabolized via the trichloromethyl radical (Fig. 6.9) producing chloroform (**7**), dichlorocarbene (**8**), and phosgene (**9**), which is also a metabolite of chloroform. Capture of oxygen yields the long-lived peroxy radical (**10**) which is held responsible for liver damage.

　　Carbon tetrachloride causes liver cancer in animals and there is a possibility that this extends to humans.

1,1,1-Trichloroethane.　This is an important industrial solvent (Section 6.3). It occurs in air at about 1 ppm and has been found in food and water samples. It does not appear to be metabolized in the manner of Fig. 6.9, being largely exhaled from the lungs unchanged.

　　Deaths from acute occupational exposure have been recorded and solvent abusers are also at risk of heart failure. There is no evidence of long-term chronic effects but the accepted workplace level is 10 ppm.

　　1993 saw the first prosecution of a company director by the National Rivers Authority under section 85 of the Water Resources Act. The pollution incident occurred in the Severn–Trent region following the release of chemicals including 1,1,1-trichloroethane into the river, which led to a major fish kill. The level of trichloroethane reached 50 times the 2 mg/l recommended by the WHO, that is 99 mg/l, and a fine of £7000 was imposed together with mandatory preventive measures (NRA, 1994).

6.3.8   Detergents

Detergents are a group of synthetic compounds within the general class of surfactants. They were first produced during the First World War in Germany but the period after the Second World War saw a dramatic escalation in their use. The principal products were based on cheap and readily available petrochemicals. The dramatic displacement of soaps by detergents is illustrated in Table 6.25.

A typical propylene tetramer (Fig. 6.10, **11**) was combined with benzene in an acid-catalysed reaction. The position of the double bond and also the number of polymerized alkene molecules varies because it is not economic to separate single substances, nor is a mixture of alkyl groups any less effective for detergency.

Types of detergent.   The sulphonate (**12**), like soaps of the stearate group (**13**), dissolves in water and becomes effective as its anion. These compounds are therefore known as anionic detergents. The sulphonates have the

Table 6.25   The displacement of soap by detergents in the USA (Davidsohn and Milwidsky, 1987)

Year	Soap usage (10^3 t)	Detergent usage (10^3 t)
1940	1410	4.5
1950	1340	655
1960	583	1645
1972	587	4448
1982	545	5090

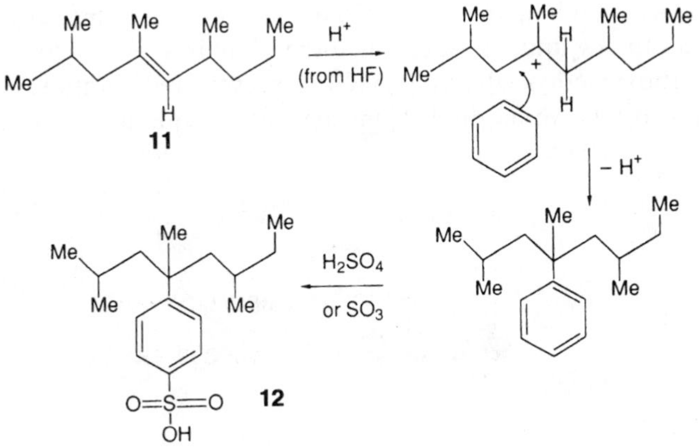

Figure 6.10   Synthesis of an alkylbenzene sulphonate.

$$CH_3—(CH_2)_{16}—C\lessgtr^O_{O^-Na^+}$$

A soap - sodium stearate

13

$$CH_3—(CH_2)_{14}—CH_2—\overset{Me}{\underset{Me}{\overset{|}{\underset{|}{N^+}}}}—Me \quad \overset{Cl^-}{}$$

Hexadecyl trimethylammonium chloride
used for home laundry and fabric conditioning

14

$$C_8H_{17}—\langle\ \rangle—O—CH_2—CH_2(—O—CH_2—CH_2)_n—O—CH_2—CH_2OH$$

15

hexaethoxyethylene glycol monoether of *p*-octyl phenol (when *n*=4)

advantage that, unlike the stearates, they are not precipitated in hard water as calcium or magnesium salts. Both types depend for their action on removing organic material from soiled surfaces on the residence of the hydrophilic head in the aqueous and the lipophilic tail in the organic phase (Fig. 6.11). This greatly reduces the surface tension, allowing the non-aqueous materials to be wetted and washed out. Above a certain critical concentration the surfactant will support water insolubles within aggregates or micelles, which present the heads to the water while the tails form an internal clump.

Anionic detergents still occupy a dominant place in the market, but the use of detergents is widespread in industry and the home (Karsa *et al.*, 1991). This necessitates the formulation of many specialist types, including cationic (**14**) and non-ionic detergents (**15**).

Builders.   In the original preparation of sulphonates, sulphuric acid in the residues was neutralized with alkali so forming sodium sulphate. This was retained so as to bulk up the product and make it easier to measure out. This practice continues and, in addition, other builders are added, including sodium carboxymethylcellulose (CMC) to prevent redeposition of dirt and phosphates to remove hardness from the water. Equation 6.16 shows how

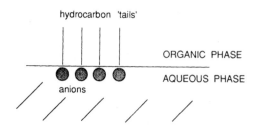

Figure 6.11   Phase interaction of an anionic surfactant.

one of the phosphate formulations, tetrasodium pyrophosphate, sequesters Mg^{2+} so softening the water without precipitation:

$$2Na^+(Na_2P_2O_7)^{2-} + MgCl_2 \longrightarrow 2NaCl + 2Na^+(MgP_2O_7)^{2-} \qquad (6.16)$$

Pollution by detergents.   The upsurge in the use of anionic sulphonates based on propylene led to concern about persistence and foaming when they were released to water bodies. The traditional soaps are degradable by the effluent bacteria but the branched chains of the sulphonate are resistant. Consequently, by the early 1960s industry began to formulate these detergents from degradable straight-chain polymers of ethylene, following the synthetic route shown above (Fig. 6.10).

The production of sodium tripolyphosphate as an additive increased 100-fold between 1947 and 1970 and levels in surface waters increased in proportion. Lake Constance, situated in a populated area on the German–Swiss border, has no outflow and phosphate rose from a natural level of 0.2 mg (as P)/m^3 to 3 mg/m^3 by 1953 (Davidsohn and Milwidsky, 1987): today this has risen by an order of magnitude. This type of accumulation of phosphate has supported the growth of algae and the associated unsightly 'blooms'.

To control algal growth, alternative sequestering agents have been substituted for phosphate. One of these is the chelating agent nitriloacetic acid, which associates with Ca^{2+} and Mg^{2+} as shown in structure **16** (*cf.* Equation 6.16). Nitriloacetic acid (NTA) is, however, environmentally suspect since it also sequesters and, therefore, mobilizes heavy metals such as Pb and Cd from sediments. The most promising alternative to phosphate are zeolites, which are polymeric sodium aluminium silicates. A simplified form is the structure **17**, where the trivalent Al atoms in sharing with the tetravalent Si aquire a surplus negative charge; as a result the zeolite associates with Ca^{2+} in order to achieve electrical neutrality.

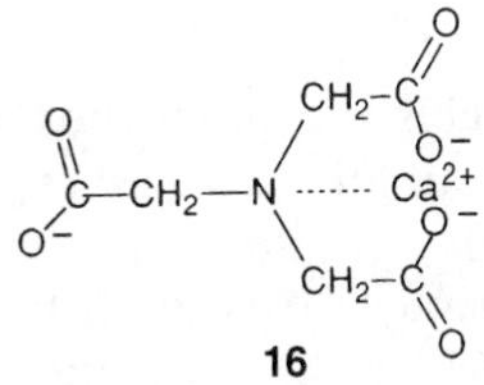

16
association of sodium nitriloacetic
acid with Ca^{2+}

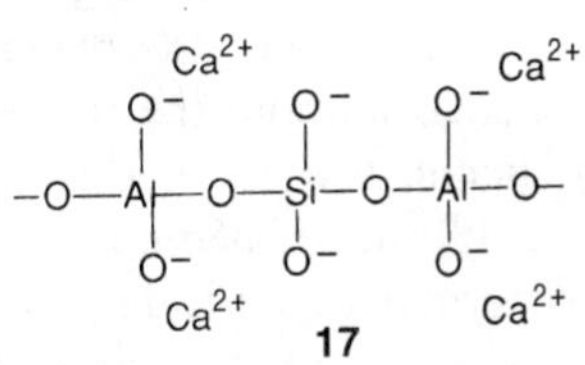

17
simplified structure of a zeolite

6.4.1   Historical note

Chlorine was first isolated by the Swedish chemist Carl Scheele in 1774 and its commercial use as a bleaching agent dates from 1789. In 1800, Cruickshank showed that it could be obtained by the electrolysis of brine and following the development of the mercury and diaphragm cells during 1883–93, this method is now employed for large-scale production. The mercury cell reactions are:

$$Na^+ + Cl^- \longrightarrow Na/mercury + Cl^\bullet$$

whence Na^+ discharged at the cathode reacts with water to produce NaOH and hydrogen. Chlorine gas is collected from the anodic discharge.

Mercury is surprisingly soluble in water (10^{-1} mg/l at 50°C) and as a result losses amount to between 100 and 200 g/t of chlorine produced. Owing to the toxicity of mercury (Section 5.3.6) and more stringent controls on its release, this has necessitated the use of diaphragm cells in industry. In this modification, the chlorine stream at the anode is separated from hydrogen and NaOH by a layer of asbestos (Section 7.7) which is vacuum-deposited on the cathode. The brine now percolates this diaphragm under close control to enter the cathode compartment.

In addition to its use in bleaching, large quantities of chlorine were required for use as an insecticide. One example of the importance of this application is the elimination of cholera in developed countries as a direct result of chlorination of water. In the UK, the first recorded death from this loathsome disease was that of William Sproat in 1832 and about 32 000 deaths occurred here in a two-year period. Cholera was pandemic during 1862–72 and came under control following identification of the bacillus by Koch in 1884. The last recorded outbreak in western Europe was that in 1892 in Hamburg when 10 000 died, but 1000 deaths in Peru were attributed to cholera in 1991.

6.4.2   Organochlorine production

Despite its use in water treatment, worldwide chlorine production was relatively modest at about 2 million tons in 1940, but this increased rapidly with the demand for organochlorine compounds for use as pesticides. Figure 6.12a shows the exponential growth in the synthesis of chemicals from petroleum and the related demand for chlorine in the USA. By 1971, petrochemicals amounted to over 90% of the total in the USA, Europe and Japan; chlorine consumption levelled out partly as a result of restrictions on polychlorinated pesticides (Section 6.4.8). Nevertheless, the major uses for the production of PVC and as an insecticide ensured continuing growth in chlorine production as well as in general chemicals (Fig. 6.12b).

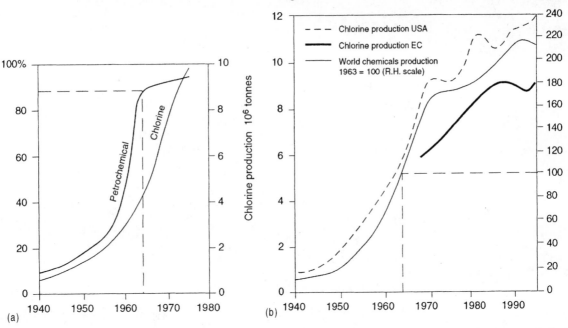

Figure 6.12   Comparative growth of (a) petrochemical and chlorine production in the USA 1940–80. (b) Comparative growth of chlorine and world chemicals 1940–1994.

The first compound produced commercially was carbon tetrachloride in 1907, followed by others, such as trichloroethane, in the 1920s. Annual production worldwide now ranges from over 10^9 kg of industrial solvents such as dichloroethane down to a few million kilograms of more specialized substances, for example hexachlorobenzene.

6.4.3   DDT (dichlorodiphenyl trichloroethane)

Apart from the ready availability of chlorine and petroleum products in the years following the Second World War, the successful use of DDT (Fig. 6.13) in controlling disease during its final stage encouraged the research and development of organochlorines as a whole. DDT was first described by Othmar Zeidler in 1874, although over 60 years elapsed before it was formulated as an insecticide by the Geigy company, following the work of Muller (1944).

The preparation of DDT depends on the condensation of chloral with chlorobenzene in sulphuric acid (Fig. 6.13). This is not a unique pathway and significant amounts of the isomer with one chlorine in an *ortho*-position are produced; also chloral produced by the chlorination of acetaldehyde is

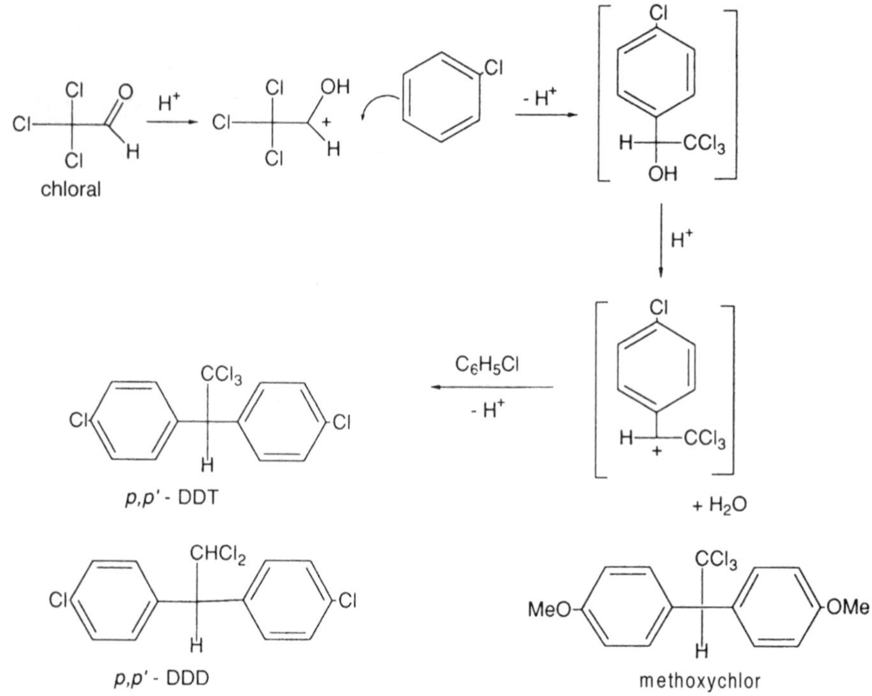

Figure 6.13 The synthesis of DDT.

contaminated by dichloroacetaldehyde and gives rise to DDD. Commercial DDT has been shown to be a mixture of 14 substances: DDT itself constitutes 65–80%; *o,p*-DDT 15–20% and DDD may be up to 4%.

6.4.4 Lindane, hexachlorocyclohexane

Lindane is obtained by the addition of three molecules of chlorine to benzene activated by UV irradiation (Fig. 6.14). In theory there are eight possible geometrical isomers, in which the chlorine atoms occupy different relative positions about the cyclohexane ring. Structure (**18**) is drawn in a way which shows the characteristic 'chair' form with six axial (a) bonds parallel to the main axis equatorial (e) bonds diposed around the belt of the molecule. The active form, the *gamma*-isomer, has three consecutive axial and equatorial substituents and forms about 15% of the mixture of reaction products, which includes five of the possible isomers. It was first described by Van de Linden in 1912 and introduced commercially by ICI in 1942.

Lindane or gammexane has similar properties to DDT and is widely used in Third World countries as the crude product is cheap to produce although

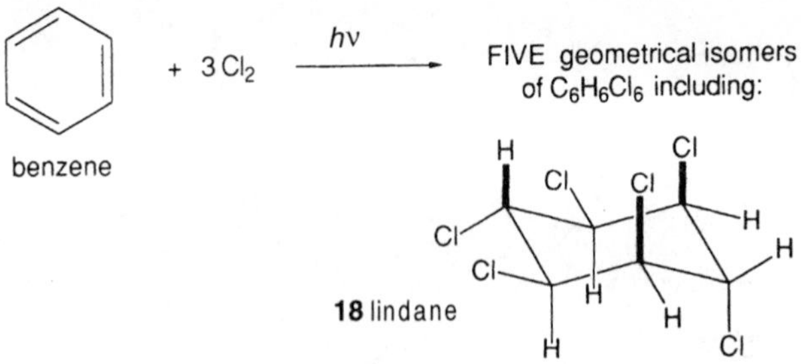

Figure 6.14   The synthesis of Lindane.

Table 6.26   Distribution of hexachlorocyclohexanes in ocean air, 1990 (Iwata *et al.*, 1993)

Ocean	Lindane mean levels	Alpha-isomer (pico. g/m^3)
North Pacific	76	520
Bay of Bengal	1100	8600
North Atlantic	66	200
South China Sea	500	810

some of the components have a musty taste. For control of pests in food crops such as the potato it is necessary to obtain pure lindane, which is tasteless, by recrystallization. It is superior to DDT in controlling soil pests.

It has been estimated that 10^7 tons of the technical mixture had been produced worldwide by 1992; India produced 20 000 tons in 1991. Sediments in Casco Bay (Maine in 1990) contained only low levels of < 0.5 ppb, but the distribution in ocean air (Table 6.26) is evidence of the continued use of the cheaper mixture in Asian countries.

A study of 35 Malaysian workers from six different plantations showed that all had experienced muscular weakness on exposure to lindane, most became dizzy and had stomach pains and nausea. The LD_{50} value in the rat is 90 mg/kg; the LC_{50} in rainbow trout is 0.06 mg/kg; lindane is carcinogenic in rats and mice.

Control levels have been set in the UK and USA at a TWA of 0.5 mg/m^3. The Codex Alimentarius gives an acceptable daily intake of 8 µg/kg.

6.4.5   Some other chlorinated pesticides

The structures of some other principal chlorinated pesticides are shown in Fig. 6.15. A number of these compounds were obtained by the Diels–Alder addition reaction of hexachlorocyclopentadiene (**19**). Combination with

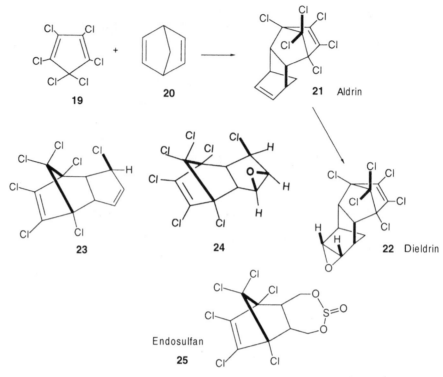

Figure 6.15   The synthesis of the Aldrin group of pesticides.

bicycloheptadiene (**20**) leads to Aldrin (**21**), which on epoxidation affords Dieldrin (**22**). A similar relationship is found between Heptachlor (**23**) and its more active epoxide (**24**).

Epoxidation is an important *in vivo* process and results in the hydroxylation and consequent solubilization of organic pollutants (Section 6.2.5). While this change is the key to degradation of toxic compounds, there are occasions when the cure is worse than the disease. Aldrin is converted into Dieldrin *in vivo*; the latter is in general more toxic to fish (Table 6.27). In 1972 following treatment of rice seed, there were heavy losses of ducks and other birds, and the compound was detected in drinking water from wells in

Table 6.27   Toxicity to fish of Aldrin and Dieldrin (96-hour static LC$_{50}$ (μg/g))

	Aldrin	Dieldrin
Striped mullet	100	23
Bluehead	12	6
Rainbow trout	36	19

California. The oral LD_{50} values in the rat lie in the range 33–100 mg/kg for both substances and the occupational exposure limits were set in the UK and the USA at a TWA of 0.25 mg/m^3. Dieldrin is a human carcinogen and was recently detected at levels above the acceptable daily intake in Cairo residents, where bread is the major source. The concen- tration of residual chemicals in food chains is a further cause for concern and has led to restrictions on the use of these compounds (Section 6.4.8).

The cyclic sulphite Endosulfan (**25**) is still in use, even in the USA, since the high level of oxygenation ensures that it is less persistent than other members of this group. Endosulfan is hydrolysed to the diol and SO_2 and is rapidly detoxified in mammals. The technical grade is a mixture of the α-isomer (66% as shown) and the β-isomer (34%); the dermal LD_{50} value in the rabbit is 359 mg/kg and, unless applied under proper control conditions, penetration of the skin is dangerous to workers in the field.

Farmers in the Sudan, who were aware of the risks to humans and animals, used Endosulfan in excess to treat maize. Subsequently, 16 pounds of the crop were fed in part to poultry, which died. The remainder was sold on to make bread for a funeral party – 31 of the guests died and several hundred were affected, notably by memory loss.

The EPA considers Endosulfan to be a teratogen and maximum levels in workplace air are set in the UK and USA at a TWA of 0.1 mg/m^3. The acceptable daily intake is 0.006 mg/kg.

6.4.6 Organochlorine herbicides

Early non-selective chemical weed control agents included oil waste and creosote (Cremlyn, 1991). A degree of selectivity was achieved by spraying cereal crops with copper sulphate solution or dilute sulphuric acid, when the large broad-leafed weeds were more susceptible.

In 1993 the first use of a truly selective agent, dinitro-*o*-cresol (**26**, Fig. 6.16) was seen, and although limited by its failure to control perennial weeds, it is still useful as a winter wash for fruit trees and is not persistent. It is, however, very poisonous to mammals with an LD_{50} of 30 mg/kg in rats.

The discovery by Kogl in 1934 that 3-indolylacetic acid (**27**) was a plant growth hormone led to the development of more active and more stable compounds for crop protection during the Second World War. Notable among these were 2,4-D (**28**), 2,4,5-T (**29**) and MCPA (**30**).

These substances were applied typically at a level of 1 kg/ha to broad-leafed weeds in grassland. They act as mimics for the natural growth hormone leading to over-production of RNA and death of weeds, which cannot then obtain sufficient nutrients from the roots to sustain the abnormal growth. Demand is decreasing because of resistance, the emergence of alternatives and the presence of toxic by-products in the formulations.

An aliphatic substance, Dalapon (**31**) is useful for the control of couch

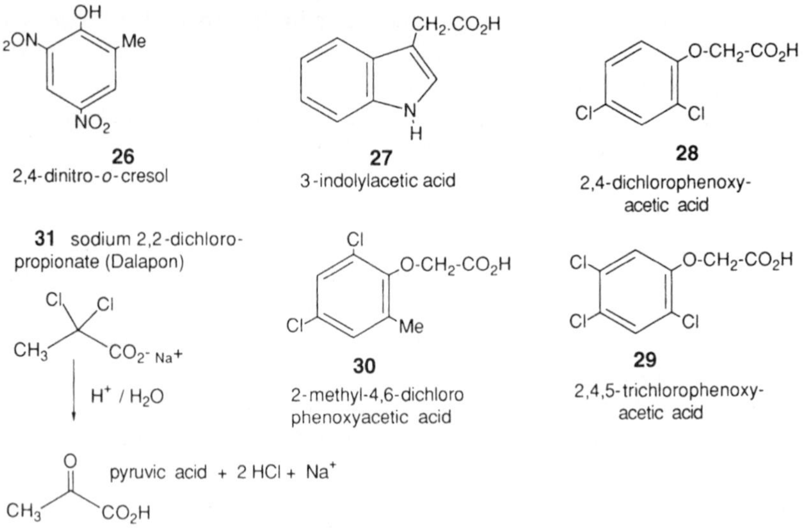

26
2,4-dinitro-o-cresol

27
3-indolylacetic acid

28
2,4-dichlorophenoxy-
acetic acid

31 sodium 2,2-dichloro-
propionate (Dalapon)

30
2-methyl-4,6-dichloro
phenoxyacetic acid

29
2,4,5-trichlorophenoxy-
acetic acid

Figure 6.16   Some important herbicides.

grass. This agent is not persistent as it is readily hyrolysed to pyruvic acid (Fig. 6.16).

Despite their cheapness and early dramatic successes, undesirable pollution has arisen from the widespread use of the organochlorine pesticides. This has led to restrictions on their use and a search for alternatives, which will be more expensive (Section 6.5, p. 307).

6.4.7   Toxic effects of insecticides

It can be appreciated that to assay the toxicity of pure individual compounds in a mixture arising from commercial synthesis may be difficult: commercial samples of DDT include 14 substances. The PCBs provide an extreme example of this; they are formed as a mixture arising from the mechanism of their synthesis rather than from impurities in starting materials (Section 6.4.10).

The oral LD_{50} of DDT in the rat is 110 mg/kg but DDT is not very toxic to humans – a human test group ingested 35 mg daily over an extended period without ill effects and the human fatal dose is estimated to be 500 mg/kg body weight. However, a typical insect toxicity would be LC_{50} of 5 µg/l for a 48 hour exposure; for rainbow trout the LC_{50} (96 hour) is 7 µg/l. It is extremely toxic to cold blooded creatures owing to their inability to limit the opening of Na^+ nerve channels (*cf*. Section 6.5.2) by the wedge-shaped DDT molecule; as a result nerve cells are continuously activated and death follows.

Penetration of the insect cuticle is favoured by high levels of molecular chlorine; this enhances the effectiveness of chlorine-rich molecules such as DDT ($C_{14}H_9Cl_5$). Unfortunately, the survivors of successive generations of pests breed ever more resistant strains and by 1950 several species of house fly and of cabbage rootfly had become resistant to DDT. This is not likely to have arisen from induced mutations but rather from the survival of a small resistant proportion of the insect population followed by the relatively rapid breeding of these more resistant survivors (Hutson and Roberts, 1987). This process of recovery is often accelerated by the action of high doses and the reduction of predator populations, which recover their numbers more slowly. By 1980 over 400 species of insects and the very rapidly reproducing mites had become resistant to DDT. Related compounds like methoxychlor acquire resistance by a crossover mechanism but this does not extend to compounds in a different cross-resistant group (Fig. 6.13) such as Aldrin (**21**) and Lindane (**18**). To overcome resistance, application rates have to rise and the search for effective alternative treatments is ongoing. It is also found that the introduction of electronegative chlorosubstituents makes organic compounds more difficult to oxidize (Section 6.4.12) and so extends their lifetime, which for DDT may extend to > 10 years.

Largely as a result of unwise excessive applications, the use of DDT and its analogues was called into question in developed countries. This followed the realization that fish and raptorial birds at the head of food chains were being poisoned owing to biomagnification along the chain.

Figure 6.17 shows relationships in the complex ecosystem of Lake Kariba, Zimbabwe, with clear biomagnification of sedimentary DDT levels to those in the filter-feeding mussel *Corbicula africana*. The relation between kapenta and algae is not clear, possibly because the short-lived plankton reflect DDT levels over a short time period while the levels in kapenta have been accumulated over a much longer period. Biomagnification is clearly shown from kapenta to the tigerfish and the cormorant. Other evidence of biomagnification is seen between fish and the crocodile at the head of the chain.

Lack of understanding of food-chain relationships has led to numerous pollution incidents which became the subject of major public concern.

Clear Lake, California. This is a body of water about 100 square miles (180 km²) in area, 70 miles (94 km) north of San Francisco and much used for recreation (Moriarty, 1988). Following complaints about clouds of gnats, the lake was first treated with DDD in 1949 and the nuisance was controlled; following recovery in the numbers, a successful reapplication was made in 1954. However, by 1957 the gnats and numerous other insect species had developed resistance to DDD. Towards the end of 1954 many dead western grebes were found, leading to a public outcry which ultimately ended the treatment. These birds were fish-eating divers and their death was attributed to bioaccumulation through plankton → plankton-consuming

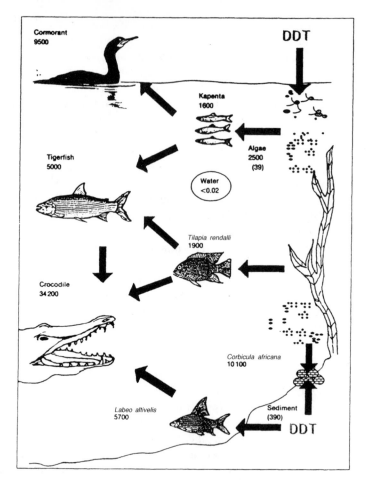

Figure 6.17   Mean levels of DDT (ng/g fat) in the Lake Kariba ecosystem (Berg *et al.*, 1992).

fish → carnivorous fish → grebes. The DDD was originally dispersed in the lake water at a concentration of only 20 ppb, but the birds had accumulated DDD in their fat at levels up to 1600 ppm: a bioaccumulation factor of 80 000.

Death of predatory birds.   A small fall in the peregrine population around 1940 (Fig. 6.18; Cremlyn, 1991) was the result of deliberate shooting in wartime to protect carrier pigeons. Post-war recovery was followed during 1950–60 by a steep and clearly abnormal decline, which was attributed to the application of organochlorines, especially Aldrin and Dieldrin, to protect cereal seeds from the wheat bulb fly. Other toxic effects were seen among pigeons, other predatory birds and foxes at the head of the food chain.

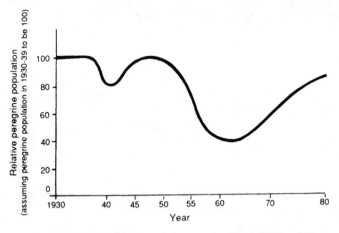

Figure 6.18 UK populations of the peregrine falcon (1930 = 100).

DDT and other organochlorines are responsible for the thinning of eggshells and the consequent reduction in hatching rates.

6.4.8 Control of pesticides

With increasing awareness of the toxic effects of chlorinated agrochemicals, limitations have been set on their use and in some instances a complete ban has been imposed. The use of DDT was first banned in the USA in 1973 and a ban on Aldrin and Dieldrin followed in 1983. In the UK, DDT was banned in 1984 and the use of Aldrin and Dieldrin for seed dressing was stopped in 1986. However, it must be emphasized that even following prohibition these substances will persist in the environment for a period dependent on temperature, rainfall and soil type; some median half-life values are given in Table 6.28.

It should also be noted that insects amount to 70% of all animals and of one million insect species, about 1% are significant pests, so alternatives must be found. In developed countries, the policy has been to reduce dependence on organochlorines and to introduce environmentally

Table 6.28 Persistence of some pesticides in soil (Goring *et at.*, 1972)

	Decay (%)	Time taken (years)
DDT	44	8
BHC	50	1
Dieldrin	49–53	3

acceptable but more expensive substitutes, such as pyrethroids (Section 6.5.1) and organophosphorus compounds (Section 6.5.2).

In Third World countries, a ban on effective control chemicals is not enforced as rigorously as in the developed countries since the magnitude of the problem posed by pests overshadows possible toxic effects. As an illustration, it should be noted that 100 million clinical cases of malaria are reported annually and of these about 1 million are fatal. A problem compounded by the fact that 50 of the 60 species of malarial mosquitoes are resistant to DDT and Dieldrin. Another instance is the distribution of the tsetse fly over 12 million square kilometres of Africa with the associated risk of trypanosomal disease in cattle and humans.

The problem of contamination of herbicides by dioxins is closely related to that arising from the oxidation of PCBs and is discussed in the following section.

6.4.9   Vinyl chloride and polyvinyl chloride (Kirk-Othmer, 1996)

In this section, and that following, pollution arising from accidental leakage of organochlorines produced for applications other than control of pests is considered.

Some 85% of synthetic chemicals are used for the manufacture of polymers. In 1985 about 7 million tonnes of vinyl chloride were produced worldwide by a route dependent on the addition of chlorine to ethylene and the subsequent thermal elimination of hydrogen chloride. To avoid wastage of hydrogen chloride, acetylene has been added to the system to trap it and so produce a balanced process:

1,2-dichloroethane

$$C_2H_4 + C_2H_2 + Cl_2 \longrightarrow 2CH_2 = CHCl \qquad (6.18)$$

More recently re-incorporation of hydrogen chloride is ensured by oxidative addition, followed by thermal cracking:

$$C_2H_4 + 2HCl + [O] \longrightarrow \underset{\underset{Cl}{|}}{CH_2} - \underset{\underset{Cl}{|}}{CH_2} + H_2O \qquad (6.19)$$

The production of polyvinylchloride (PVC) began in 1940 through the induction of polymerization of the monomer by free radicals. The reaction is conducted on an aqueous suspension in glass-lined steel kettles and a typical catalyst is benzoyl peroxide, in which the weak central bond is cleaved at 50°C to generate two initiators (32):

$$Ph \overset{O}{\underset{\|}{C}} O\text{-}O \overset{O}{\underset{\|}{C}} Ph \longrightarrow 2\ Ph \overset{O}{\underset{\|}{C}} O^{\bullet} \qquad (6.20)$$

$$\mathbf{32}$$

$$Ph \overset{O}{\underset{\|}{C}} O^{\bullet}\ CH_2{=}CH.Cl \longrightarrow \left[Ph \overset{O}{\underset{\|}{C}} O\text{-}CH_2 \overset{H}{\underset{\overset{\bullet}{C}}{\underset{|}{C}}} Cl \right] \qquad (6.21)$$

$$CH_2{=}CHCl$$

and growth continues in this way to give a degree of association of about 1000 monomer units and PVC granules with a molecular weight of about 63 000.

It is often not appreciated, especially by the media, that very few chemicals have been proved to cause cancer in humans (Section 3.2.2). However, vinyl chloride is a proven carcinogen, and by 1970 it was shown that industrial workers when exposed to it were prone to cancer. The causative agent is an oxidation product, chloroethylene oxide, which provides another example of toxicity emerging on modification in the environment (*cf.* Dieldrin, Section 6.4.5). Conversion of 10% of ethylene into PVC meets a major demand for which no suitable substitute exists and hence there has been no ban on the monomer, although control measures have been adopted and it is no longer available as a laboratory chemical.

6.4.10 Polychlorobiphenyls

Polychlorobiphenyls (PCBs) were first produced industrially in 1929 and were early on incorporated in printing inks and paints; subsequently, their use was extended to the softening of plastics and as insulators in transformers. In the USA, production peaked at 7×10^4 tons in 1970 and it is estimated that 75×10^4 tons have been manufactured overall. Their thermal stability made them attractive as insulating transformer fluids and biphenyl (**33**), the essential precursor, was readily available, as was chlorine. The chlorination (Fig. 6.19) is catalysed by iron(III) chloride but exhibits the characteristics of a free radical reaction in that a diversity of biphenyls at different chlorination levels appear shortly after reaction begins. As a result it is not economic, or even feasible, to separate indivi-duals and the mixed products are marketed under trade names such as Aroclor, Phenoclor and Clophen. The Monsanto nomenclature typically designates Arochlor 1242 as the fraction containing 42% of chlorine, b.p. 325–366°C; Aroclor 1268 contains 68% of chlorine, b.p. 435–450°C.

Of the three possible initial products, the 4-chloroisomer (**34**) pre-dominates with a significant amount of the 2-chloroisomer (**35**) and a little of the 3-chlorobiphenyl (**36**). Each of these compounds subsequently and

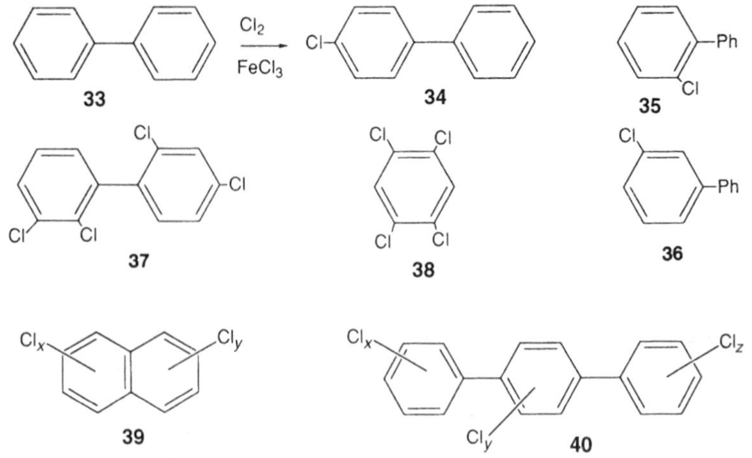

Figure 6.19 Synthesis of PCBs with the structures of some related compounds.

Table 6.29 Occurrence of PCBs in the environment

	ng/m^3		mg/kg
Air	0.1–20	Plankton	0.01–20
Water	0.1–30000	Invertebrates	0.01–10
Sediments	1.0–1000	Fish	0.01–25
		Birds' eggs	0.1–500
		Humans	0.1–10

rapidly captures chlorine in either ring A or ring B, so that substances such as 2,2′,3,4′-tetrachlorobiphenyl (**37**) and others of higher chlorine number occur in Aroclor. There are 42 possible tetrachloroisomers and also 42 hexachloroisomers, because in the latter the four unsubsituted positions permute in the same way as do the four chlorinated ones. In total, there are 209 PCBs, ranging from chlorine number one to the fully substituted decachlorobiphenyl. Hence the separation of all the pure components and evaluation of all the individual toxicities was a nearly impossible task. However, PCBs were identified as pollutants in 1966 with similar behaviour on gas chromatograms to that of DDT, and like it they are now ubiquitous in the environment. Table 6.29 shows some typical occurrences.

It should also be pointed out that other chlorinated aromatic compounds constitute an environmental risk. These include chlorobenzenes (e.g. **38**), the chloronaphthalenes (e.g. **39**) and polychloroterphenyls (e.g. **40**). The last named are found in the environment at lower levels than PCBs but levels in human fat are comparable.

At normal levels of exposure PCBs are not very toxic to humans, although in the Yusho incident their inclusion in rice bran cooking oil led to

Table 6.30 Pollution indicators and incidents involving PCBs

1950 onwards	Decline in the populations of the European otter
1950–70	Decline in seal populations in the Baltic and Dutch Wadden sea (Harrison, 1992) Levels of penta- and hexachlorobiphenyls in the range 2–15 ppb in Jordanian soils; these are similar to those found in industrialized countries (Alawi and Heidman, 1991)
1970–76	Reduction in herring gull populations on the Great Lakes Erie and Ontario, which are most at risk from anthropogenic sources
1987–88	Over 700 dolphin carcasses were found on the Atlantic coast of the USA. The deaths were attributed to brevitoxin, a neurotoxin produced by the Florida red tide organism (*Ptyodiscus brevis*) but stress from the ingestion of organochlorines was a likely factor (Kuel *et al.*, 1991). Blubber levels in ppm:

	max.	mean
PCB	195	40
DDE	80	8
Dieldrin	3.3	1

1991	Rural air samples from Ulm, Germany, typically contain 170 pg/m^3 of PCBs similar to those found elsewhere in Europe and over the Great Lakes. Air levels in Reunion, South Indian Ocean, were as high as 125 pg/m^3. It is evident that the principal input to the northern hemisphere disperses slowly into the southern hemisphere (Ballschmitell and Wittlinger, 1991).
1992	Milk from five species of seals was sampled from feeding regions in the Arctic, Antarctic, California and Australia (Bacon *et al.*, 1992). All contained organochlorines and PCBs, which were principally a common pair of hexachloroisomers and a sole heptachloroisomer. Typical levels (ppm) in the milk were: Australia 44 Arctic 232 California 370 Antarctic 10

serious consequences. Residents in South Japan became affected by a peculiar skin disease which first broke out in March 1968; it became notorious as 'Yusho disease' and another outbreak occurred in Taiwan in March 1979 when about 2000 people were affected.

In 1973, a community of several thousand in Michigan was also seriously affected by the inclusion of polybromobiphenyl in cattle feed. This product consists mainly of penta- and hexabromobiphenyls, is used as a fire retardant and is marketed as ' Firemaster'.

Table 6.30 summarizes some observations and incidents which establish that pollution by PCBs is serious and widespread. Once the distribution of PCBs and the similarity of toxic effects to those of DDT were established, their use declined and was banned for all but closed systems by the Dow company in 1971. Production worldwide had practically ceased by 1977 but the problem of destruction of the residues remained.

The incident at Binghampton. There have been numerous examples of pollution arising from transformer fires, but that in the Binghampton State office building on 5 February, 1981 rates as the most serious problem in terms of damage to the building, state records and human exposure (Schechter *et al.*, 1985). The transformer contained Aroclor 1254 and tetrachlorobenzene (**38**), the combustion products entered the ventilation system and hence were distributed throughout. Some hundreds of workers engaged in the clean-up were exposed to risk before the full extent of contamination was discovered and two years after the fire the building was still unusable. Table 6.31 lists the principal findings.

This incident graphically illustrates the toxic mix which results when organochlorines are burnt below the oxygen levels and temperatures needed for safe incineration (Section 6.4.13). The formation of a PCDF during PCB combustion probably involves the hydroxyl radical; the mechanism of the reaction is shown in Fig. 6.20. 3,3′,4,4′-Tetrachlorobiphenyl (**41**) is taken as an example and the formation of the free radical (**42**) from it follows a similar course to that shown in the conversion of methane to PAN (Section 6.2.3). The next step gives the phenoxy radical (**43**) and is followed by attack within the same molecule with closure of the furan ring and loss of a hydrogen atom. The product, 2,3,7,8-tetrachlorodibenzofuran (**44**), is the most toxic of this group and compares with 2,3,7,8-tetrachlorodibenzodioxin in the dioxin group. The TCDFs were implicated in the rice oil incidents and further examples of their toxicity are discussed in Section 6.4.11, covering pollution by herbicides.

It should also be noted that PCB fires generate chlorobiphenylenes such as (**45**) by dimerization of the radical intermediate (**42**).

Table 6.31 Pollution of the Binghampton Office Building

Initial level[a] (ppb)		Human adipose levels PCDD + PCDF (ppt)[b]	
10% of soot equiv. to	100×10^6	Highest recorded	8400
of which PCDF	2.1×10^6	Arithmetic mean	2400
PCDD	$(1–2) \times 10^4$	Mean from four people	
		not exposed	1150
Early PCB in air	$80\,\mu g/m^3$		

[a] General standard of detection 1 ppt.
[b] PCDD, polychlorodibenzodioxins; PCDF, polychlorodibenzofurans.

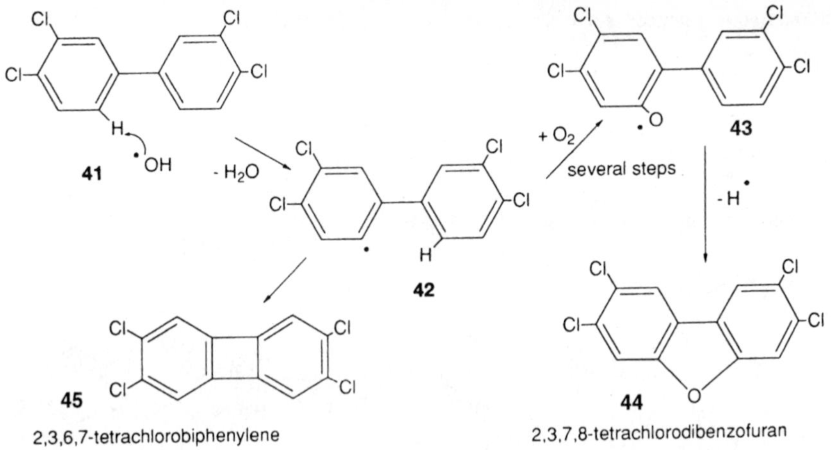

41

42

43

+ O₂

several steps

- H·

- H·

- H₂O

45   2,3,6,7-tetrachlorobiphenylene

44   2,3,7,8-tetrachlorodibenzofuran

Figure 6.20   The formation of a PCDF from a PCB.

6.4.11   Toxic substances in herbicides

An alternative title to 'toxic substances in herbicides' (Kamrin and Rodgers, 1985) would be 'the dioxin problem'. In the late 1960s, chemists became aware of the presence of the dioxin analogues of the PCDFs (e.g. **44**); of these the most dangerous is the symmetrical 2,3,7,8-tetrachlorodibenzo-dioxin (**48**). It was realized that this compound was formed as a by-product whenever the temperature of a reaction mixture containing 2,4,5-trichloro-phenol exceeded 220°C. This explains (Fig. 6.21) the presence of the dioxin in formulations of 2,4,5-trichlorophenoxyacetic acid.

In the required reaction, the hydroxide ion displaces chloride from the tetrachlorobenzene (**38**) to form the chlorophenolate ion (**46**). At higher temperatures, which may exist locally within the mix or which may be reached if controls fail, then the attack of one phenolate ion on another begins to compete. The first product is the diphenyl ether (**47**) but this rapidly cyclizes in a manner formally analogous to the last step in PCDF formation to give TCDD (**48**). It is estimated that 30 lb TCDD is produced annually worldwide.

There is no obvious route from PCBs to dioxins but they may still be formed in transformer fires, as 1,2,4,5-tetrachlorobenzene is often added to keep the mix fluid. Clearly replacement of chlorine by oxygen during combustion is all that is required to generate dioxins in the manner shown in Fig. 6.21.

The formation of the dioxin (**48**) occurred widely and as a result it became incorporated in industrial wastes that were frequently handled without knowledge of the risk. The formation of highly toxic dibenzodioxins and dibenzofurans during incineration is discussed in Section 6.4.13. The

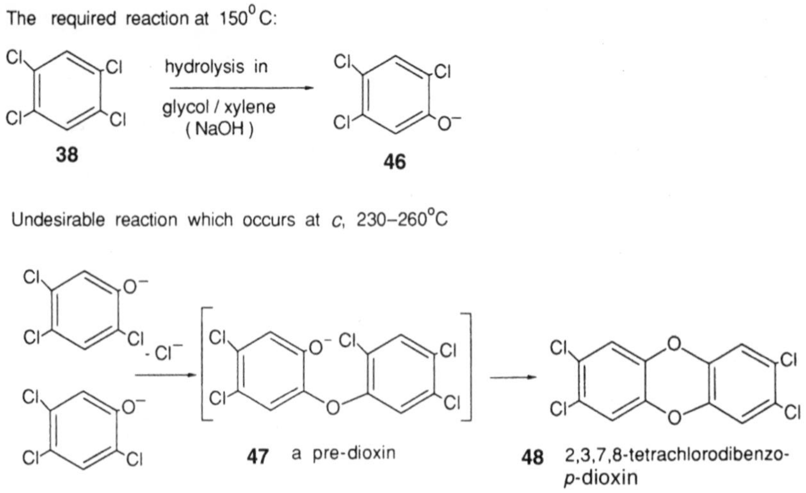

The required reaction at 150°C:

38 → **46**

Undesirable reaction which occurs at *c*, 230–260°C

47   a pre-dioxin   **48**   2,3,7,8-tetrachlorodibenzo-*p*-dioxin

Figure 6.21   The formation of dioxin during the preparation of 2,4,5-T.

incident at Séveso and the Agent Orange issue (Table 6.32) are of first importance and will now be treated as case histories. A more extensive history of issues related to dioxin and 'dioxin-like' compounds has been published (Webster and Commoner, 1994). The latter group of compounds includes PCDFs, PCBs, polyhalodiphenyl ethers and naphthalenes.

Agent orange.   This was the major material used for defoliation by the USAF during the Vietnam War (Young and Reggiani, 1988); others used early on were codenamed Agents White and Purple (Fig. 6.22). The mixture of herbicides in Agent Orange contained equal amounts of 2,4-D and 2,4,5-T (Fig. 6.16) and variable amounts of dioxins, depending on the source and the degree of control during preparation. The levels of 2,3,7,8-TCDD ranged between 2 and 50 µg/g. This was not appreciated when the programme began in January 1962, but opposition in the USA grew until following an investigation by the National Institute of Health in 1969 the last mission was flown in May 1970.

In fact the preparation of 2,4-D does not present the same risk as that from 2,4,5-T, as the side reaction of 2,4-dichlorophenol can only lead to dioxins of chlorine number less than four; this group is not dangerously toxic. It is worthy of note that the risk is now defined as a toxic equivalent factor (TEF) expressed as a fraction of 2,3,7,8-TCDD taken as unity. On this scale other dangerous individual compounds are 1,2,3,7,8-penta-chlorodibenzdioxin (TEF = 0.5) and 2,3,4,7,8-pentachlorodibenzfuran (TEF = 0.5). On this scale 2,3,7,8-TCDF has a TEF value of 0.1.

Concern about dioxins is largely based on the very small amounts required to produce birth defects in small mammals (Table 6.33). Guinea pigs fed 2,3,7,8-TCDD in corn oil at a dose of 6 mg/kg suffered lethality of

Table 6.32   Some dioxin pollution incidents

1949	A 'kettle' used for trichlorophenol production at the Monsanto plant in West Virginia blew and exposed workers were affected by chloracne, a severe and persistent form of acne known to be typical of contact with 2,3,7,8-TCDD (Kamrin and Rodgers, 1985)
1976	On 10th July overheating a reactor at the ICMESA plant at Séveso near Milan led to the release of a cloud of solvent containing trichlorophenol and TCDD. About 1800 ha were contaminated, and birds, animals and vegetation were seriously affected (Pocchiari *et al.*, 1983).
1977	Massive quantities of 'Agent Orange', an equal mixture of 2,4-D and 2,4,5-T, were sprayed from aircraft in the 1960s to defoliate supply trails used by the Vietnamese during the war with the USA. In the summer of 1977, 8 million litres of unused herbicide were destroyed by incineration at sea. It was estimated to contain 23 kg 2,3,7,8-TCDD (Young and Reggiani, 1988).
1981	Local authorities in the vicinity of Chesterfield, UK were concerned at the possible pollution of water supplies arising from historic dumping of chlorophenolic wastes by the Coke and Coalite Company.
1982	Deaths of horses and birds at a horse arena in eastern Missouri were attributed to the spraying of dioxin-containing waste oil to keep down dust. Humans were affected and the source was identified as a chemical plant in Verona, Missouri that had produced trichlorophenol for synthesis of the antibacterial agent hexachlorophene during 1970–1. Fish is nearby streams were also found to contain 2,3,7,8-TCDD (Kleopfer and Kirchmer, 1985). All the streets in the small town of Times Beach, MO were sprayed with the waste oil and it had to be closed down at a cost of $33 million. In 1991, the total cost of clean-up in Missouri was $80 million.
1982	A former dumpsite at Love Canal, Niagara, USA was found to be polluted. The canal was excavated in 1894 with a view to linking the upper and lower Niagara rivers but was never completed and it was used to dispose of chlorophenolic wastes produced by the Hooker Corporation. The site was subsequently developed as a residential area and later when the risks were realized, storm sewer sediments were found to contain up to 600 ppb of 2,3,7,8-TCDD (Kamrin and Rodgers, 1985; *cf.* p. 369).
1991–6	An investigation in Derbyshire, by the Ministry of Agriculture Fisheries and Food (1992), revealed the highest levels of dioxins in cows' milk ever found in the UK. The river Doe Lea passes through the grazing land near Bolsover and subsequent sediment samples from this stream were shown to contain dioxins at levels in the range 13–64 ppb TEQ. After four years, these levels had declined to within the range 0.47–1.2 ppb TEQ, probably by transport downstream. These problems were controlled by closure of an incinerator operated by the Coalite Chemical Company and through their installation under an Integrated Pollution Control authorization of an entrapment lagoon and biological treatment and filtration plant (National Rivers Authority, 1995). In February 1996 the Company was fined £500 000 for polluting the area around its works at Bolsover.
1994	There is concern about dioxins which are formed from residual cutting oils during the sintering of iron and steel: levels in the flue gasses at a German plant were in the range 24–67 ng TEQ/m^3. The gaseous flux from such a unit is typically 1×10^6 m^3/hour and it is likely that British Steel emissions, at about 590 g TEQ/year, are the largest source of dioxins in the UK (ENDS, 1995).

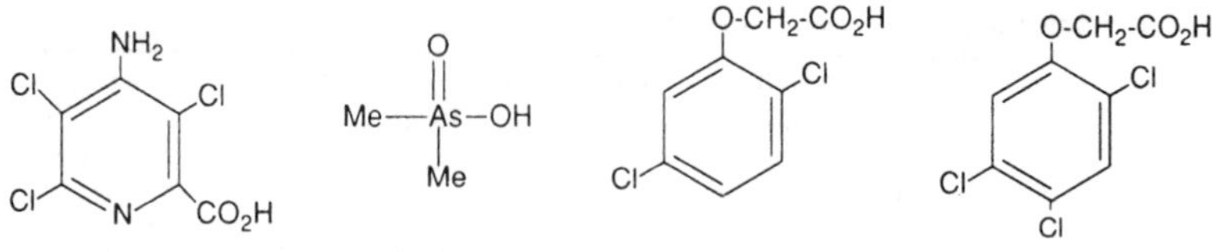

Figure 6.22 Herbicides used in Vietnam: all operations.

Table 6.33 Teratogenetic effects of 2,3,7,8-TCDD on mice and rats (Meyers, 1988)

	Dose (µg/kg)	Form administered	Effect[a]
Rat	0.5	Oral	Internal bleeding
	>1.0	Subcutaneous	Kidney damage
Mouse	1–3	Subcutaneous	Kidney damage
Hamster	Also suspectible at low µg/kg doses		

[a] Determined by the ED_{50}, the dose required to produce a positive effect in 50% of the sample.

Table 6.34 Comparative toxicities (*cf.* Table 3.2, p. 67)

Material	Probable lethal dose in humans (mg/kg)
Cacodylic acid	500–5000
Picloram	500–5000
2,4-D	50–500
2,4,5-T	50–500
DDT	50–500
Strychnine	<5
Ethanol	5000–15000

five from eight animals. One death from such a group occurred at 1 mg/kg. LD_{50} values from ingestion of polluted soil samples were about one quarter those found in corn oil. Comparative toxicities of various compounds in humans are given in Table 6.34.

After the Vietnam War US veterans initiated actions against the Government claiming adverse affects on their health and that of their families as a result of contact with dioxins in Agent Orange. These actions were prompted in the first instance by a Colombia documentary report broadcast on television in 1978. The situation was complicated by the ban on 2,4,5-T by the recently established (1971) EPA, although at the time there was no evidence of long-term injury to humans. There followed a long and intensive screening of groups who had been exposed to Agent Orange with control groups who had not. This work was made possible by the high sensitivity of the methods of gas chromatography (Section 4.4) and included studies of residual dioxin in adipose tissue and blood. The outcome was that

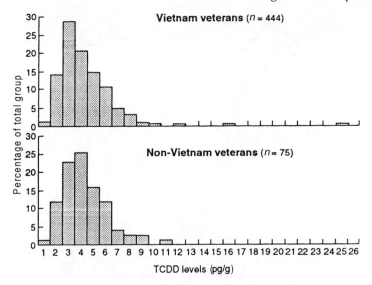

Figure 6.23 Levels in serum of 2,3,7,8-TCDD in Vietnam veterans (444) and of a control group (75) (Lathrop, 1988).

no distinction could be drawn between the two groups and one such finding is shown in Fig. 6.23. It is evident from this study that both the distribution pattern and the dioxin levels are essentially the same.

However, in 1984 seven manufacturers of Agent Orange eventually paid $180 million in an out-of-court settlement, but there was no admission of responsibility, no case law and no Federal compensation.

The Séveso incident (10 July 1976). In contrast to the use of defoliants in Vietnam, which affected an area of 1.3×10^6 hectares over a period of 7 years, the Séveso pollution was local to the ICMESA factory, which then ceased production (Pocchiari *et al.*, 1983).

In the first half of 1976, production of 2,4,5-trichlorophenol had risen to 142 000 kg and a typical reactor batch contained (values in kg):

sodium trichlorophenate	2000
ethylene glycol	1000
ethylene diglycol	1300
sodium hydroxide	360
sodium chloride	540

It was estimated that of a total load of 5200 kg there remained 2900 kg containing 600 kg of chlorophenolate. TCDD release was put in the range 13–60 kg. The analytical detection limit of the dioxin was $0.75 \, \mu g/m^2$ and it was found at a maximum limit in soil at $20 \times 10^3 \, \mu g/m^2$.

On the day, with lack of temperature control, solvents under pressure blew out the safety valve and a cloud of reagents rose to a height of 50 m

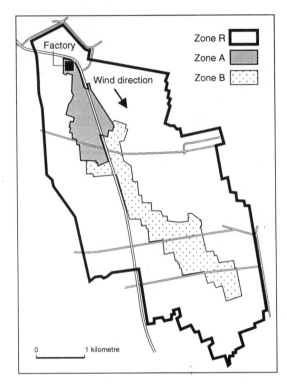

Figure 6.24   The zones affected by the ICMESA dioxin leakage. Zone A, 110 ha, population 730, TCDD in soil 400 ppt; zone B, the area subject to area-specific regulation; zone R, 1430 ha, subject to area-specific regulation (from Pocchairi *et al.*, 1983).

and was carried on a south-easterly wind over the lightly populated area (Fig. 6.24).

The background level of TCDD outside the three control zones was 1–2 ppt. Unfortunately, the problem was exacerbated by a delay in enforcing control measures. Sixteen days elapsed before the population was evacuated from the highly polluted zone A and the opportunity to fix chemicals where the plume had settled on long grasses and plants was lost when a heavy rainstorm carried the material into surface soil. Ideally the procedure would be to spray the grasses with an aqueous dispersion of vinyl acetate; this sets to a polymer so trapping the pollutants, which may then be removed by harvesting the grass or other crop.

Some homes in zone A were reclaimed, floors and walls in the school were cleaned daily until surface levels fell below $10 \, \text{mg/m}^3$. To prevent the export of chemicals, zone B was regulated by a ban on the breeding of animals or the planting of vegetables. Children and pregnant women were required to evacuate the area by day and there was to be no procreation. Milan was 70 km to the south and there was concern to protect this densely populated area from the risk of exposure of the dioxin. The cost of the clean-up was borne by the Swiss parent company Hofmann La Roche.

A model operation was mounted to remove and incinerate the 2500 kg of residual chemicals in the plant and over 5 t of equipment was decontaminated by Hofmann La Roche specialist workers. In a major disposal, the less exposed artefacts and buildings were committed to a closely monitored landfill.

Toxic effects of dioxins

There follows a very brief summary of the extensive data shortly to be published by the US EPA (EPA, 1994).

1. Dioxin release is very largely the result of industrial practice.
2. The main route to humans is the ingestion of food containing minute quantities of dioxin-like compounds originally deposited from the air onto plants.
3. The average human body burden is 40–60 ppt PCDF, PCDD and PCB, although in 10% of individuals levels may be three times this.
4. Humans fall in the middle range of animal sensitivity rather than at either extreme: it follows that the extremely high teratogenicity found in small mammals (Table 3.2, p. 67) is not a reliable guide to human experience.
5. Adverse effects, including cancer, may not be detectable until exposures exceed background by an order of magnitude or more.
6. The biological response depends upon the binding of dioxin-like substances to a cellular protein, the 'Ah receptor', which then translocates the chemicals to the nucleus where they bind to DNA. This is similar to the action of toxic aromatic hydrocarbons (Section 6.2.5) and leads to changes in gene expression (Schecter, 1994). It follows that dioxin-like compounds are likely to initiate many biochemical mechanisms leading to a range of toxic effects, including liver damage, birth defects and cancer.

In the case studies mentioned above involving Vietnam veterans and a control group, there was no evidence of toxic effects. A continued study of individuals exposed at Séveso failed to detect any abnormalities in the 5 years following the incident, where especial attention was paid to the abortion rate.

Over this period there were 12 680 live births from 14 690 pregnancies – 4.9% of women of childbearing age were pregnant, which was lower than the national average but this was attributed to social rather than biological factors; there was a general decline throughout the Lombardy region. The abortion rate of 19% was high but just within a range worldwide of 10–20%, while 3.5–6.7% of children had birth defects also within the accepted range of the WHO classification.

Carcinogenicity.   There is only one record of direct toxicity testing of TCDD on humans. This unethical study involved inmates of Holmesberg prison in Pennsylvania and the effect on most of the individuals was not recorded. Evaluation, therefore, depends upon epidemiology: the analysis of disease in statistically significant exposed populations and control groups.

Media comment frequently implies that dioxins are carcinogenic to humans and there was a report in 1979 by Hardell and Sandstrom that raised the possibility of induction of relatively rare soft tissue cancer. This was subsequently discounted by the Evatt Commission adjudicating cases in Australia and also by Sir Richard Doll.

Studies of US veterans showed that four deaths could be attributed to cancer out of 44 000 individuals who had been in Vietnam. A rate not significantly different from deaths from the same cause in a control group of 78 000 other veterans who had not been exposed.

In the zone B of Séveso, soil levels of TCDD sampled to a depth of 7 cm, were up to $5 \mu g/m^3$. A group of adults who had been resident here was compared with a control group from the low toxicity zone R. A statistically significant increase in two specialized cancers was claimed while uterine and breast cancers were reduced in the test group. There were no statistically significant increase in cancers amongst the 0–19 year age group resident in zone B.

On this controversial topic, the EPA concluded in 1985 that 'TCDD is a proven animal carcinogen and a probable human carcinogen'. An important later study was published by the National Institute of Occupational Safety and Health (NIOSH, 1991), which reviewed a cohort of over 5000 people, from 12 different sites, involved in the production of TCDD-containing chemicals. A total of 265 deaths from cancer were recorded against 229.9 expected, a significant increase of 15% mortality. The EPA verdict reached in 1985 is unchanged in 1995 following the NIOSH report and considerable other research.

Risk assessment.   Aside from a possible link to cancer there is evidence that at some levels TCDD, through its interaction with DNA, induces adverse health effects in humans. These include low sperm levels and immunosuppression and occur at levels close to those which affect laboratory animals.

The situation is complicated by the fact that humans are inevitably exposed to contact with the other dioxin-like toxins which also interact with the Ah receptor.

One method of assessing the risk of cancer assumes that there is no completely safe dose because induction occurs through a sequence of irreversible stages. The slope of the plot of effect against dose is linear even at low levels. Using this model and data from rat tumour induction, the EPA arrived at an acceptable daily dose TCDD of 0.006 pg/kg per day. The average citizen of an industrialized country is exposed to a level in the much higher range of 1–3 pg/kg per day.

The EPA requirement is disputed, for the US Food and Drug Administration has proposed a level, based on the same model, which is an order of magnitude higher. Other authorities have evaluated an aceeptable level based on the theory that there exists a threshold dose below which there is no risk of cancer. Using this model, Canada and the WHO both propose an acceptable daily dose of 10 pg/day, more than three orders of magnitude higher than that of the EPA and in excess of present average exposure.

Serious economic pressure will be experienced by any community which attempts to achieve the most stringent of these controls. As discussed above (Section 1.2), the cost of control increases exponentially as the level of control rises (Fig. 1.1). Up to about 20% control, the cost was less than $20/unit; at 65% control the cost rose sharply to $110/unit. As control approaches completeness, the cost becomes unsustainable; in the illustration it became $500/unit for 95% control.

The extremely stringent USEPA requirement for TCDD control falls in the upper 90% level and for it to be attained the inflated costs will have to be met by the wide range of industries which handle related chemicals. These include municipal and hazardous waste incineration, iron, steel and magnesium production, the recycling, degreasing and finishing of metals, paper pulp bleaching and other industrial users of chlorine.

6.4.12 Metabolism of chloroaromatic compounds

Pollutants of this group are susceptible to oxidation by P-450 cytochrome enzymes leading to the formation of arene oxides and subsequent excretion (p. 271). The PCBs are subject to this reaction but also undergo oxidation by direct hydroxylation (Fig. 6.25) (Preston *et al.*, 1983). Electronegative chlorosubstituents reduce the rate of oxidation, which depends on abstraction of electrons, and PCBs of higher chlorine number tend to accumulate in human tissues (Borlakoglu and Dills, 1991).

6.4.13 Disposal of organochlorine compounds

These substances are largely classified as hazardous or toxic wastes and as such must be disposed of by incineration and not directly to landfill.

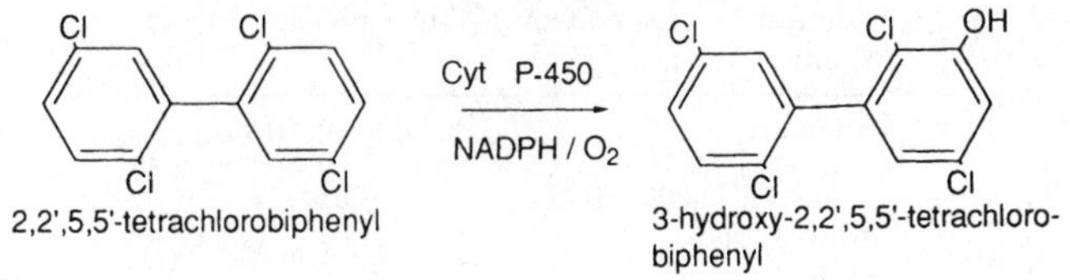

Cyt P-450
NADPH / O₂

2,2',5,5'-tetrachlorobiphenyl

3-hydroxy-2,2',5,5'-tetrachloro-biphenyl

Figure 6.25 Direct hydroxylation of a PCB.

Organochlorines are difficult to incinerate efficiently because of their low flammability and because of the range of reactions which can occur during combustion.

Toxic waste incinerators are operated 'in house' by major industrial producers or else toxic waste is transported to 'merchant incinerators'. In the UK, over 50 sites are registered with the HMIP who monitor their efficiency at the time of commissioning. Conditions for the incineration of PCBs require exposure to temperatures in the range 1000–1200°C with oxygen in 3% excess during a residence time of 1.5–2.0 s: in the USA, the EPA require an efficiency of 99.9999%. This is barely credible as it implies an accuracy of measurement of 0.0001% for a process which is difficult to analyse owing to the additional uncertainties of feed rate, positioning of the sampler and the extent to which the products are taken up on emitted particulates. With the eyes of the world on it, the incinerator ship Vulcan disposed of 10 000 t of Agent Orange with an efficiency of 99.9%, which left 0.1% or 10 t unaccounted for.

In the USA, the problem has been highlighted by objections by citizens' groups and by state legislatures to the incineration of chemicals from disposal of chemical weapons (Silton, 1993), including organophosphorus nerve agents (p. 317).

Incomplete combustion of PCBs leads to the formation of PCDDs and PCDFs by a mechanism which is similar to that shown in Fig. 6.20. In areas which are efficiently monitored, such as southern Ontario, the normal level in air of the TCDDs is about 10^{-5} ng/m^3. The average adult inhales 20 m^3/day, which at this level is $20 \times 10^{-5} \times 10^{-9} = 0.2$ pg/day; the intake from food will be greater by more than an order of magnitude. Annual emissions in such an environment have been estimated (Table 6.35) (Paterson *et al.*, 1990).

Municipal solid waste incinerators (MSW) are regarded as the principal producers of PCDD/PCDF (Oppelt, 1987). Complex reactions leading to unexpected products can occur within incinerators; for example, in the range 600–800°C, chloroform gives rise to hexachlorobenzene. Domestic waste contains aromatic precursors such as lignin and polystyrene; PVC provides a source of chlorine, and synthesis of PCDDs has been shown to occur near 600°C. These lower temperatures follow the addition of too high a proportion of vegetable matter to municipal solid waste. It was shown in a

Table 6.35 Estimated PCDD emissions in southern Ontario

Source	PCDD (g/year)
Industrial incinerators	65
Automobiles	> 67
Wood for fuel	1260
Municipal incinerators	4522

Quebec study (Finkelstein *et al.*, 1990) that unsatisfactory emissions could be brought within proper limits by changing the configuration, the air supply, and the controls: PCDD/PCDF emissions of up to 4000 ng/m^3 were reduced to 50 ng/m^3. A typical output is 30 000 m^3/hour, which operating inefficiently over a working year of 5000 hours amounts to:

$$4 \times 10^3 \times 3 \times 10^4 \times 5 \times 10^3 \text{ ng/yr} = 600 \text{ g/year}$$

6.4.14   Cremation or burial of wastes

Wastes can be either incinerated or buried (Ayres, 1987). In the UK at 1987 prices, landfill cost £5/t and incineration £15/t but the latter alternative is favoured by rising land values and shortage of suitable sites. Incineration reduces the bulk of domestic waste to one-fifth, an important factor when landfill amounts to 0.5 t per head per year. Although the bulk for disposal to landfill is reduced, the fly ash from municipal incinerators is much higher in PCDD/PCDF than the raw waste, a disadvantage brought about by synthesis within the incinerator. In a test group of six municipal units, the levels of TCDDs ranged between 3 and 2600 ng/g.

6.4.15   Use of biodegrading microorganisms

TCDD is attacked by light and by microorganisms (Section 2.6.1) and has a half-life of about 1 year in soil. Attempts to dispose of chlorinated waste by microorganisms has been researched (Bennett and Olmstead, 1992), but it is difficult to divert them from their natural metabolism. In addition, as discussed above (Section 6.4.12), highly chlorinated compounds are resistant.

Chlorinated lignin (Section 6.2.1) and chlorophenols in the bleaching effluent from paper mills have been successfully degraded by the white rot fungus *Phanerochaete chrysosporium* (Pellinen *et al.*, 1988).

6.5
Natural, organophosphorus and carbamate pesticides

In advanced countries, the use of organochlorine pesticides has declined and some are banned unless 'no suitable safe alternative exists'. Acceptable alternatives are commonly found among the group to be discussed, but unfortunately they cost about three times as much to produce; consequently, in Third World countries, DDT, Aldrin and Dieldrin continue in large-scale use.

6.5.1   Naturally occurring pesticides

Although chemists are often accused of introducing dangerous substances into the environment, very many toxic compounds are produced

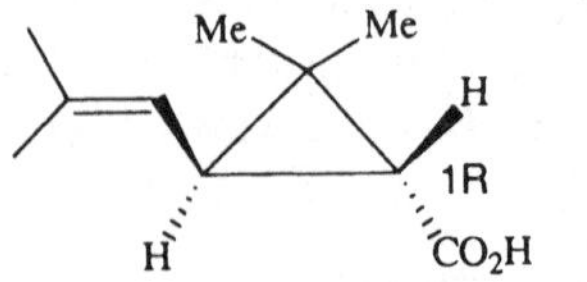

49 (1*R*)-chrysanthemic acid

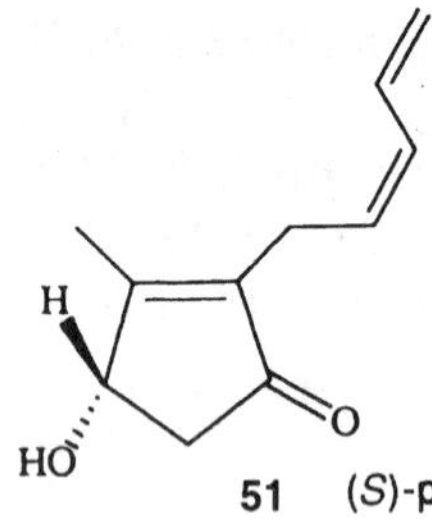

51 (*S*)-pyrethrolone

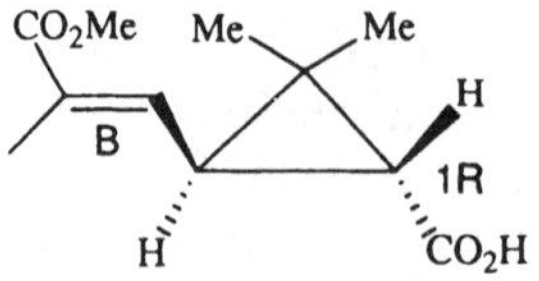

50
(1*R*)-pyrethric acid
(+) *trans*

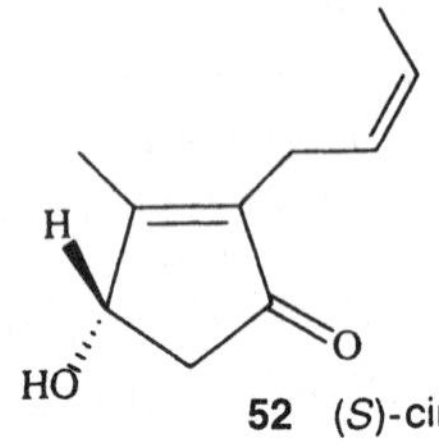

52 (*S*)-cinerolone

Figure 6.26 Principal constituents of the pyrethrins.

by plants and animals (Section 3.1). The natural pesticides provide an example of this.

Pyrethrins. These are obtained largely from the plant *Chrysanthemum cinerariaefolium.* The insecticidal activity of the flowers was known to the Chinese in the first century AD, but commercial production of pyrethrum did not begin until 1850 and the first evidence of their structure was obtained in 1920 by Ruzicka and Staudinger. About half the current world production comes from Kenya; other major sources are Russia and Japan. In 1983 users spent \$630 × 10^6 to treat 52 × 10^6 ha.

Each flower head contains 2–4 mg (*c.* 2%) of a mixture of active components; these are extracted with light petroleum. The four principal compounds are esters formed by combination of either of the acids, chrysanthemic (**49**) and pyrethric (**50**), with one or other of the alcohols, pyrethrolone (**51**) and cinerolone (**52**) (Fig. 6.26).

Structure–activity relationships. Key requirements relating structure to activity (Elliot and Janes, 1978) are: (i) a bulky group like *gem*-dimethyl two places removed from the ester function, and (ii) unsaturated centres at the two extremities of the molecule.

The activity is also related to the absolute configuration. Chrysanthemic acid, with its two asymmetric centres, can exist in four enantiomeric forms.

One of these has been shown above and the remaining three are structures (**53–55**).

53 (–) *trans* (S)

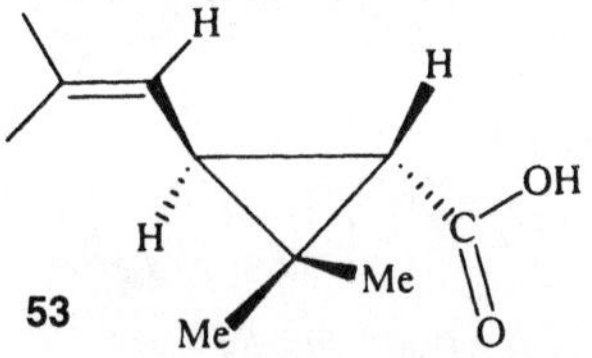

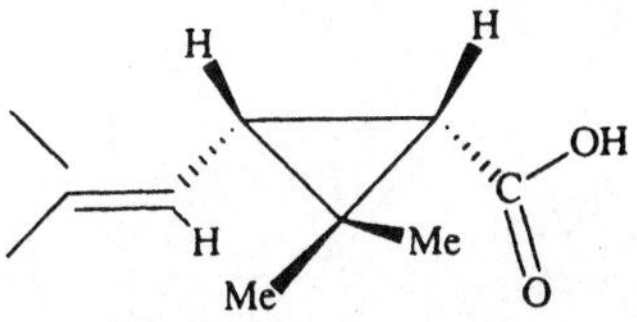

54 (–) *cis* (S)

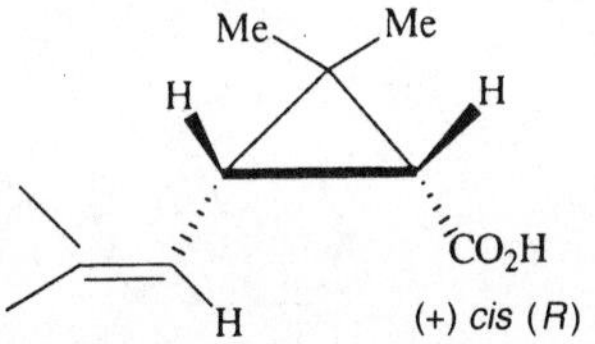

55 (+) *cis* (R)

The *cis*- and *trans*-esters of the dextrorotatory (+) (*R*)-acids are 20–50 times as active as those of the corresponding laevorotatory (−) (*S*)-isomers. Evidently when the key structural elements relate to this (*R*) configuration they can interact with enzymes of the target species.

The highly active pyrethrin 1(*R*) (**56**) is the pyrethrolyl ester of chrysanthemic acid; it meets the structural criteria for activity and has an LD$_{50}$ of 0.3 µg/insect. The (+)-*trans* ester is slightly more toxic to mammals than the (+)-*cis* analogue, with an LD$_{50}$ of 5 mg/kg in rats by the intravenous route. The result of the interaction with the target is similar to that which follows exposure to DDT and death follows from hyperexcitation of the nervous system.

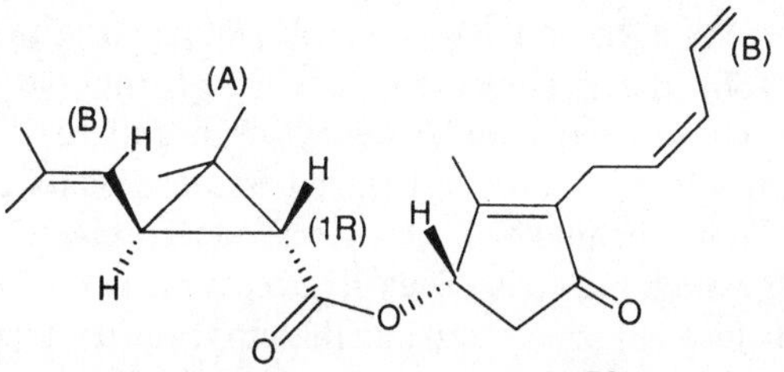

56

The advantages of the pyrethrins lie in their unsaturated aliphatic structures, which undergo rapid photolytically induced oxidation; this means that they have short half-lives. However, it also means that they are commonly mixed with small amounts of other pesticides, such as rotenone and resmethrin, to ensure that insects do not recover after they have been knocked down.

Synthetic analogues of pyrethrin. These have been extensively researched and a major contribution was made by Elliot and Janes at the Rothamsted Research Station (Fig. 6.27). The first of these compounds was Allethrin

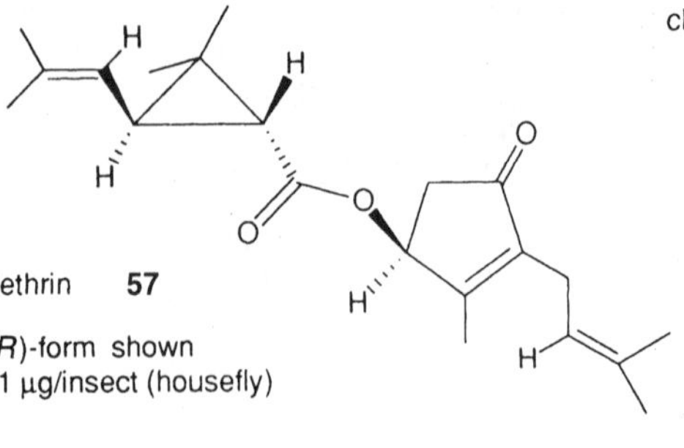

Allethrin **57**

(1*R*)-form shown
0.1 μg/insect (housefly)

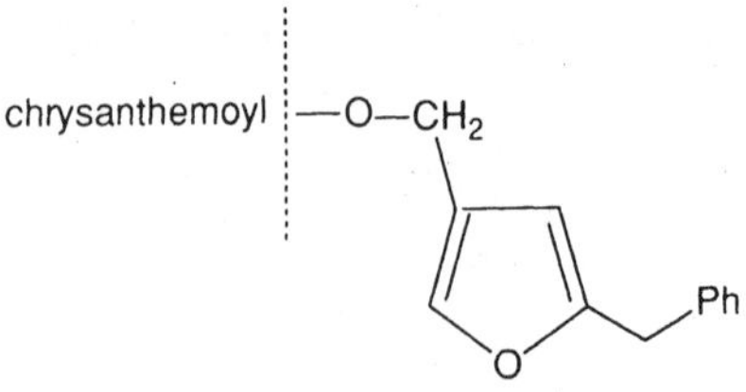

Bioresmethrin **58**
LD_{50} 0.005 μg/ insect

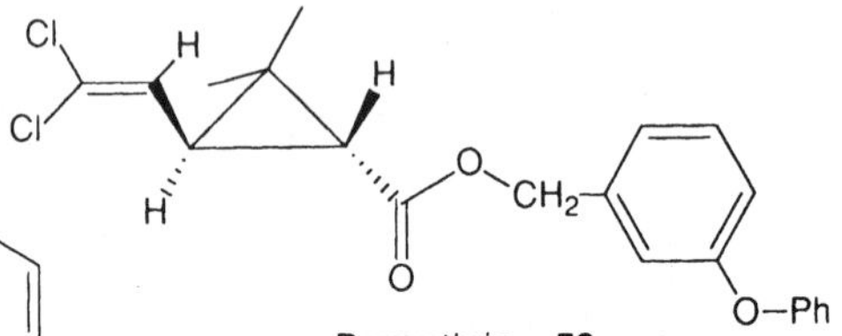

Permethrin **59**

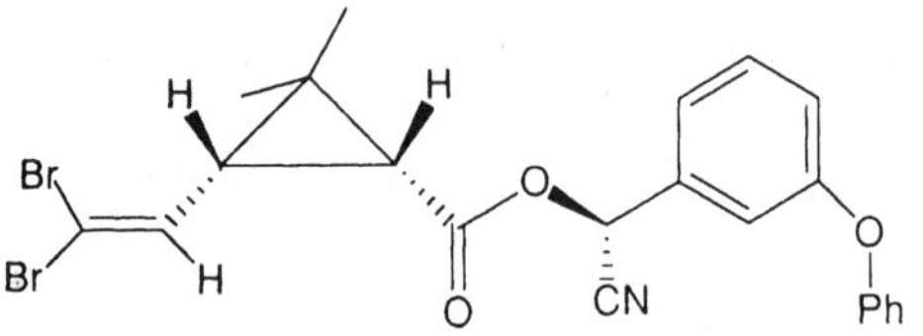

Deltamethrin **60**

Figure 6.27 Some synthetic pyrethroids.

(**57**) prepared from a racemic mixture of (*R*, *S*)-chrysanthemic acid. In Bioresmethrin (**58**) the cyclopentenone ring of the natural product is replaced by the structurally simpler benzylfuran and yet the compound is extremely potent, although still short-lived because of its photosensitivity. Some of the thinking behind the commercial development is illustrated by Permethrin (**59**), which includes the easily accessible phenyl ether group and the inclusion of two chlorine atoms in the chrysanthemic acid part. These electronegative substituents were seen (Section 6.4.7) to create too long a lifetime in polychlorocompounds; here they have the desirable effect of usefully extending the lifetime of an overly sensitive substance. Deltamethrine (**60**) is another example. The toxicity towards mammals remains low, although fish are more susceptible to poisoning as shown in Table 6.36, which gives the LD_{50} values for fish and rats.

Permethrin is supplied as a mixture (*R*, *S*) with geometrical isomer ratio in the range 40–50 : 60–50; it is marketed as an insecticide under several tradenames, for example Ambush (ICI), Cooper powder (Wellcome Foundation Ltd), Permit (Pan Britannica Ltd), as Rentokil Musk control and in mixtures with other substances. Deltamethrin is a single isomer.

Table 6.36 Toxicity data for some pyrethrins

Compound	Rat LD$_{50}$ (mg/kg)	Fish LC$_{50}$ (mg/l)	Comment
Pyrethrin	600–900[a]	114 (catfish)	Toxic to bees
Allethrin	1100	17 (trout)	
Bioresmethrin	7000	0.5 (guppy)	Toxic to bees
Permethrin[b]	400–4000	0.009 (trout)	
Deltamethrin[c]	1000	0.0012 (sunfish)	harmless to bees outdoors

[a] Depends on sex, age and *cis/trans* ratio.
[b] Half-life in soil < 38 days.
[c] Biodegrades in soil within 1–2 weeks, in plants in 10 days.

When the complexity of these structures is compared with that of DDT, it can be appreciated that their synthesis is more costly than the route to DDT, depending as this does on cheap, readily available starting materials (Section 6.4.3).

Other naturally occurring pesticides.   Although they are less important than the pyrethrins, mention should be made of the rotenoids obtained from the roots of *Derris elliptica*, of which rotenone (**61**) is a typical component (Crombie, 1980). Although toxic to fish and many insects, the rotenoids are not harmful to mammals; the rotenone LD$_{50}$ to rats is 135 mg/kg. Its principal use today is in horticulture for the control of aphids, caterpillars, sawflies and wasps. The symptoms of insect poisoning by rotenoids differ from the hyperactivity associated with DDT and the pyrethrins.

Rotenone **61**

(S) (-) -Nicotine **62**
b.p. 247° [α]$_d$ 161°
sol. in water

Also of interest is nicotine (**62**), first shown to be the active principle of tobacco extracts in 1828. The naturally occurring active isomer shown is the (−)(S) form. Its use has declined rapidly because of its high mammalian toxicity. The heart rate of adult humans responds to as little as 2 mg of the bitartrate, equivalent to 0.5 mg nicotine. Serious poisonings were frequent and in extreme cases lead to death from respiratory failure; the estimated fatal dose in a human adult is about 100 mg.

Nicotine becomes bound to the red cells in the blood of smokers and reaches levels in the range 0.4–2.0 µg/g. It is linked to heart disease and may be excreted in human milk at a level of 0.5 µg/g. The action of the molecule is critically linked to its dimensions, leading to its interference with cholinesterase in a manner which is discussed in the next section.

6.5.2   Organophosphorus pesticides

Organophosphorus pesticides (Fest and Schmidt, 1982) of commercial value were developed from work on nerve poisons which was carried out during the Second World War by Saunders in Cambridge and Schrader in Germany. The simplest of these compounds is tetraethyl pyrophosphate (**63**), which is highly toxic to mammals, as is the warfare agent Sarin (**64**).

63 TEPP   **64** Sarin   **65** choline

The mechanism of action.   The toxic action takes place at the synapses, or junctions between nerve cells, where impulses are transmitted chemically through the acetylation of choline (**65**). When functioning normally this reaction is rapidly reversed through the action of cholinesterase, so that stimulation ceases and the site is ready to receive a new impulse. The nerve poisons act by inhibiting cholinesterase, which leads to the failure of nervous transmission, convulsions and death. The action of nicotine is similar in that its critical dimensions are sufficiently close to those of acetylcholine for this to be displaced at the synapses, so disrupting the nervous system.

The simplified scheme (Fig. 6.28) shows how amino acid residues on the peptidic surface of the enzyme interact in a stereo-dependent way with acetylcholine. The process can for clarity be separated into steps:

1. The positively charged end of the substrate is attracted to a carboxylate ion on the surface.
2. A specific acidic group of the enzyme forms a hydrogen bond to the acetoxy group so exposing it to nucleophilic attack.
3. The acetyl group is thereby transferred to a serine residue on the enzyme and choline is regenerated (Equation 6.22).

$$\text{Enzyme---CH}_2\text{-OH} \quad \xrightarrow{-\text{H}^+} \quad \text{Enzyme-CH}_2\text{-O} \overset{O}{\underset{}{\|}} \text{CH}_3$$

$$\text{(6.22)}$$

$$+ \text{ choline}$$

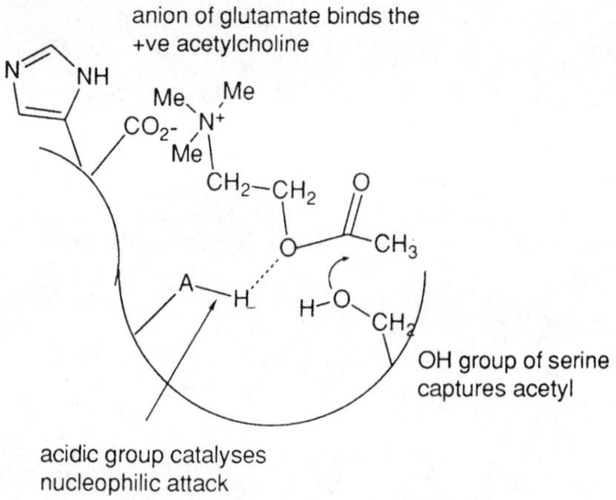

Figure 6.28   Interaction of acetylcholine and cholinesterase.

Sarin and related phosphorus compounds interfere with the action of the enzyme by converting the serine into a type of phosphorus ester. Unlike O-acetylation, this reaction (Equation 6.23) is only slowly reversed and so the inserted group blocks the normal transmission of impulses.

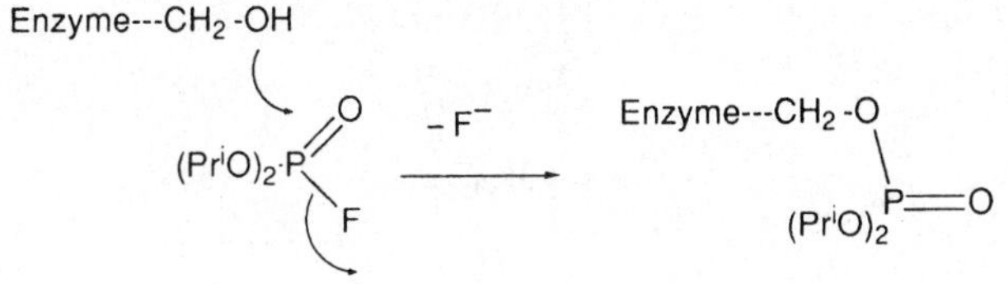

$$(6.23)$$

Some individual compounds

This section concludes with mention of carbamate pesticides. They possess an ester function and react in a very similar manner (Equation 6.24) with the insertion of a blocking group. Those singled out for comment have been chosen to illustrate a range of properties from among those of greatest commercial importance (Fig. 6.29).

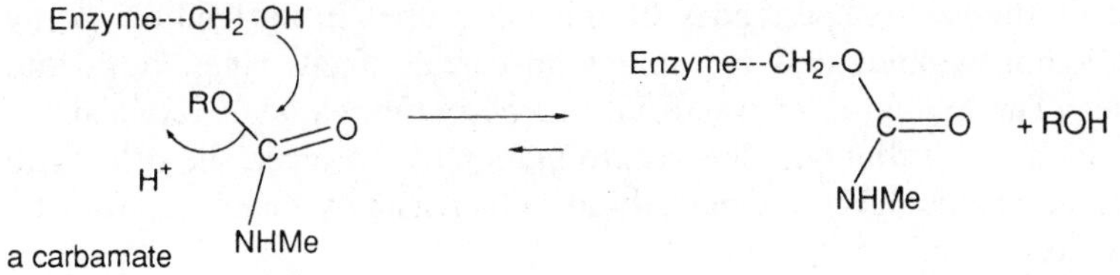

$$(6.24)$$

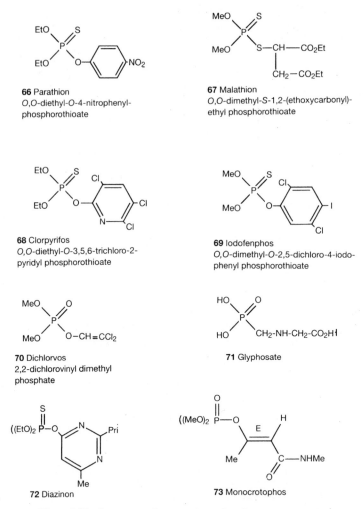

66 Parathion
O,O-diethyl-O-4-nitrophenyl-phosphorothioate

67 Malathion
O,O-dimethyl-S-1,2-(ethoxycarbonyl)-ethyl phosphorothioate

68 Clorpyrifos
O,O-diethyl-O-3,5,6-trichloro-2-pyridyl phosphorothioate

69 Iodofenphos
O,O-dimethyl-O-2,5-dichloro-4-iodo-phenyl phosphorothioate

70 Dichlorvos
2,2-dichlorovinyl dimethyl phosphate

71 Glyphosate

72 Diazinon

73 Monocrotophos

Figure 6.29   Structures of some organophosphorus compounds.

Parathion.   The ethyl ester Parathion (**66**) was discovered by Shrader in 1946 and was introduced in 1952 by ICI Plant Protection Ltd. Parathion was one of the earliest pesticides of this class used in agriculture, being effective against aphids and red spider on cereals, fruit, vines, hops and vegetables. The methyl ester was found to be preferable as it retained the activity against pests but with less risk to mammals; however, the risk is still considerable (Table 6.37) and Parathion is increasingly being replaced by less toxic compounds.

Table 6.37 Toxicity data for some organophosphorus pesticides

Compound	Oral LD$_{50}$ rat (mg/kg)	LC$_{50}$ fish (mg/l)	Persistence	Comment
Parathion (Et)	14–24	2.7 trout	2–8 days on plant, longer elsewhere	Dimethyl safer MRL[b] 0.2 fruit and vegetables
Malathion	2800[a]	0.1 bluegill	Excreted in <24 h	MRL 8.0 cereals 0.5 fruit and vegetables
Clorpyrifos	130–160	0.003	$t_{1/2}$ soil 60–120 day	MRL 0.3 citrus 0.2 vegetables
Iodofenphos	2100	0.1 trout	–	Systemic
Dichlorvos	60–80	0.17 bluegill	$t_{1/2}$ water 20–80 h $t_{1/2}$ mammals 25 min	MRL 2.0 wheat, 0.1 fruit and vegetables
Monocrotophos	15	5.0 trout	–	–
Diazinon	250–320	2.6 trout	–	Strongly adsorbed onto soil
Glyphosate	5600	120 sunfish	Low	Rapidly excreted by mammals

[a] The oxyphosphate, Malaoxon, has an LD$_{50}$ of only 90 mg/kg.
[b] EC limit (μg/g).

Parathion is synthesized by the reaction:

$$(EtO)_2 \cdot P \overset{S}{\underset{Cl}{\diagup}} + Na^+ \; {}^-O\!-\!\langle \rangle\!-\!NO_2 \longrightarrow (EtO)_2 \cdot P \overset{S}{\underset{O-\langle \rangle-NO_2}{\diagup}} + NaCl \qquad (6.25)$$

which resembles the preparation of a carboxylic ester from an alkoxide and the acid chloride. Since Parathion is an ester of thiophosphoric acid, it is rapidly detoxified in the environment by hydrolysis:

$$(EtO)_2 \cdot P \overset{S}{\underset{O-Ar}{\diagup}} \xrightarrow{H_2O} (EtO)_2 \cdot P \overset{S\,(O)}{\underset{OH}{\diagup}} + \underset{nitrophenol}{Ar\text{-}OH} \qquad (6.26)$$

The thioesters do not react so readily with cholinesterase and are not pesticides, but the mixed function oxidases (MFO) of insects convert them into the toxic oxyesters and so encompass their own destruction:

$$\underset{non\text{-}toxic}{(RO)_2 \cdot P \overset{S}{\underset{O-R'}{\diagup}}} \xrightarrow{MFO} \underset{toxic}{(RO)_2 \cdot P \overset{O}{\underset{O-R'}{\diagup}}} \qquad (6.27)$$

Parathion and other phosphorothioates are known as proinsecticides, they are only toxic to those animals whose MFO can effect the change into the oxyphosphate (Equation 6.27).

The USEPA has set a limit in drinking water of 9 μg/l for the ethyl ester, and a limit of 0.2 μg/l in workplace air is effective in the UK and USA.

Malathion. Unlike Parathion, Malathion (**67**, also introduced in 1952) is not toxic to mammals (Table 6.37); consequently it is used for the protection of a wide range of crops including cotton, fruit, potatoes, rice, vegetables, stored grain and it is also effective against mammalian parasites. However, the toxicity of impurities and transformation products may exceed that of Malathion itself, for example:

74 Isomalathion
LD_{50} rat 113 mg/kg

75 Malaoxon
LD_{50} rat 158 mg/kg

The maximum permitted level of Malathion in drinking water in the UK is $7 \, \mu g/l$ and in the EC it is $0.1 \, \mu g/l$. The workplace limit in air in the UK and the USA is a TWA of $10 \, mg/m^3$. Malathion is marketed under the cloak of 29 different trade names.

Chlorpyrifos. Chlorpyrifos (**68**) was introduced in 1965 for use in major crops such as citrus fruits, coffee, cotton, maize, and also against the mosquito. The dose rate is in the range $200-1200 \, g/ha$ and in soil it persists for only 60–120 days, being cleaved by hydrolysis into phosphate and trichloropyridine-2-ol.

Chloropyrifos is a major import into South America, where uncontrolled use in banana plantations has led to river pollution. In Costa Rica, cucumbers grown downstream of these sources included 8 ppm Chlorpyrifos and the International Water Tribunal ruled that proper controls be introduced. In the workplace in the USA and UK the maximum permitted TWA in air is $0.2 \, mg/m^3$.

Iodofenphos. Like the above compounds, Iodofenphos (**69**) is a contact insecticide but it also has systemic properties. That is to say it is translocated in plants and persists long enough to make the host itself toxic to insects that eat or suck it, but not to those that merely alight on it. It was introduced in 1969 and has an important application in poultry houses.

Dichlorvos. This is a low molecular weight compound (**70**) and is very volatile, which leads to its incorporation into slow release insecticidal strips for the control of flies, moths and mosquitoes. It is also effective in plants against sucking insects and as an anthelmintic. Although appreciably toxic to mammals, it is not persistent, being rapidly hydrolysed, with a half-life in human blood of 11 minutes.

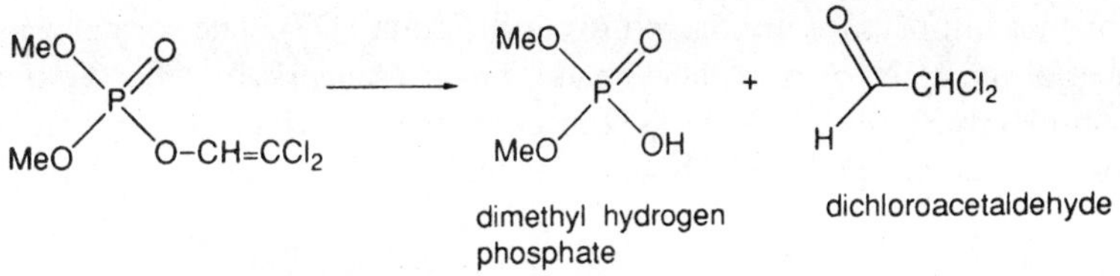

MeO—P(=O)(OMe)—O—CH=CCl₂ ⟶ dimethyl hydrogen phosphate + dichloroacetaldehyde   (6.28)

Structures shown: starting dimethyl phosphate ester with O–CH=CCl₂ group reacts to give dimethyl hydrogen phosphate (MeO, MeO, P=O, OH) plus dichloroacetaldehyde (O=CH–CHCl₂).

It is further degraded by the action of MFO to Cl^-, phosphoric acid and CO_2.

Monocrotophos.   Monocrotophos (**73**) is an insecticide and acaricide which is very toxic if swallowed, it also penetrates the skin. India uses large quantities for control of cotton pests; in 1988–9, the 655 tons imported amounted to 45% of the total pesticides. There have been numerous incidents of poisoning arising from its misuse and in the Parana state of Brazil in 1990, 412 incidents were recorded, of these 107 (25%) were attributed to monocrotophos and 93 (22%) to Parathion.

The limit in workplace air in the USA is a TWA of $0.25\,mg/m^3$.

Diazinon.   The use of diazinon (**72**) is mainly to control pests that have become resistant to organochlorines, for instance it is used in South Africa against locusts in place of BHC. It is also employed to control sucking and chewing insects on fruit, vines, vegetables and cereals.

Diazinon is strongly adsorbed onto soil. The principal metabolities are diethyl thiophosphate and diethyl phosphate; on storage it gives the more toxic Sulfotep.

Glyphosate.   The activity of glyphosate (**71**) was first reported in 1971. It is used as a herbicide and is widely effective at doses of 0.3 kg/ha for annual weeds and up to 2.2 kg/ha for perennials; in soil it acts as an insecticide. Common trade names are Tumbleweed and Touchdown; there are over 20 others.

The toxic action differs from the thiophosphate esters and depends upon the fact that glyphosate is a zwitterion and a modified glycine. It acts as a glycine mimic and becomes accepted into peptides where it blocks normal development. Glyphosate is not carcinogenic and is excreted unchanged by mammals; the half-life in soil is less than 60 days. The maximum level permitted in drinking water in the UK is 0.2 mg/l.

Disposal of chemical warfare agents

Because of their extreme toxicity, the organophosphorus nerve poisons are of major importance and their safe disposal is a matter of concern (Yang, 1995).

Two of first importance are Sarin (**76**) and Tabun (**77**). The former was used in the fatal terrorist attack made on passengers on the Tokyo underground on 20 March 1995. These compounds, which are esters of phosphorus acids, are best hydrolysed and solubilized in alkaline solutions:

$$
\begin{array}{cc}
\text{Pr}^i\!-\!\text{O}\!-\!\overset{\displaystyle\text{O}}{\overset{\|}{\underset{\underset{\text{Me}}{|}}{\text{P}}}}\!-\!\text{F} & \text{Me}_2\text{N}\!-\!\overset{\displaystyle\text{O}}{\overset{\|}{\underset{\underset{\text{OEt}}{|}}{\text{P}}}}\!-\!\text{CN}\\[2mm]
\textbf{76}\ \text{Sarin} & \textbf{77}\ \text{Tabun}
\end{array}
$$

For example sarin is treated with excess sodium hydroxide in aqueous alcohol at ambient temperature:

$$
\text{Pr}^i\!-\!\text{O}\!-\!\overset{\text{O}}{\overset{\|}{\underset{\underset{\text{Me}}{|}}{\text{P}}}}\!-\!\text{F} \longrightarrow \text{Pr}^i\!-\!\text{O}\!-\!\overset{\text{O}}{\overset{\|}{\underset{\underset{\text{Me}}{|}}{\text{P}}}}\!-\!\text{OH} \xrightarrow{\text{Na}^{\oplus}\text{OH}^{\ominus}} \text{Pr}^i\!-\!\text{O}\!-\!\overset{\text{O}}{\overset{\|}{\underset{\underset{\text{Me}}{|}}{\text{P}}}}\!-\!\text{O}^{\ominus}\ \text{Na}^{\oplus}
$$

HO$^{\ominus}$

Sodium salt of methyl phosphonic acid

(6.29)

Similar alkaline conditions can be used to dispose of mustard gas (**78**).

Cl.CH$_2$.CH$_2$ — S — CH$_2$ ... CH$_2$—Cl   in alkali →   Cl.CH$_2$.CH$_2$ — S$^{\oplus}$ — CH$_2$ + Cl$^{\ominus}$

Mustard gas **78**

HO.CH$_2$.CH$_2$ — S — CH$_2$.CH$_2$.OH $\underset{+\text{OH}^{\ominus}}{\overset{-\text{Cl}^{\ominus}}{\longleftarrow}}$ [Cl.CH$_2$.CH$_2$— S — CH$_2$.CH$_2$.OH]

Thiodiglycol

(6.30)

6.5.3   Carbamate pesticides

There are currently about 40 of these carbamate pesticides; they have a substantial share in the market and many of them are systemic in plants. Their activity against cholinesterase (Section 6.5.2) was first suspected in the naturally occurring alkaloid physostigmine (**79**; Fig. 6.30). This is, however, an ionizable base and derived ions cannot penetrate the sheath which protects the nervous system of insects. Those compounds which have come into use are, therefore, not ionizable and have the general structure (**80**).

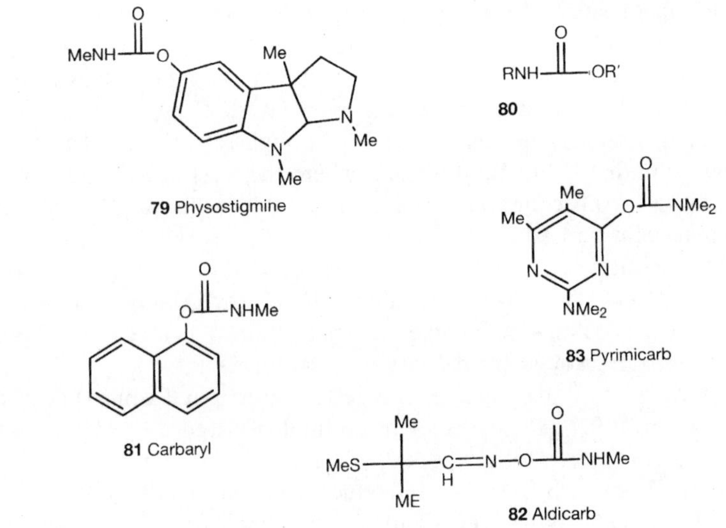

Figure 6.30 Carbamates and carbamate pesticides.

Notice that this resemblance to acetylcholine favours it in competition for the sites on cholinesterase (Fig. 6.28).

Carbaryl. Carbaryl (**81**), a cheap and widely used compound, was introduced as early as 1956 by the Union Carbide Company. It is synthesized by the reaction of α-naphthol with phosgene, which is essentially a dibasic acid chloride. This means that after the first esterification step, an amide link can be established by reaction with the amine in the second step:

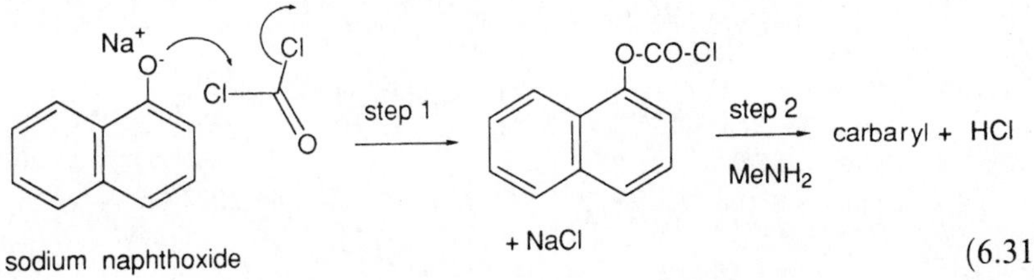

$$(6.31)$$

Carbaryl is effective against pests in cotton, fruit and vegetables and can be used as a substitute for DDT. The mammalian toxicity is moderate, for the rat the LD_{50} is 500 mg/kg while 200 µg/g in the diet showed no effect over 2 years.

Aldicarb. The best known of the carbamoyl oximes, Aldicarb (**82**) was indroduced by the Union Carbide Company in 1965 for use against aphids, nematodes, flea beetles, thrips and whitefly. It is rapidly metabolized to the sulphoxide and sulphone, which retain the activity. Aldicarb is highly toxic to mammals, with an oral LD_{50} in the rat of 1 mg/kg and is dangerous to fish, game, wild birds and animals.

The water solubility is high at 6 g/l, although it was not detected in groundwater until 1979 in Long Island, where the level in half the samples exceeded 7 µg/l. Subsequent sampling of potable wells in 34 US states showed that eight had Aldicarb residues.

In 1985, uncontrolled spraying of water melons led to the largest recorded outbreak of food-borne pesticide illness in North America. Levels of 0.2 ppm may cause illness and some melons contained as much as 2.7 ppm: 1170 cases were registered from Alaska to California.

In the UK the drinking water level is restricted to 7 µg/l while that of the EC is set lower at 0.1 µg/l; the USA has a limit of 3 µg/l.

Pirimicarb. Pirimicarb (**83**) was introduced in 1968 by ICI after it was discovered that carbamates which had a pyrimidine carrier group were less toxic to mammals: for Pirimicarb the oral LD_{50} in rats is about 150 mg/kg. It is more expensive than carbaryl but is effective against aphids including those which have become resistant to organophosphorus insecticides.

Atrazine and Simazine. Mention must also be made of two triazine relatives of Pyrimicarb, Atrazine (**84**) and Simazine (**85**). Atrazine was introduced in 1958 by Ciba-Geigy AG and is used as a pre-emergence weed killer in corn and for the culture of sugar cane and pineapples; it has a half-life in soil of 30 days or less, depending on soil type. The LD_{50} in rat, mouse and rabbit lies in the range 750–3080 mg/kg and the LC_{50} in fish in the range 4.3–8.8 mg/l.

84 Atrazine
water sol. 70 µg/g at 25°C
m.p. 175°C

85 Simazine
water sol. 5 µg/g at 20°C
m.p. 227°C

Simazine was first reported in 1956 by A. Gast *et al.* and was also introduced by Ciba-Geigy AG. It has systemic action against annual grasses and broad-leaved weeds in fruit, vines, olives, some vegetables, maize, tea, coffee, cocoa, etc. It is decomposed by UV light and has a half-life in soil of about 75 days.

Simazine has an LD_{50} of over 5 g/kg in the rat and the LC_{50} in rainbow trout is over 100 mg/l.

Table 6.38 Insecticides whose use is banned in the USA (Magnus Francis, 1994)

Pesticide	Year	Reason
Aramite	1955	Carcinogen
Aminotriazole	1959	Carcinogen
Alkylmercury fungicides	1970	Chronic toxicity
DDT and DDD	1971	Bioaccumulation
Strychnine and sodium fluoroacetate	1972	Acute toxicity
Aldrin and Dieldrin	1974	Carcinogens, bioaccumulation
Heptachlor and Chlordane	1976	Carcinogens, bioaccumulation
Strobane	1976	Carcinogen
Mirex	1976	Bioaccumulation
Leptophos	1976	Delayed neurotoxicity
DBCP	1977	Male sterility
Chloranil	1977	Carcinogen
Chlordimeform	1977	Bladder carcinogen
Chlordecone (kepone)	1978	Neurotoxicity
BHC	1978	Off-flavours
2,4,5-T and silvex	1979	Dioxin contamination
Endrin[a]	1979	Fish toxicity
Chlorobenzilate	1979	Carcinogen, male sterilant
Pyriminil	1980	Pancreatic poison
Nitrofen[b]	1980	Teratogen
Aldicarb	1981	Water pollutant
EDB	1983	Carcinogen
Toxaphene	1983	Carcinogen, fish toxicity
Dicofol	1986	Bioaccumulation
Dinoseb	1986	Teratogen
Alachlor	1987	Carcinogen, water pollutant
Chlordane	1987	Indoor air pollutant
Parathion-ethyl[c]	1991	Acute toxicity

[a] Banned east of the Mississippi only.
[b] Technically, nitrofen was *withdrawn* by the manufacturer.
[c] Most uses banned by US EPA.

Despite their low toxicity, both Simazine and Atrazine have their water solubility enhanced by the action of soil organisms, which leads to the replacement of the chloro-substituent by hydroxyl. Water solubility is also increased by the action of UV light, which results in dealkylation. As a result, these metabolites have polluted supplies in the Thames Valley and consequently both Atrazine and Simazine have been placed on the red list in the UK.

Annual usage of Atrazine in the 14 mid-western states of the USA is about 21×10^6 kg, and after spring application groundwater levels may rise in June to 10 times the EPA limit of 3 µg/l. Under similar conditions, the Simazine level in groundwater in the cornbelt can rise to 7 µg/l, compared with an EPA limit of 1 µg/l.

Some EC countries have banned these triazines because of their penetration of water supplies. Levels in the chalk aquifers in the UK lie in the range 20–400 ng/l and the present approved level for combined s-triazines in drinking water is 30 µg/l.

For a survey of the numerous other carbamates and other specialized pesticides the reader is referred to Cremlyn (1991). Table 6.38 lists those insecticides that have been banned in the USA.

6.6 Odours

Odours play an important role in people's lives and are a very complex sensation. With regard to pollution, however, the main considerations are unpleasant odours, their sources and control. The odours which cause the most complaints are mainly associated with livestock production, industries rendering animal wastes, waste disposal (especially sewage sludge and animal manure disposal onto land) and some industrial processes (Elsom, 1987; Clarke, 1992). However, even 'pleasant' odours such as those from perfumeries can be a nuisance when present in excessive quantities, in the wrong place or at the wrong time. Although odours are usually a local problem, they can also be carried several kilometres downwind and therefore offend a relatively large number of people. The 'Bedford smell' (alkane thiols and SO_x) from the brick kilns near the town of Bedford is a good example of this (Owen, 1981) (Section 5.4.3).

In the UK in 1989–90, there were 9800 official complaints of smell from industrial operations and 3800 related to agricultural practices (e.g. manure spreading) (Murley, 1995). In the US, 10% of the population (25 million people) considered odours a problem which required some form of control (Elsom, 1987).

6.6.1 Important properties of odours

Certain properties of odours are important in characterizing their impact (Murley, 1995):

1. Intensity – the strength of the odour sensation which depends on the concentrations of odour substances present
2. Odour character – properties which enable substances to be detected (those odours which a recognizable character are considered to be the most annoying)
3. Hedonic tone – scale of pleasantness and unpleasantness, in the context of a specific situation
4. Frequency/duration of odour releases – important determinants of nuisance.

It is important to note that many odours have very low threshold concentrations at which they can be detected, for example allyl mercaptan, which has a disagreeable garlic-like odour and a threshold of 0.00005 mg/l (or 0.05 ppb). In contrast, SO_2 has a threshold of 0.009 mg/l and NH_3 a threshold of 0.37 mg/l (Cheremisinoff, 1992). Odours are usually mixtures of gaseous substances which stimulate the olfactory senses. They are often present at low concentrations near to limit of detection and are, therefore, difficult to analyse accurately (GC-MS is the most suitable chemical method). One way of measuring an odour is to determine the number of times that it needs to be diluted with clean air until 50% of the people on a test panel can no longer detect it (Clarke, 1992).

Odour fatigue can also occur when a subject becomes acclimatized to an odour and no longer notices it. This can be dangerous in the case of toxic gases such as H_2S because their odour provides a warning of the presence of a toxicity hazard.

Some of the important malodorous substances caused by sewage treatment and other major causes of odour nuisance include H_2S, skatole (C_9H_9N), ethane thiol (C_2H_5SH) and methane thiol (CH_3SH), cadavarine and various amines. In the case of sewers, the major odour problems occur when the sewage does not move fast enough and anaerobic conditions develop. This is often worst where garbage grinders are used, which add a lot of putrescible solid material to the sewage. The most unpleasant odours from industrial processes are those from substances containing sulphur and nitrogen (alkane thiols and amines, respectively) (Elsom, 1987).

6.6.2   Methods of odour control

Odours can be controlled by (Murley, 1995):

1. Process and plant modifications.
2. Dilution by the atmosphere.
3. Odour modification by adding other chemicals to alter the response to the original odour (masking or counteracting), e.g. air wicks or aerosol fragrances for indoor use. Masking is controversial because it involves adding more chemicals to the atmosphere and, therefore, is the opposite of cleaning-up.

6.6.3   Methods of odour treatment

Odours can be can be treated in a number of ways (Murley, 1995; Clarke, 1992):

1. Incineration by raising the emission temperature of a process sufficiently and for long enough to completely oxidize the malodorous organic compounds in the emissions. This can be done by using thermal afterburners, but the fuel cost can be prohibitive unless heat recovery, catalysts or another furnace is used. Temperatures need to be at least 650–800°C but can be lower (650°C) with catalysts and good mixing with air to ensure complete oxidation.

2. Absorption of malodours compounds in reagent solutions (commonly water).

3. Adsorption on activated carbon (commonly used in domestic and catering kitchens).

4. Chemical oxidation with ozone, or hydrogen peroxide.

5. Biological treatment – passing the odorous air through columns with microorganisms growing on the packing or through peat or soil.

References   AEA (1995) 7th Report by Bower, J. S., Broughton, G. F. S., Willis, P. G. and Clark, H., *Air Pollution in the United Kingdom: 1993/4*. AEA Technology, Culham.

Alawi, M. A. and Heidman, W. A. (1991) Analysis of PCBs in environmental samples from the Jordan valley. *Toxicol. Environ. Chem.*, **33**, 93–99.

Ayres, D. C. (1987) Cremation or burial. *Chem. in Britain*, **23**, 41–44.

Bacon, C. E., Jarman, W. M. and Costa, D. P. (1992) Organochlorine and PCB levels in pinniped milk from the Arctic, the Antarctic, California and Australia. *Chemosphere*, **24**, 779–791.

Ballschmitter, K. and Wittlinger, R. (1991) Interhemisphere exchange of hexachlorohexanes, hexachlorobenzene, PCBs and DDT in the lower troposphere. *Environ. Sci. Technol.*, **25**, 1103–1111.

Bennett, G. F. and Olmstead, K. P. (1992) Micro-organisms get to work. *Chem. in Britain*, **28**, 133–137.

Berg, H., Kiibus, M. and Kautsky, N. (1992) DDT and other insecticides in the Lake Kariba ecosystem, Zimbabwe. *Ambio*, **21**, 444–450.

Beevers, S., Bell, S., Brangwyn, M. and Rice, J. (1994) *1st Report of the London Air Quality Network*. London Boroughs Association, London.

Blake, D. R. and Rowland, F. S. (1995) Urban air leakage of liquified petroleum gas and its impact on Mexico City air quality. *Science*, **269**, 53–56.

Boubel, R. W., Fox, D. L., Turner, D. B. and Stern, A. C. (1994) *Fundamentals of Air Pollution* (3rd edn). Academic Press, San Diego.

Borlakoglu, J. T. and Dills, R. R. (1991) PCBs in human tissue. *Chem. in Britain*, **27**, 815–818.

BP (1991) *Statistical Review of World Energy*. British Petroleum, London.

BP (1995) *Statistical Review of World Energy*. British Petroleum, London.

Bridgman, H. (1990) *Global Air Pollution*. Bellhaven, London and Newcastle.

Carslaw, D. C. and Fricker, N. (1995) Natural gas vehicles, *Chem. Ind.*, **15**, 593–596.

Cass, G. R., Hildeman, L. M., Mazurek, M. A. and Simone, L. T. (1993) Mathematical modelling of urban organic aerosols: properties measured by high resolution GC. *Environ. Sci. Technol.*, **27**, 2045–2055.

Central Pollution Control Board (1992) *National Ambient Air Quality Standards of India*. Delhi.

Central Pollution Control Board (1995) *Ambient Air Quality Survey of Major Traffic Intersections in Delhi*. Delhi.

Cerutti, C., Sandroni, S., Froussou, M., Asimakopoulos, D. N. and Helmis, C. G. (1989) *Air Quality in the Greater Athens Area*. European Commission, 12218.

Cerutti, C., Noriega, A. and Sandroni, S. (1990) *Air Quality in the Greater Madrid Area*. European Commission, 13999.

Cheremisinoff, P. N. (1992) *Industrial Odour Control*. Butterworth-Heinemann London.

Clarke, A. G. (1992) In *Understanding Our Environment*, (2nd edn), Ch. 2 (ed. R. M. Harrison). Royal Society of Chemistry, Cambridge.

Cremlyn, R. J. (1991) *Agrochemicals, Preparation and Mode of Action*. Wiley, Chichester.

Crombie, L. (1980) Chemistry and biosynthesis of natural pyrethrins. *Pesticide Sci.*, **11**, 102–118.

Davidsohn, A. S. and Milwidsky, B. (1987) *Synthetic Detergents* (7th edn) Longman, Harlow.

DOE (1993) *Urban Air Quality in the United Kingdom*, pp. 164–169. HMSO, London.

DOSE (1994) *Dictionary of Substances and their Effects* (eds M. L. Richardson and S. Gangolli). Royal Society of Chemistry, Cambridge.

Elliot, M. and Janes, N. F. (1978) Synthetic pyrethroids – a new class of insecticide. *Chem. Soc. Rev.*, **7**, 473.

Elsom, D. (1987) *Atmospheric Pollution*. Blackwell, Oxford.

ENDS (Environmental Data Services) (1995) *ENDS Report*, **240**, 3–5.

EPA (1994) *Public Review Draft of Wealth Assessment Document for 2,3,7,8-Tetrachloro-p-dioxin (TCDD) and Related Compounds*, EPA/600/BP-92/001a, 001b and 001c. EPA, Cincinnati.

EUROSTAT (1995) *Basic Statistics of the European Union*, (32nd edn).

EUROSTAT (1995) *Europe in Figures*. Office of Official Publications, Luxembourg.

Ferris, B. and Wiederkehr, P. (1995) Technical options for reducing vehicle emissions, *Chem. Industry*, 597–600.

Fest, C. and Schmid, K. J. (1982) *Chemistry of Organophosphorus Pesticides* (2nd edn). Springer, Berlin.

Finkelstein, A., Klicius, R. and Hay, D. (1990) The national incinerator testing and evaluation programme: air pollution control technology assessment results. In *Emissions From Combustion Processes*, (eds R. Clement and R. Kagal). Lewis, Chelsea, MI.

Glowa, J. R. (1990) Behavioural toxicology of solvents. *Drug Devel. Res.*, **20**, 411–428.

Goring, C. A. I. and Hamaker, J. W. (1972) *Organic Chemicals in the Soil Environment*. Dekker, New York.

Hannigan, M. P., Cass, G. R., Lafleur. A. L. Longwell, J. P. and Thilly, W. G. (1994) Bacterial mutagenicity of urban organic aerosol sources in comparison to atmospheric samples, *Environ. Sci. Technol.*, **28**, 2014–2024.

Harrison, R. (ed.) (1992) *Understanding Our Environment*. Royal Society of Chemistry, Cambridge.

Harvey, R. G. (ed.) (1985) PAH and carcinogenesis. *Am. Chem. Soc. Symp. Series*, **283**, 1985.

Higgins, B. G. and Britton, J. R. (1995) Geographical and social class effects on asthma mortality in England and Wales, *Resp. Med.*, **89**, 341–346.

Hutson, D. H. and Roberts, D. R. (1987) *Progr. Pesticide Biochem. Toxicol.*, **6**.

International Agency for Research on Cancer (IARC) (1983a) PAH. *Part 1. Chemical, Environmental and Experimental Data*, **32**. WHO, Lyons.

International Agency for Research on Cancer (IARC) (1983b) *Part 1. Polycyclic aromatic hydrocarbons*. **32**, 41–45.

Iwata, H., Tanabe, S., Sakai, N. and Tatsukawa, R. (1993) Distribution of persistent organochlorines in the oceanic air and surface seawater and the role of oceans on their global transport and fate. *Environ. Sci Technol.*, **27**, 1080–1098.

Jerina, D. M. and Daly, J. W. (1974) Arene oxides: a new aspect of drug metabolism. *Science*, **185**, 573–582.

Jones, M.R. (1995) Hybrid vehicles – the best of both worlds? *Chem. Industry*, **15**, 589–592.

Kamrin, M. A. and Rodgers, P. W. (1985) *Dioxins in the Environment: Hemisphere*. McGraw-Hill, Washington, DC.

Karsa, D. R., Goode, J. M. and Donnelly, P. J. (eds) (1991) *Surfactants Applications Directory*. Blackie, Glasgow.

Kennaway, E. (1995) The identification of a carcinogenic compound in coal tar. *Br. Med. J.*, **2**, 749–752.

Kirk-Othmer (1994) *Encyclopedia of Chemical Technology* (4th edn). Wiley, New York.

Kleopfer, R. D. and Kirchmer, C. J. (1985) Quality assurance plan for 2,3,7,8-tetrachloro-dibenzo-p-dioxin monitoring in Missouri. In *Chlorinated Dioxins and Dibenzofurans in the Total Environment*, (eds L. H. Keith, C. Rappe and G. Chaudhary). Butterworth, Boston.

Kuel, D. W., Haebler, R. and Potter, C. (1991) Chemical residues in dolphins from the US Atlantic coast including bottlenose obtained during the 1987/8 mass mortality. *Chemosphere*, **22**, 1071–1084.

Lathrop, G. D. (1988) In *Agent Orange and its Associated Dioxin*, (eds A. L. Young and G. M. Reggiani). Elsevier, Amsterdam.

Lewis R. J. (ed.) (1992) *Sax's Dangerous Properties of Industrial Materials* (8th edn). Van Nostrand, New York.

Magnus Francis, B. (1994) *Toxic Substances in the Environment*, p. 254. Wiley, New York.

Martin, H. and Woodcock, D. (1983) *The Scientific Principles of Crop Production* (7th edn). Arnold, London.

Martin, D. E. (1990) The environment effects of oil and gas production. In *Energy and the Environment*, (ed. J. Dunderdale). Royal Society of Chemistry, Cambridge.

McCabe, R. W. and Kisenyi, J. M. (1995) Advances in automotive catalyst technology. *Chem. Industry*, **15**, 605–608.

Meyers, V. K. (ed.) (1988) *Teratogens: Chemicals which Cause Birth Defects.* Elsevier, Amsterdam.

Ministry of Agriculture, Fisheries and Food (1992) *Report of Studies on Dioxins in Derbyshire.* HMSO, London.

Montreal Protocol Assessment (1991) *Report of the Solvents, Coatings and Adhesives Technical Options Committee.*

Moriarty, F. (1988) *Ecotoxicology: the Study of Pollutants in Ecosystems.* Academic Press, London.

Muller, P., Lauger, P. and Martin, H. (1994) The constitution and toxic effects of botanical and new synthetic insecticides. *Helv. Chim. Acta,* **27**, 892–928.

Murley, L. (ed.) (1995) *1995 Pollution Handbook.* National Society for Clean Air and Environmental Protection, Brighton.

NRA (National Rivers Authority) (1994) *Water Pollution Incidents in England and Wales.* HMSO, London.

NRA (National Rivers Authority) (1995) *Dioxins and the river Doe Lea.* NY-12/95-0. 1k-E-AREX. HMSO, London.

NIOSH (1991) *Cancer Mortality Study of US Chemical Workers.* NIOSH, Cincinatti, Ohio

OECD (1995) *Motor Vehicle Pollution: Reduction Strategies Beyond 2010.* OECD, Paris.

Oppelt, T. E. (1987) *J. Air Pollution Control Assoc.,* **37**, 558–586.

Osborne, M. O. and Crosby, N. T. (1987) *Benzopyrenes.* p. 198. Cambridge University Press, Cambridge.

Owen, K. (1981) *The Times* (London), January 23.

Paterson, S., Shiu, W. Y., Mackay, D. and Phyper, J. D. (1990) Dioxins from combustion processes: environment fate and exposure. In *Emissions from Combustion Processes.* (ed. R. Clement and R. Kagel). Lewis, Boca Raton, FL.

Pellinen, J., Yin, C. F., Joyce, T. W. and Chang, H. M. (1988) Treatment of chlorine bleaching effluent using a white-rot fungus. *J. Biotecchnol.,* **8**, 67–75.

Pocchiari, F., Di Domenico, A., and Silano, V. (1983) Environmental impact of the accidental release of tetrachlorodibenzo-*p*-dioxin (TCDD) at Séveso (Italy). In *Accidental Exposure to Dioxins,* (eds F. Coulston and F. Pocchiari). Academic Press, New York.

Pope, C. A. Schwartz, J. and Ransom, M. R. (1992) Daily mortality and PM_{10} pollution in the Utah valley. *Archiv. Environ. Health,* **47**, 211–217.

Pottier, A. C. W., Taylor, J. C., Norman, C. L., Meyer, L. C., Anderson, H. R. and Ramsey, J. D. (1992) *Trends in Deaths Associated with Abuse of Volatile Substances 1971–1990.* St Georges Hospital, London.

Preston, B. D., Miller, J. A., and Miller, E. C. (1983) Non-arene oxide aromatic ring hydroxylation of 2,2′,5,5′-tetrachlorobiphenyl. *J. Biol. Chem.,* **258**, 8304–8311.

Rodricks, J. V. (1994) *Calculated Risks,* Cambridge University Press, Cambridge.

Rogge, W. F., Hildemann, L. M., Mazurek, M. A., Cass, G. R. and Simoneit, B. R. (1994) Sources of fine organic aerosols 6. Cigarette smoke in the urban atmosphere. *Environ. Sci. Technol.,* **28**, 1375–1388.

RCEP (Royal Commission on Environmental Pollution) (1985) 11th Report, *Managing Waste: the duty of care.* HMSO, London.

(RCEP) Royal Commission on Environmental Pollution (1991) 15th Report, *Emissions from Heavy Duty Diesel Vehicles.* HMSO, London.

Schechter, A. (eds) (1994) *Dioxins and Health,* p. 253. Plenum Press, New York.

Schechter, A., Tiernan, T. O., Taylor, M. L., Van Ness, G. F., Garett, J. H., Wagel, D. J. Gitlitz, G. and Bogdasarian, M. (1985) Biological markers after exposure to polychlorinated dibenzo-*p*-dioxins, dibenzofurans, biphenyls and biphenylenes. Part 1: Findings using fat biopsies to estimate exposure. In *Chlorinated Dioxins and Dibenzofurans in the Total Environment,* (eds L. H. Keith, C. Rappe and G. Chaudhary). Butterworth, Boston.

Silton, T. (1993) Out of the frying pan: chemical weapons incineration in the USA. *Ecologist,* **23**, 18–25.

Suess, M. J. (1976) The environmental load and cycle of PAHs. *Sci. Total Environ.,* **6**, 239.

Timberlake, L. and Thomas, L. (1990) *When the bough breaks.* Earthscan Publications, London.

UN (1994) *Statistical Yearbook,* (39th edn). UN, New York.

UNEP/WHO (1993) *Megacities of the World.* UNEP/WHO, Washington, DC.

Urban Air Review Group (1993a) *1st Report.* DOE, Bradford.

Urban Air Review Group (1993b) *2nd Report*. DOE Bradford.
USEPA (1994a) *National Air Quality and Emissions Trends Report, 1993*. USEPA, Research Triangle Park, North Carolina.
USEPA (1994b) *Public review draft of Health Assessment Document for 2,3,7,8-tetrachloro-p-dioxin (TCDD) and Related Compounds*, EPA/600/BP-92/001a, 001b and 001c, USEPA, Cincinatti
Verschueren, K. (1996) *Handbook of Environmental Data of Organic Chemicals* (3rd edn). Van Nostrand, New York.
Wayne, R. P. (1991) *Chemistry of Atmospheres*. Oxford University Press, Oxford.
Webster, T. and Commoner, B. (1994) Overview, the dioxin debate. In *Dioxins and Health* (ed. A. Schecter). Plenum, New York.
WHO (1987) *Occupational Exposure Limits for Airbourne Toxic Substances*. International Labour Office, Geneva.
WHO (1987) *Air Quality Guidelines for Europe*. WHO, Copenhagen.
WHO (1989) *Aldrin and Dieldrin*. WHO, Copenhagen.
Yang, Y.-C. (1995) *Chem. Industry*, Ist May, 334–337.
Young A. L. and Reggiani, G. M. (1988) *Agent Orange and its Associated Dioxin*. Elsevier, Amsterdam.

Hydrocarbons

Coughtrey, P. J., Martin, M. H. and Unworth, M. H. (eds) (1987) *Pollution Transport and Fate*. Oxford University Press, Oxford.
Ferguson, J. E. (1982) *Inorganic Chemistry and the Earth*. Pergamon, Oxford.
Henderson, B. and Sellers, B. (1984) *Pollution of Our Atmosphere*. Adam Hilger, Bristol.
Myerson, J. (1991) *Mad Car Disease*, Greenpeace, London; *Pollution Handbook* (1991) National Society for Clean Air.
Vo-Dinh, T. (ed.) (1989) *Chemical Analysis of PAH*, p. 8. Wiley, New York.
Wolf, S.M. (1988) *Pollution Law Handbook*. Quorum, New York.

Organic solvents

Brandrop, J. and Immergut, E. H. (eds) (1989) *Polymer Handbook* (3rd edn). Wiley, New York.
Fawell, J. K. and Hunt, S. (1988) *Environmental Toxicology, Organic Pollutants*, Ellis Horwood, Chichester.
International Agency for Research on Cancer (IARC) (1989) *Some Organic Solvents, Resin Monomers, Pigments and Occupational Exposures in Paint Manufacture*, Vol. 47, Lyon.
Russell, J. (1993) Butane abuse: the neglected killer. *New Scientist*, **1859**, 21–23.
Sheftel, V. O. (1990) *Toxic Properties of Polymers and Additives*, RAPRA, Shrewsbury.
Snyder, R. (ed.) (1987) *Ethel Browning's Toxicity and Metabolism of Industrial Solvents*, Vol. (1) *Hydrocarbons*. Elsevier, Amsterdam.

Organochlorines

Brown, P. (1992) Who cares about malaria? *New Scientist*, **1845**, 37–41.
British Crop Protection Council (1986) Biotechnology, Crop Improvement and Production, *Monograph*, **34**.
British Medical Association Guide to Pesticides, Chemicals and Health (1992). Arnold, London.
Hayes, W. J. and Laws, E. R. (1991) *Handbook of Pesticide Toxicology*. Academic Press, San Diego.
Ragsdale, N. N. and Menzer, R. E. (1989) Carcinogenicity and pesticides, *Am. Chem. Soc., Symp. Series*, **414**, Washington, DC.

Organophosphorus and other pesticides

Kidd, H. and James, D. R. (1991) *The Pesticide Index* (2nd edn). Royal Society of Chemistry, Cambridge.

Matsumura, F. (1985) *Toxicology of Pesticides* (2nd edn). Plenum, New York.

Ministry of Agriculture, Fisheries and Food (1990) *Pesticides*. HMSO, London.

Watterson, A. (1988) *Pesticide Users Health and Safety Handbook*. Gower, Aldershot.

Indoor pollution 7

A list of common indoor pollutants and their sources was given in Table 2.8 (p. 26). Especial concern is now felt over this type of pollution because levels in the home and workplace often exceed those outdoors (National Academy, 1981; Bloemen and Burn, 1993) and because individuals spend 90% of their time indoors.

7.1
Volatile organic compounds

Indoor air contains a complex mixture of volatile organic compounds (VOCs) derived in large part from the use of propellants (Section 6.3.3) to dispense cleaning compounds, deodorizers and foodstuffs. To this source must be added gases from cooking and heating fuels, refrigerants, adhesives, dry-cleaning solvents, textiles, floor coverings, dyestuffs and pesticides. Analysis of the complex mixture of VOCs requires refined techniques including capillary GLC (Bayer and Black, 1987).

It has been shown (Rogge *et al.*, 1993) that natural gas space heaters and top burner gas cookers release indoor and outdoor PM_{10}. The emission rate for space heaters is in the range 6–300 ng/kJ and for gas cookers it is 240–620 ng/kJ. Although they are not present originally in the fuel, GC/MS evaluation established the presence of high-molecular-weight hydrocarbons such as nonadecane ($C_{19}H_{40}$, m.p. 32°C) and hentriacontane ($C_{31}H_{64}$, m.p. 68°C). These alkanes were emitted over the range 1.5–164 pg/kJ.

Other emitted particles were mainly organic and included PAH, aza-arenes such as phenanthridine (**1**, 48 pg/kJ) and thia-arenes such as benzo-[b]-naphthothiophene (**2**, 325 pg/kJ).

1 phenanthridine   **2** benzo-[b]-naphthothiophene

1,1,1-Trichloroethane (TCE) is used in dry cleaning and both tri- and tetra-chloroethylene contribute to the input to indoor pollution from dry cleaning (Section 6.3.5). It has been shown (Guo *et al.*, 1990) that wool retains 96% of tetrachloroethylene after airing at 20°C for 1 hour and that 5 hours at 40°C is needed to reduce the residual cleaner to 1%. The permissible level of TCE in the UK is to be revised from the present 8-hour TWA of 50 ppm (335 mg/m^3); in the USA, the current limit is 25 ppm (170 mg/m^3). The oral LD_{50} in the rat is 2600 mg/kg. Alkanes and alkylated hydrocarbons also occur, together with carbonyl compounds formed by oxidation (Section 6.2.3, p. 258).

The health effects of hydrocarbons have been researched (Mehlman, 1992) in the context of motor fuels but the conclusions reached are valid also for the evaluation of the risk they present to workers in the industry and by infiltration into buildings. In the UK, a detailed study (Crump, 1995) of 174 homes in the Avon district showed that outdoor emissions of toluene and benzene were largely responsible for levels indoors. For toluene, the mean levels were 12 μg/m^3 outdoors and 42 μg/m^3 indoors, whilst for benzene they were 5 μg/m^3 outdoors and 8 μg/m^3 indoors. In the UK an expert panel has recommended that benzene levels indoors be limited to a maximum annual average of 5 ppb (16 μg/m^3).

About 20 alkylbenzenes are present in petrol but the principal compounds are benzene itself (2.3–6.0%), toluene (4.6%) and the isomeric xylenes (9.9%). A number of these compounds are carcinogenic in animals, but benzene presents the greatest risk. The USA uses 11 billion gallons of benzene annually and over 200 thousand people are occupationally exposed to it, including those in printing and publishing. Benzene is a human carcinogen and induces a range of leukaemias; the relative risk of death from exposure to it has been put at 3.75–25 per 1000 workers.

Human cancers have also been linked to 1,3-butadiene and numerous less specific studies have linked exposure to petrol vapour with cancer in rats and mice.

Exposure to benzene, alkylbenzenes and oxygenated solvents occurs in factories manufacturing and spraying paints. A study was made of 196 workers from two manufacturing units and 25 paint-spraying sites in Taipei City where toluene (50%) and the xylenes (24%) were the major components in air samples (Wang and Chen, 1993) (Table 7.1). It was concluded that exposure to the median level of mixed solvents would lead to acute and chronic neurological symptoms.

An example of extensive exposure to solvents is to be found in the Camacari Petroleum Complex of 54 companies which employ 50 000 people in Brazil. Although they use benzene, *n*-hexane, haloalkanes and alcohols, which are all known to attack the central and peripheral nervous systems, many workers are unaware of the risk (Kato *et al.*, 1993) The EC has established EURONEST, a programme to study the effects of chemicals on the nervous system (Gilioli, 1993).

Table 7.1 Concentration of solvents in the air samples of paint works

Compound	Range (ppm)	Median (ppm)	TLV[a]	No. exceeding TLV
Xylenes	0–365	18	100	7
Toluene	0–540	10	100	2
Benzene	0–20	1	100	3
Acetone	0–124	3	750	0

[a] Recommended by American Conference of Government Industrial Hygienists; all samples were 8-hour time weighted averages.

Table 7.2 Substances detected in samples collected overnight at domestic premises

Compound	Median level ($\mu g/m^3$)		Detection limit ($\mu g/m^3$)
	Indoor	Outdoor	
1,1,1-Trichloroethane	8.5	4.2	0.16
p-Xylene	9.5	5.7	0.06
Limonene	4.5	Nil	0.12
Benzene	4.2	2.7	0.09
Carbon tetrachloride	0.8	0.7	0.17
Tetrachloroethylene	1.2	1.0	0.05

Researchers in Los Angeles County (Michael *et al.*, 1990) were unable to correlate indoor and outdoor levels with one another or with a central sampling site. They sampled 43 households using Tenax cartridges (Section 4.4.5, p. 99) and GC analysis. Twenty individual VOCs were identified and most were quantified. Typical median levels for samples collected overnight are given in Table 7.2.

7.1.1 Oxidized forms of VOCs

Formaldehyde is the most important of the oxidized forms of VOC which occur indoors and it arises chiefly from the outgassing of urea–formaldehyde foam insulation and from glues used to formulate particle board. Mobile homes are especially at risk from these releases, while the slow deterioration of furnishings such as carpets, wallpapers and ceiling tiles may contribute in a moist acidic environment. Formaldehyde has also been shown to be formed when ozone reacts with furnishings (Section 7.2).

In a study (Krysanowski *et al.*, 1990) of 202 households in Arizona, formaldehyde was found at levels which averaged 26 ppb. In kitchens, levels > 90 ppb were found, rising to a maximum of 140 ppb; typically indoor levels are in the range 50–300 ppb. It was shown that asthma and chronic bronchitis were more common amongst children in houses with higher formaldehyde levels but there was no clear correlation for adults. Formaldehyde triggers attacks of asthma, symptoms of exposure include

Table 7.3 Levels of aldehydes in and around houses of New Jersey

Compound[a]	Mean(ppb)		75%ile (ppb)	
	Outdoor	Indoor	Outdoor	Indoor
Formaldehyde	12.5	54.6	20.4	67.2
Acetaldehyde	2.6	2.9	2.5	3.3
Propionaldehyde	1.3	1.1	1.5	1.4
n-Hexaldehyde	0.6	1.3	0.6	1.4
Benzaldehyde	0.25	0.38	0.43	0.52

[a] The remaining four aldehydes detected were in the mean concentration range of 0.91–0.27 ppb.

eye irritation, headache and sleeplessness; it is a probable human carcinogen (USEPA, 1987).

Formaldehyde has a b.p. of $-19.5°C$ and water solubility at $20°C$ of $550 \, g/l$. The LC_{50} (96 hours) of rainbow trout is $440 \, mg/l$ and the LC_{50} (4 hours, inhaled) of the mouse is $0.48 \, mg/l$. In the USA, the threshold limit value is $0.3 \, ppm$ $(0.37 \, mg/m^3)$.

Spider plants, *Chlorophytum comosum*, were once claimed to purge formaldehyde from the air but this has not been substantiated.

Oxidation of hydrocarbons gives rise indoors to lower levels of other carbonyl compounds. A study in New Jersey (Zhang *et al.*, 1994) in which 36 samples were obtained showed that nine aldehydes were present in houses (Table 7.3). The average person's exposure to formaldehyde was estimated as $873 \, \mu g/day$ and the total aldehydes as $1005 \, \mu g/day$. The high indoor/outdoor ratio of seven for formaldehyde indicated that there was significant emission from furnishings and construction materials. The samples were collected in the afternoon when outdoor levels peaked; however, the concentration patterns for formaldehyde and ozone were very similar throughout the day, which indicates that formation of the aldehyde was photochemically induced.

Table 7.4 gives the exposure limits set in the UK for exposure to carbonyl compounds. No limits exist at present for benzaldehyde; the LD_{50}

Table 7.4 Exposure limits for carbonyl compounds

Compound	UK (8-hour TWA)[a]		UK short-term limit (STEL)	
	ppm	mg/m^3	ppm	mg/m^3
Formaldehyde	2	2.5	2	2.5
Acetaldehyde[b]	100	180	150	270
Acetone	750	1780	1500	3560
Methyl ethyl ketone	200	590	300	885

[a] The Health and Safety Executive (1993).
[b] The ACGIH now sets the 8-hour TWA at 25 ppm and the STEL at 45 ppm. Acetaldehyde like formaldehyde is a suspected human carcinogen.

in the guinea pig is 1000 mg/kg and the LC_{50} (96 hours) for rainbow trout is 11 mg/l.

Ozone reacts with materials in carpets and with other oxidizable compounds in air including NO, CO and carbonyl compounds. These reactions, while producing other pollutants, also reduce ambient indoor levels of ozone to within 20–80% of those outdoors. The reaction with indoor NO produces the peroxide (Equation 5.6, p. 160) and the reaction with carpets (Weschler et al., 1992) liberates aldehydes. Carpet emissions include 4-phenylcyclohexene (**3**), which is typical of new materials, residual styrene monomer (**5**) and 4-vinylcyclohexene (**4**), which is a residue from latex and adhesives. Levels of ozone in excess of 37 ppb were frequently exceeded indoors and there are significant reactions when the level indoors is in the range 28–44 ppb; when the concentration of the alkenes (**3–5**) falls, new irritants are produced. Formaldehyde was detected at 8.1 ppb and acetaldehyde at 4.6 ppb. Higher aldehydes with 5–10 carbon atoms were found indoors in air in the range 0.1–1.4 ppb and probably arose from the action of ozone on the material of carpet fibre. It should be noted that ozone will also react *inter alia* with room fresheners, such as limonene (**6**) or α-pinene (**7**), and many products of styrene, latex, aromatic based detergents, polyvinyl resins, etc.

Ph-CH=CH$_2$

5 styrene monomer

3 4-phenyl-
cyclohexene

4 4-vinylcyclohexene

6 limonene
(*R*-form)

7 α-pinene
(−)-form

In the USA, 150 million citizens live with an ozone level above the 0.12 ppm level accepted as the NAAQS (1986) (Table 6.9, p. 262). It has been shown that there exists a strong relation between the high summer ozone concentrations and emergency hospital visits by asthmatics. A similar link was found for Southern Ontario but not in Vancouver, which has lower pollution levels (Burnett et al., 1994)

In Toronto, the mean daily hourly maximum for ozone over the period 1983–8 was 50 ppb with a 95%ile of 90 ppb, some levels there and in Ottawa and Windsor reached 150 ppb.

Table 7.5   Ozone and nitrogen oxides within an office building

	Ozone (ppb)[a]		NO (ppb)[a]		NO$_2$ (ppt)[a]	
	Outdoor	Indoor	Outdoor	Indoor	Outdoor	Indoor
October 92	89	32	138	127	81	72
March 93	53	29	209	199	88	86
August 93	85	35	84	79	71	65

[a] 95% concentrations from monthly measurements.

Acids have a synergistic efffect on asthma and other respiratory infections and it has been shown that they are increased in socially deprived children the day following raised ozone levels.

In a Burbank, California office building with air-exchange rates in the range 0.4–0.6 of the initial volume per hour indoor ozone was consistently lower than outdoors (Weschler *et al.*, 1994), while NO and NO$_2$ levels were similar both indoors and out (Table 7.5). This is further evidence of the reactions of ozone indoors and of the lack of protection afforded by the building from conditions outside.

The reaction of ozone with the oxides of nitrogen to form nitrous and nitric acids is discussed in the following section.

7.3
Oxides of nitrogen

The nitrogen oxides arise from ineffectively vented gas cookers and heating systems. In these circumstances, levels frequently exceed the air quality standards for outdoors (p. 262) and may fall in the range 20–300 ppm. Levels in the range 2–6 mg/m^3 (1–3 ppm) are said to have adverse effects on the lung, with loss of lung cilia and a susceptibility to viral infection.

In an Austrian study, involving 558 primary schoolchildren at four different locations, 24-hourly averages of 21.6, 15.3, 32.2 and 14.9 µg/m^3 were found for NO$_2$ (Frischer *et al.*, 1993). All these figures are within the standard of 100 µg/m^3.

Following the initial fast reaction of NO with ozone, the NO$_2$ formed can react further (Equation 7.1) to give the nitrate radical, but this step is very much slower:

$$O_3 + NO_2 \longrightarrow O{=}N\big\langle{\overset{\displaystyle O\bullet}{\underset{\displaystyle O}{}}} + O_2 \qquad (7.1)$$

(NO and NO$_2$ both have odd electrons and proper description of their bonding can only be given in terms of molecular orbitals. A clear account is given by Brady and Holum, 1987). If there is limited air changing and high humidity, the nitrate radical formed in Equation 7.1 can produce nitric acid:

$$(7.2)$$

which may also arise through capture of a hydrogen atom from organic substrates:

$$R : H + {}^{\bullet}NO_3 \longrightarrow HNO_3 + R^{\bullet} \qquad (7.3)$$

A possible source of nitrous acid under these conditions is the dispropor- tionation of NO_2 in aqueous surface films:

$$2NO_2 + H_2O \text{ (surface)} \longrightarrow HO{-}NO \text{ (aq.)} + H^+NO_3^- \qquad (7.4)$$

These nitrogen oxyacids also act as irritants towards asthma sufferers.

A study in Albuquerque showed that levels of nitrous acid correlated with those of NO_2 is the range 5–15% (Spengler *et al.*, 1993). In Los Angeles, the half-life of nitrous acid is about 120 minutes; it will be longer indoors as there is no photodegradation. Higher levels of nitrous acid were found indoors in both summer and winter in Massachusetts and Boston.

Table 7.6 shows pollution levels arising from the use of gas cookers in non-smoking homes in the Albuquerque study (Spengler *et al.*, 1993). The sources of sulphate here are odorants such as propane-2-thiol $(CH_3-CH(SH)-CH_3)$, which is incorporated in the fuel gas at a con- centration of 0.7 lb/million ft^3.

These levels were evaluated over the whole house during 24 hours.

7.4 Carbon monoxide

Carbon monoxide is prevalent in underground garages and is also produced indoors by gas heaters and cookers. In recent years there have been a

Table 7.6 Air pollutants arising from the use of gas cookers

Pollutant	Mean	Minimum	Maximum
NO_2 (ppb)	60	24	115
HNO_2 (ppb)	4.7	1.8	8.0
NH_3 (ppb)	20	14	30
NO_3^- (nmol/m^3)	8.7	0	23
SO_4^{2-} (nmol/m^3	11.5	0	56
NH_4^+ (nmol/m^3)	12.8	0.5	53
Time cooker in use (min)	130	35	426
Total floor area (ft^2)	940	348	1512

number of fatalities arising from accumulation of CO overnight in student flats and Spanish holiday accommodation. Outside levels are generated largely by vehicles, which were responsible for 77% of USA emissions in 1993, with significant contributions from fixed industrial sources, wood-burning and incineration. The NAAQS for the 8-hourly average is 9 ppm(10 mg/m^3) and the hourly average is 35 ppm (40 mg/m^3); generally levels up to 100 ppm can be tolerated for several hours without ill effect. At about 1000 ppm for 1 hour, headache and impaired performance will be experienced and 2000 ppm over an hour can be fatal.

Those suffering from angina or peripheral vascular disease are most at risk from CO poisoning. This is because CO interferes with the provision of oxygen in the bloodstream by displacing oxygen from its complex with haemoglobin (Hb):

$$HbO_2 + CO \longrightarrow HbCO + O_2 \qquad (7.5)$$

The affinity of haemoglobin for CO is at least two orders of magnitude greater than that for oxygen, the HbCO complex, therefore, continues to circulate and so blocks oxygen transfer for an extended period, which throws a strain on the heart. Nitrous oxide behaves in a similar fashion.

Carbon monoxide is oxidized only slowly by ozone and other oxidants; nevertheless, it gives rise to CO_2 indoors, which can accumulate in sealed office buildings to a level of 2000 ppm – about six times the level outdoors (Fig. 5.2, p. 166). Symptoms such as fatigue may be experience when CO_2 levels reach 800 ppm.

7.5
Other gaseous oxides

Carbon dioxide levels may reach 2000 µg/g, five times that outdoors, but SO_2 is usually lower indoors, at < 20 ng/g.

7.6
Cigarette smoking

Cigarette smoke contributes to levels of CO_2 but smoking is especially dangerous for those with heart disease because combustion of tobacco in this way, with limited oxygen available, introduces CO directly into the bloodstream. Among the very many products of combustion (Table 7.7) nitrosamines (9, 10), benzo-[a]-pyrene (Section 6.2.5) and nicotine (8) are highly poisonous. Nicotine also throws an additional strain on the heart by stimulating the basal metabolic rate.

8 nicotine (s-form)

9 N'-nitrosonor-nicotine

10 N-nitroso-pyrrolidine

Table 7.7 Some toxic components of cigarette smoke (IARC, 1986)

Component of smoke	Concentration per cigarette (400–500 mg)
Total particles	15–40 mg
Nicotine (**8**)[a]	1.0–2.3 mg
Acetaldehyde	0.5–1.2 mg
Acetone	100–250 µg
Benzene	20–50 µg
Formaldehyde	70–100 µg
N'-Nitrosonornicotine (**9**)	200–3000 ng
N-Nitrosopyrrolidine (**10**)	0–110 ng
Benzo-[a]-pyrene[b]	20–40 ng

[a] The TWA for nicotine is $0.5\,mg/m^3$ and the STEL is $1.5\,mg/m^3$.
[b] Benzo-[a]-pyrene is a human carcinogen.

7.6.1   N-*Nitrosocompounds*

The nitrosamines are members of a formidable group over half of which are known to cause cancer in animals. In addition to those (**9, 10**) found in tobacco smoke, the following occur in snuff (oral tobacco), notably:

N-nitrosodimethylamine at	37 µg/kg
N-nitrosopyrrolidine at	120 µg/kg
N-nitrosopropylamine at	3980 µg/kg

Nitrite is added during tobacco processing and will react with amines, probably through nitrous anhydride (**11**):

$$2\ HO-N{=}O \longrightarrow O{=}N-O-N{=}O\ +\ H_2O \qquad (7.6)$$

11

then

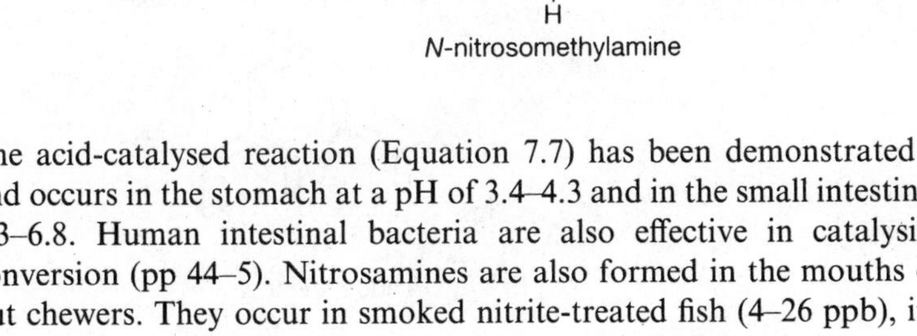

N-nitrosomethylamine

The acid-catalysed reaction (Equation 7.7) has been demonstrated in rats and occurs in the stomach at a pH of 3.4–4.3 and in the small intestine at pH 6.3–6.8. Human intestinal bacteria are also effective in catalysing this conversion (pp 44–5). Nitrosamines are also formed in the mouths of betel nut chewers. They occur in smoked nitrite-treated fish (4–26 ppb), in many meat products (1–5 ppb) and as traces in milk, cheese and beer.

Mice, rats and hamsters developed tumours after one week on a daily diet containing 0.4 mg/kg. In the rat, dimethylnitrosamine is metabolized to formaldehyde.

Dimethylnitrosamine was manufactured in the USA at the rate of 5×10^5 kg/year as the precursor of the rocket fuel dimethylhydrazine. Levels in air of up to 36 000 ng/m^3 were recorded on a factory site and 1000 ng/m^3 was measured in the adjacent neighbourhood of Boston, MA; production ceased in 1976.

7.6.2   Polycyclic aromatic hydrocarbons

Nicotine is a useful tracer for tobacco smoke and meets the criteria:

1. it is unique to tobacco smoke
2. it is easy to detect in air
3. it occurs in proportion to environmental tobacco smoke (ETS) components which are injurious to human health.

It is possible to determine nicotine and PAH simultaneously by adsorption on XAD-4 resin, solvent extraction and GC–MS analysis (Chuang et al., 1990). The levels of PAH are an order of magnitude less than that of nicotine in tobacco smoke but nevertheless they are clearly derived from cigarettes (Table 7.8).

There is a correlation between mutagenic substances indoors with somewhat higher levels outdoors. The sources of street dust in Tokyo were identified as car exhausts and diesel combustion (Takada et al., 1990); wear of asphalt was a minor contributor. PAHs were found in Tokyo Bay sediments. In residential areas the most significant were dibenzothiophene (**12**), 2-methylphenanthrene (**13**), fluoroanthene (**14**), pyrene and chrysene (Fig. 6.6, p. 268).

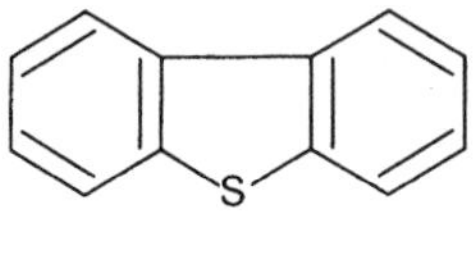

12 dibenzothiophene

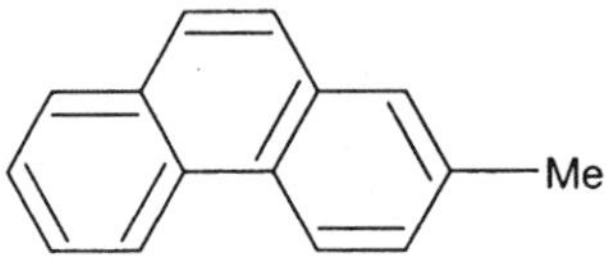

13 2-methylphenanthene

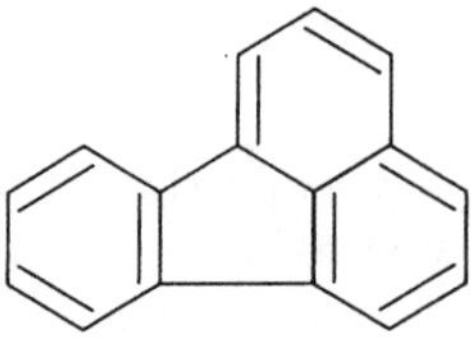

14 fluoanthene

Table 7.8 Relative occurrence of PAH in homes of smokers and non-smokers

Compound	Level (ng/m^3) in the home of	
	Smoker	Non-smoker
Phenanthrene	220	75
Pyrene	18	5
Anthracene	15	Neglible

Other data on PAH levels in water, food and tobacco smoke are given in Section 6.2.5.

It is important to distinguish the two kinds of environmental tobacco smoke, namely mainstream, which is inhaled after filtering through unsmoked tobacco, and the more toxic unfiltered sidestream smoke. The latter is generated most of the time the cigarette is burning and leads to exposure of non-smokers, or passive smoking. A number of successful legal actions have been brought by heavily exposed non-smokers whose health had been adversely affected by smokers in their workplace.

Apart from the effect on the heart, tobacco smoking exacerbates bronchial infection and causes 30–40% of all deaths from cancer. The higher incidence of cancer in smokers arises from inhalation of PAH and nitrosoamines (Yoshikawa *et al.*, 1994). There are genetic variations in cytochrome P-4501 Al, which converts smoke carcinogens into active DNA-binding metabolites or inactive metabolites, depending on the individual. The active derivatives are oxygenated by aryl hydrocarbon hydroxylase (AHH); hydroxyl radicals play a part in a reaction which may be catalysed by iron(III). It is known that workers in the iron industry and those exposed to iron-containing crocidolite asbestos are more at risk from cancer.

7.6.3 Other mutagens

Lung cancer is the leading cause of cancer death among women in Hong Kong, but it has been estimated that in almost two-thirds of these cases this is *not* the result of a history of smoking (Tewes *et al.*, 1990) Nevertheless, the incidence rate of 27 per 100 000 (in 1982) ranks as one of the highest in the world and is attributed in part to the drinking of infusions of green tea. One cup produces mutagenic activity equivalent to 0.5 mg benz-[a]-pyrene; it is suspected that this toxicity is due to the extraction of flavonols such as quercetin (**15**, X = OH) and kaempferol (**15**, X = H).

15 kaempferol (X = H)
quercetin (X = OH)

The majority (80%) of quercetin is excreted but a small proportion may be metabolized and so enter the lung.

Other mutagenic products which arise from smoking are formed by the pyrolysis of tryptophan (**16**) (Manabe and Wada, 1990) and two of them, Trp-P-1 (3-amino-1,4-dimethyl-5*H*-pyrido-[4,3-b]-indole; **17**, X = Me) and Trp-P-2 (3-amino-1-methyl-5*H*-pyrido-[4,3-b]-indole; **17**, X = H), occurred at higher levels in air indoors than outdoors. They can be isolated by HPLC (Section 4.5.2) following detection by fluorescence at 399 nm initiated by irradiation at 266 nm.

16 tryptophan

17 Trp-P-1 (X = Me)
Trp-P-2 (X = H)

Smoking also increases the risk to workers in industries which generate fine particles. For example, Indian workers crushing sandstone at Chandigarh were held to be more at risk of chromosome damage from inhalation of particles, which had taken up mutagens on the surface. For comment on pollution by mineral particles out-of-doors see Section 5.6.

7.7
Asbestos fibres   Inhalation of separate dry fibres in a confined air space is the major hazard from asbestos (Section 5.6). When they are bonded they constitute a relatively low hazard because there will be few loose fibres. However, damaged thermal insulation (e.g. during alterations or demolition) and accumulated fibres from the manufacture or wear of asbestos or other mineral materials, such as brake shoe pad dust, constitute a major hazard within buildings. On the streets, fibre dust from wear of brake shoe materials presents a risk which may be carried indoors.

Asbestos structures include single fibres, bundles, matrices and clusters; a concentration of fewer than 10^5 structures/litre in water is acceptable (Hardy *et al.*, 1992). In exposed situations, such as abraded asbestos cement, levels can rise to over 10 million/litre.

In New York city, samples were taken in 886 buildings chosen from 16 different categories ranging from tall offices to petrol stations and including hotels, churches, family houses, factories and warehouses. These were analysed and 68% were found to contain asbestos (Lundy and Barer, 1992). Family dwellings were less affected than were flats with lifts, and it was estimated that there was surface exposure of $326 \times 10^6 \, \text{ft}^2$ in the whole city; this was mostly as thermal insulation. Appropriate safety masks and skin protection should always be worn when handling mineral fibres.

The group name *asbestos* is applied to a range of naturally occurring fibrous magnesium silicate minerals with the approximate formula $Mg_3P(Si_2O_5)(OH)_4$. Three common types are used:

chrysolite (white asbestos): least hazardous to health
crocidolite (blue asbestos): most hazardous
amosite (brown asbestos): second most hazardous.

7.7.1   Analysis

Unlike most pollutants, mineral fibres are easily seen with a microscope and can be counted on air filters. Identification is usually by X-ray diffraction spectrometry (XRD) . Mineral pollutants are difficult to detect in soils and sediments owing to the abundance of similar minerals, although scanning electron microscopy with linked X-ray analysis can be used to identify the particles.

The US EPA took 387 airborne asbestos samples in 49 Government buildings (Chesson, 1990). They found that levels indoors were much the same as those outdoors. Fibres, bundles, clusters and matrices were identified and counted under a magnification of 20 000. Taking the sample mean value from buildings with no asbestos as unity, the outdoors rated as 1.95, within buildings with asbestos in good condition as 2.95 and within those with damaged asbestos as 3.65.

7.7.2   Toxicity

Other sources of asbestos include ship breaking, waste disposal and, more insidiously, the exposure of workers' families from particles carried on clothing.

Asbestos fibres can cause lung and bowel cancer as well as non-cancerous lung diseases. Approximately 50% of the inhaled fibres are cleared from the

Table 7.9   Mortality amongst asbestos workers

Group	Asbestos exposure	Cigarette smoker	Death rate	Mortality ratio
Control[a]	No	No	38.3	1.00
Asbestos workers	Yes	No	208.2	5.44
Control	No	Yes	60.3	1.57
Asbestos workers	Yes	Yes	334.2	8.73

[a] 3219 workers not exposed to asbestos dust or fumes or vapours.

lungs and swallowed, which then exposes the throat and digestive system to their hazardous effects. Water from asbestos–cement pipes also poses a further source of digestive tract exposure. Respiratory diseases include asbestosis (a pneumonia-like condition), bronchial cancer and mesothelioma, which has a latency period of 20–30 years. There appears to be a synergistic reaction between cigarette smoke and asbestos in that the onset of disease is more pronounced in heavy smokers than in non-smokers (Table 7.9).

An asbestos factory in Armley, Leeds was closed in 1958 but while operating had contaminated neighbouring residential streets and dwellings. In October 1995, a sufferer and a dependant of a mesothelioma victim were awarded substantial damages by the High Court in the UK, although the illness was not diagnosed until 30 years after exposure. Manufacture continued elsewhere in the UK through the 1960s and, unless overturned on appeal, the judgment must result in many similar legal actions.

A retrospective mortality cohort study for the period 1972–86 has been made of workers in Tianjin (Cheng and Kong, 1992). In a group of 662 male and 510 females working with chrysolite asbestos, a significant excess mortality from lung cancer was found and a time lapse of over 19 years may occur before death from lung cancer. In the early years of this period, the mean concentration of dust in workplace air was as high as 167 mg/m^3 but this was reduced to <20 mg/m^3 by 1986.

The widespread use of asbestos in schools and other public buildings for thermal insulation has resulted in large numbers of people being potentially exposed to these particles. The transport and dumping of building wastes has also contributed to the load of airborne asbestos. In the USA, it has been estimated that the major emissions of asbestos to air are from its mining and milling (700 t/year), product manufacture (100 t/year) and from landfills (18 t/year). The current USA limits in workplace air are 2 fibres/cm^3, but stricter limits of 0.5 or 0.2 fibres/cm^3 are proposed (Harte et al., 1991).

Non-asbestos mineral fibres such as glass fibre and rockwool are increasingly used as substitutes (Brown, 1992); while they are not as hazardous as asbestos, excessive exposure and inhalation of particles could still cause disease.

In developing countries hazards in the workplace are worsened by:

- low awareness of risk by workers and managers
- inadequate facilities
- high unemployment and pressure to take high-risk jobs
- importation of hazardous waste and hazardous industries from developed countries.

Asbestosis is on the increase and is found in 5% of Thai workers, where a sample of 1013 people was taken from 24 factories (Levy and Seplow, 1992); the figure for India is 7% and for China 10%.

The major producers are the former Soviet Union, Canada and South Africa. In the last, there was a strong correlation between miners and asbestosis and mesothelioma. Natural outcrops have been used for building and villagers in Turkey, where asbestos was used for stucco, were found to have abnormalities on chest X-ray examination. In Brazil in 1986, one-third of new houses and two-thirds of new factories had asbestos-cement roofs.

The International Labour Office produced a convention to protect workers in 1991 but so far only 11 countries have ratified it.

7.7.3 Other particulate pollutants

Respirable fine particles ($<5\,\mu m$) of several metals or their compounds, including Be, Ni and Cr, can constitute serious hazards to health through their effects on the respiratory system. However, these mainly affect occupationally exposed workers or possibly people living near to industrial plants using or producing the metals or their compounds. These metallic pollutants are dealt with in Section 5.3.

Many ubiquitous particles which are commonly suspended in air can cause respiratory illness in susceptible people. These particles and fibres include pollen, mould spores, faeces and pieces of exoskeleton of the house dust mite (*Dermatophagoides pteronyssinus* and *D. farinae*), animal hairs, skin fragments (dander) and bird feathers; all of which can cause allergic conditions such as asthma and hay fever. It is known that guanine (**18**) is a waste product of arthropods, insects, birds and some mammals.

18 guanine

Table 7.10 Comparative enzymatic activity of some allergens

Allergen	Enzymatic activity (mU/mg)[a]
Female cat, urinary protein	20
Female cat, dander	7.7
Dog dander	44
Female dog, urinary protein	92
Cockroach	287
Locust	1299

[a] 1 enzyme unit = 1000 mU and converts 1 μmol substrate/minute during an experiment.

Guanine is not itself allergenic but its proportion in house dust is a pointer to the total contribution from these organisms. Typical allergens which induce a response in the antibodies of susceptable individuals are the serine and cysteine proteinase enzymes. Protease and acid phosphatase activity also correlate with allergenic potential (Berrens, 1991). The enzymatic activity of some highly allergenic materials has been determined (Table 7.10).

There is considerable interest in indoor allergens nowadays owing to the occurrence of the 'sick office' syndrome in buildings with large air-conditioned offices. Likewise, asthma incidence is increasing in the general population and may be due, at least in part, to the increasing amount of carpeting used in houses, which provides a suitable habitat for the house dust mite. However, increasing concentrations of air pollutant gases, including NO_x, ozone and others, may also be responsible (Sections 5.1.10 and 5.2.2).

7.8
Lead

Reference has already been made to the sources and toxicology of environmental Pb (pp 198, 215). Lead is toxic to the nervous system, to red blood cells and to the cardiovascular system. Lead also leads to anaemia, renal failure and neuropsychological impairment. By the end of the Roman Empire in AD 50 the total Pb burden was 39 Mt; by 1980 this had reached 241 Mt. By 1989, environmental Pb was being deposited at a rate of 4 Mt/year, of which 1.3 Mt/year was in the USA. Modern Americans have a body burden which is 1000 times higher than that of pre-industrial North American Indians and Pb toxicity is one of the most common diseases of childhood. Indoor pollution by Pb arises from exhaust emissions entering through windows, resuspension of particles trampled into carpets, deterioration of old Pb painted surfaces and by solution from Pb plumbing in old houses. The US EPA set the blood level in children at 10 μg/dl in 1991 and it has been found that 17% of all preschool children in the USA have elevated blood levels. In poor areas this may reach 68% of this group of children.

Table 7.11 Toxic effects seen at rising levels of lead in blood (Mushak, 1992)

Effect	Lead in bloodstream (µg/dl)
Death	120
Severe CNS effects	80
Overt anaemia	60
Prenatal CNS effects	< 10

In a classical Chicago study of inner city children through the 1970s (Schwarz and Levin, 1991), toxic effects were seen once the blood level exceeded 30 µg/dl. Children with high Pb levels in their home stood ten times the risk of poisoning in Chicago in the summer. The foetus, older adults and lead workers are those most at risk. Table 7.11 gives the principal toxic levels.

Lifestyle and environmental factors were shown to determine blood levels in a Swiss population (Berode *et al.*, 1991). It was found that 15–30% of body Pb was from inhalation of particles and 70–85% was ingested in food and water. Smoking and taking alcohol contribute to raised blood levels.

The average in Swiss men was 0.63 µmol/l (13.1 µg/dl) and that in women was 0.44 µmol/l (9.2 µg/dl). Similar figures were obtained for the UK and Belgium.

The expectation that blood lead levels would improve following the reduced sales of leaded petrol has been borne out by a number of investigations. Thus, in the USA a drop of 30–40% was observed in parallel

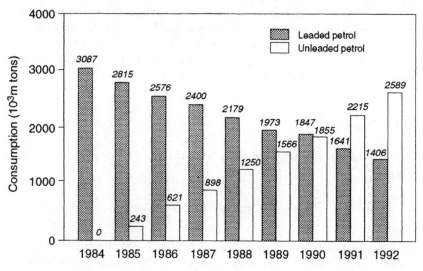

Figure 7.1 Evolution of petrol consumption in Switzerland (Wietlisbach, 1995).

with the phasing in of lead-free petrol. A study in a Swiss region with a population of 770 000 showed that a reduction in the mean blood Pb levels in men fell by 43% from 0.63 µmol/l (13.1 µg/dl) to 0.36 µmol/l (7.5 µg/dl) over the years 1984–93. In women, the fall was 36%, from 0.44 µmol/l (9.2 µg/dl) to 0.28 µmol/l (5.8 µg/dl). Figure 7.1 shows the proportional reduction in sales of leaded petrol over this period.

There was a significant increase in blood lead level for men and women in the 65–74 age group. The increase in total fuel consumption from 3087×10^3 t in 1984 to 3995×10^3 t in 1992 points to an increase of about 29% in traffic over the period.

7.9 Radon

Notwithstanding the hazards posed by occasional failure of control of nuclear power stations, the penetration of buildings by chemically inert radon gas presents a far more serious risk to human health (National Academy, 1981). As is so often found, the risk to humans was first in evidence amongst industrial workers. Increased incidence of lung cancer in uranium miners in western USA and Czechoslovakia was attributed to their inhaling ^{222}Rn.

The decay of ^{238}U produces ^{222}Rn (Fig. 5.12, p. 221) which emanates from radioactive minerals in building foundations. Because it has a half-life of only 3.8 days, the risk is only significant when the gas can rise quickly into the immediate subsoil, conditions which obtain above fissured granite. In the UK, high levels occur in Cornwall, Devon, Derbyshire, Northamptonshire and in Scotland; New Yorkers living in basements and lower floors may also be affected. The pressure within buildings is normally lower than without and so Rn is drawn in via cavities and cracks in walls and through construction joints; it has seven times the density of air and so basements and ground floor rooms are the most exposed. The indoor pressure may be as much as 25–50 pA below that outdoors and soil gas entry may reach 20 litres/minute. This pressure difference can also lead to the entry of volatile chemicals from nearby waste tips. In one such incident in Los Angeles, vinyl chloride was detected indoors (Nazarov and Teichman, 1990).

7.9.1 Cancer risk

The worst cancer risk from Rn arises from the first four *solid* decay products of radon:

$$^{222}\text{Rn} \longrightarrow {}^{218}\text{Po} \xrightarrow[\alpha\,3.05\,\text{m}]{} {}^{214}\text{Pb} \xrightarrow[\beta\gamma\,27\,\text{m}]{} {}^{214}\text{Bi} \xrightarrow[\beta\gamma\,20\,\text{m}]{} {}^{214}\text{Po} \xrightarrow[\alpha,\gamma]{1.6\times10^4\text{s}} {}^{210}\text{Pb}$$

These products are dangerous because they associate with moisture and dust and become deposited in the lung, so exposing the bronchial epithelium to α-particles from ^{218}Po and ^{214}Po. Other nuclides of Rn exist but they do not present the same risk as that from ^{222}Rn; for example, ^{220}Rn formed by decay of ^{232}Th has too short a half-life to allow survival above ground.

Units.   The ^{222}Rn dose is measured in Becquerels (Bq), which are defined as 1 disintegration/s (dps), a tiny fraction of the Curie (Ci) which is equivalent to 3.7×10^{10} dps (Section 5.5.3). For practical purposes, the pico-Curie is preferred (note that 1 pCi/l $=$ 37 Bq/m^3).

Average exposure indoors to Rn in the USA is 1 pCi/l (Berger, 1990) but levels can rise into the 10–100 pCi/l range. In 70 000 homes the dose exceeds 800 Bq/m^3 (22 pCi/l) and the experience of 75% of these full-time occupants is four work-level months, or the limit for uranium miners: several thousand cancer cases result each year. At the extreme, exposure to Rn is equivalent to heavy cigarette smoking. Table 7.12 gives the incidence of death from lung cancer in smokers and non-smokers in the USA and the effect of Rn exposure.

There is evidence of synergism between smoking and exposure to Rn (Nazaroff and Teichman, 1993).

The median level of Rn outdoors in New York City is:

$$180 \, \text{pCi/m}^3 = 180 \times 10^{-12} \times 3.7 \times 10^{10} \, \text{Bq/m}^3 = 6.7 \, \text{Bq/m}^3$$

The median level in the basements and ground floor of buildings in New York is 30 Bq/m^3, evidence of greater risk indoors.

There is a risk from Rn in the water supply. In the USA the concentration range in water entering domestic premises is 58–137 pCi/l (Valentine and Stearns, 1994). The USEPA requires a limit of 300 pCi/l for a combination of ^{222}Rn, ^{226}Rn and ^{228}Rn; the activity will decline rapidly because they all have short half-lives.

It has been claimed (Cohen and Colditz, 1994) that there is a negative correlation between cancer mortality and average Rn levels in the USA and that there is no evidence of cancer induction by low levels of radiation, which contrasts with the positive predictions based on the no-threshold theory, i.e. a minute dose can induce cancer (pp. 304–5). The latter leads to an upper estimate of 10 000 cases per year caused by radon in homes in the

Table 7.12 Annual deaths from lung cancer in the USA (1986)

	Population ($\times 10^3$)	Lung cancer all causes	Deaths/year Rn attributed
Male			
Never smoked	63 900	1 900	200
Heavy smoker[a]	9 000	22 900	3100
Female			
Never smoked	81 300	3 100	300
Heavy smoker	5 300	7 900	1100

[a] > 25/day.

USA. Haynes (1988; see Gardner, 1992) also found a negative relationship in the UK, where the geographical distribution of cancer cases does not follow the pattern of high radon exposure.

In the UK, 2309 dwellings showed an average Rn concentration of $20\,Bq/m^3$. The highest averages of $110\,Bq/m^3$ and $74\,Bq/m^3$ were found in Cornwall and Devon, respectively. The National Radiological Protection Board advised that remedial action be taken when levels exceed $400\,Bq/m^3$. This degree of exposure corresponds to a 2% risk of death from lung cancer in non-smokers, a relatively small proportion of the risk of lung cancer run by smokers and put at 20% overall.

7.9.2   Remedial measures

These include sealing cracks in structures and laying building-weight polythene over floors. A more expensive measure is the construction underfloor of a low pressure sump packed with porous material and vented outdoors by an extractor fan.

References

Bayer, C. W. and Black, M. S. (1987) Capillary chromatographic analysis of VOCs in the indoor environment. *J. Chromatog., Sci.*, **25**, 60–63.

Berger, R. S. (1990) The carcinogenicity of radon. *Environ. Sci. Technol.*, **24**, 30–31.

Berode, M., Wietlisbach, V., Rickenbach, M. and Guillemin, M. P. (1991) Lifestyle and environmental factors as determinants of blood lead levels in Swiss population. *Environ. Res.*, **55**, 1–17.

Berrens, L. (1991) Estimation of the allergen content of house dust samples by enzymatic assay. *Environ. Res.*, **56**, 68–77.

Bloemen, H. J. Th. and Burn, J. (1993) *Chemical analysis of VOCs in the environment.* pp. 5–9. Blackie, London.

Brady, J. E. and Holum, J. R. (1987) *Elements of General and Biological Chemistry* (7th edn).

Brown, R. C., Hoskins, J. A. and Young, J. (1992) Not allowing the dust to settle. *Chem. in Brit.*, **28**, 910–915.

Burnett, R. T., Dales, R. E., Raizenne, M. E., Krewski, D., Summers, P. W., Roberts, G. R., Raad-Young, M., Dann, T. and Brook, J. (1994) Effects of low ambient levels of ozone and sulphates on the frequency of respiratory admissions to Ontario hospitals. *Environ. Res.*, **65**, 172–194.

Cheng, W. N. and Kong, J. (1992) A retrospective mortality cohort study of chrysolite asbestos product workers in Tienjin 1972–1987. *Environ. Res.*, **59**, 271–278.

Chesson J., Hatfield, J., Schultz, B., Dutrow, E. and Blake, J. (1990) Airborne asbestos in public buildings. *Environ. Res.*, **51**, 100–107.

Chuang, J. C., Kuhlman, M. R. and Wilson, N. K. (1990) Evaluation of methods for simultaneous collection and determination of nicotine and PAH in indoor air. *Environ. Sci. Technol.*, **24**, 661–665.

Cohen, B. L. and Colditz, G. A. (1994) Tests of the linear no-threshold theory for lung cancer induced by exposure to radon. *Environ. Res.*, **64**, 65–89.

Crump, D. R. (1995). In *Volatile Organic Compounds in the Atmosphere* (eds R. E. Hester and R. M. Harrison). 110–111. Royal Society of Chemistry, Cambridge.

Frischer, T., Studnicka, M., Beere, E. and Neumann, M. (1993) The effects of ambient nitrogen dioxide on lung function in primary schoolchildren. *Environ. Res.*, **62**, 179–188.

Gilioli, R. (1993) : EURONEST: a concerted action of the European Community for the study of organic solvents neurotoxicity. *Environ. Res.*, **62**, 89–98.

Guo, Z., Tichenor, B. A., Mason, M. A. and Plunkett, C. M. (1990) The temperature dependance of emission of perchloroethylene from dry cleaning fabrics. *Environ. Res.*, **52**, 107–115.

Hardy, R. J., Highsmith, V. R., Costa, D. L. and Krewer, J. A. (1992) Indoor asbestos concentrations associated with the use of asbestos contaminated tap water in portable home humidifiers. *Environ. Sci. Technol.*, **26**, 680–689.

Harte, J., Holden, C., Schneider, R. and Shirley, C. (1991) *Toxics A to Z*. University of California Press, Berkeley, CA.

Haynes, R. M. (1988) The distribution of domestic radon concentrations and lung cancer mortality in England and Wales. *Radiat. Protect. Dosim.*, **25**, 93.

Health and Safety Executive (1993). Occupational Exposure Limits. *HSE EH 40/93*. HMSO, London.

International Agency for Research on Cancer (IARC) (1986) Evaluation of the carcinogenic Risk of Chemicals to Humans, Vol. 38, *Tobacco Smoking*. WHO Lyons.

Kato, M., Rocha, M. L. R., Carvalho, A. B. Chaves, M. E. C., Rana, M. C. M. and Oliveira, F. C. (1993) Occupational exposure to neurotoxicants: preliminary survey of five industries in the Camacari petrochemical complex of Brazil. *Environ. Res.*, **61**, 133–139.

Krysanowski, M., Quackenboss, J. J. and Lebowitz, M. D. (1990) Chronic respiratory effects of indoor formaldehyde exposure. *Environ. Res.*, **52**, 117–125.

Levy, B. S. and Seplow, A. (1992) Asbestos-related hazards in developing countries. *Environ. Res.*, **59**, 167–174.

Lundy, P. and Barer, M. (1992) Asbestos-containing materials in New York City buildings. *Environ. Res.*, **58**, 15–24.

Manabe, S. and Wada, O. (1990) Carcinogenic tryptophan pyrolysis products in cigarette smoke condensate and cigarette polluted indoor air. *Environ. Pollution*, **64**, 121–132.

Mehlman, M. A. (1992) Dangerous and cancer-causing properties of products and chemicals in the oil refining and petrochemical industry. *Environ. Res.*, **59**, 238–249.

Michael, L. C., Pellizzari, E. D., Perritt, R. L., Hartwell, T. D., Westerdahl, D. and Nelson, W. C. (1990) Comparison of indoor, backyard and centralised air monitoring strategies for assessing personal exposure to VOCs. *Environ. Sci. Technol.*, **24**, 996–1003.

Mushak, P. (1992) Defining lead as the premier environmental health issue for children in America: criteria and their quantitative application. *Environ. Res.*, **59**, 281–309.

National Academy (1981) *Indoor Pollutants*. National Academic Press, Washington DC.

Nazarov, W. W. and Teichman, K. (1990). Indoor Radon. *Environ. Sci. Technol.*, **24**, 774–782.

Rogge, W. F., Hildeman, L. M., Mazurek, M. A., Cass, G. R. and Simonelt, B. R. T. (1993). Sources of fine organic aerosol. 5. Natural gas home appliances. *Environ. Sci. Technol.*, **27**, 2736–2744.

Schwartz, J. and Levin, R. (1991) The risk of lead toxicity in homes with lead paint hazard. *Environ. Res.*, **54**, 1–7.

Spengler, J. D., Brauer, M., Samat, J. M. and Lambert, W. E. (1993) Nitrous acid in Albuquerque, New Mexico homes. *Environ. Sci. Technol.*, **27**, 841–845.

Takada, H., Onda, T. and Ogura, N. (1990) Determination of PAH and their source materials by capillary GC. *Environ. Sci. Technol.*, **24**, 1179–1186.

Tewes, F. J., Koo, L. C., Meisgen, T. J. and Rylander, R. (1990) Lung cancer risk and mutagenicity of tea. *Environ. Res.*, **52**, 23–33.

United States Environmental Protection Agency (1987) Report of the Federal Panel on Formaldehyde.

Valentine, R. L. and Stearns, S. W. (1994) Radon release from water distribution system deposits. *Environ. Sci. Technol.*, **28**, 534–7.

Wang, J.-D. and Chen, J.-D. (1993) Acute and chronic neurological symptoms among paint workers exposed to mixtures of organic solvents. *Environ. Res.*, **61**, 107–116.

Weschler, C. J., Hodgson, A. T. and Wooley, J. D. (1992) Indoor chemistry: ozone, VOC and carpets. *Environ. Sci. Technol.*, **26**, 2771–2377.

Weschler, C. J., Shields, H. C. and Nalk, D. V. (1994) Indoor chemistry involving ozone, NO and NO_2 as evidenced by 14 months of measurement at a site in southern California. *Environ. Sci. Technol.*, **28**, 2120–2132.

Wietlisbach, V., Rickenbach, M., Beroda, M. and Guillemin, M. (1995) Time trends and determinants of blood levels in a Swiss populated over a transition period (1984–1993) from leaded to unleaded gasoline. *Environ. Res.*, **68**, 82–90.

Yoshikawa, M., Arashidani, K., Kawamoto, T. and Kodama, Y. (1994) Aryl hydrocarbon hydroxylase activity in human lung tissue in relation to cigarette smoking and lung cancer. *Environ. Res.*, **65**, 1–11.

Zhang, J., He, Q. and Lloy, P. J. (1994) Characteristics of aldehydes: concentrations, sources and exposures for indoor and outdoor residential microenvironments. *Environ. Sci. Technol.*, **28**, 146–152.

Further reading   Lee, B. S. and Delbert, J. E. (1991). VOC in the indoor environment. In *Organic Chemistry of the Atmosphere*. CRC Press. Boca Raton, FL.

Part Three
Wastes and Other Multipollutant Situations

Wastes and their disposal 8

8.1
Introduction

Wastes and their disposal are the cause of a great deal of environmental pollution. The pollutants involved come from a wide range of sources including 150-year-old heaps of Pb–Zn mine waste in remote upland areas, highly toxic wastes from the chemical industry and nuclear power stations and household waste. Industrial manufacturing has expanded in the twentieth century and since the Second World War there has been a pronounced trend towards increasingly complex products either containing highly toxic chemicals, such as PCBs in electrical equipment, or involving toxic materials in their manufacture. With increasing affluence in the more technologically developed countries more municipal solid waste and waste water are being produced and these need treatment and disposal. Even in the home, there is a much wider range of chemicals in everyday use (Table 2.8 and Chapter 7) as well as many more people pursuing hobbies which involve the use and disposal of chemicals (car engine oil, decorating and DIY products, photography, drycell batteries, etc.)

Industrial wastes, especially hazardous wastes, can contain a wide range of chemicals some of which may be unstable. Others are highly stable and difficult to denature in order to reduce their hazard to organisms or structures. Wastes can include many of the groups of pollutant substances featured in this book. The main considerations in this chapter are:

1. the amounts of waste produced;
2. the nature of the wastes;
3. the options available for treatment or disposal;
4. the environmental safety of these wastes management options.

8.2
Amounts of waste produced

In 1990, OECD countries produced 9×10^6 t municipal waste, 1.5×10^9 t industrial wastes (including 300×10^6 t hazardous waste and 7×10^9 t other wastes (including fossil fuel ash from electricity generation, agricultural wastes, mine wastes, demolition debris and sewage sludge) (OECD, 1991).

8.2.1 Industrial wastes

The OECD group, which includes the 24 most technologically advanced countries in the world, together produce 68% of the world's industrial waste and 89.6% of the hazardous and special wastes. North America alone produces 57% of the OECD's industrial waste and 54% of its municipal waste.

Total OECD industrial waste $= 1.43 \times 10^9$ t
Total world industrial waste $= 2.1 \times 10^9$ t

Eastern Europe industrial waste $= 0.52 \times 10^9$ t
Eastern Europe hazardous and special waste $= 19 \times 10^6$ t
(OECD, 1991).

8.2.2 Municipal wastes

Municipal wastes comprise the waste from domestic houses, offices and commercial properties but not industrial or hazardous wastes. In the USA, municipal wastes on average comprise 41–50% paper; 9–18% glass; 10–18% yard (garden) waste; 10–17% miscellaneous; 9% metals; 8–12% food waste; and 7% plastic. Table 8.1 compares the municipal waste production from various countries.

**8.3
Methods of disposal of
municipal wastes**

8.3.1 Landfilling

Landfilling involves placing the waste in compacted layers in a lined pit or a mound (sanitary landfills) with appropriate leachate and landfill gas (mainly CH_4 and CO_2) control. The percentage of municipal solid wastes disposed of by landfill in selected countries is:

USA   60%
EC   70%
Japan   38%

Landfilling is not strictly a disposal method, but really one of containment or indefinite waste storage. Many landfills are used for the disposal of both

Table 8.1 Municipal waste produced per capita in different countries (kg/cap)(OECD, 1991)

	Mid 1970s	1980	Mid 1980s	Late 1980s
N. America	633	687	734	826
Japan	341	355	344	394
OECD (Europe)	277	323	346	336
OECD (all countries)	407	436	493	513

municipal and hazardous waste. The latter are 'co-disposed' with the municipal waste. This utilizes the absorptive properties of the bulky municipal waste and the microbial degradation reactions which occur wherever putrescible material is buried. There is now a move towards exploiting the microbial methanogenic processes in landfills and using them as bioreactors to both dispose of waste and produce methane which can be recovered and used as a source of energy.

The advantages of landfilling are that it is relatively inexpensive, methane production can be exploited from municipal wastes and the wastes can be recycled or treated at a later date. One particular benefit is that in some countries, such as the UK, landfilled municipal wastes can be used for the co-disposal of some types of hazardous waste. The disadvantages are that there is a danger of leakage and pollution of groundwater, a danger of explosions from methane and the problem that there are fewer and fewer suitable sites available. Modern landfill sites need to be located in impermeable strata or sealed with an appropriate membrane before the wastes are deposited and should have a system for managing the leachate. Older landfills worked more on the principle of waste attenuation, where it was accepted that leachate would leak from the site but it was intended that the rate and volumes of leachate involved would not cause any acute environmental problems. It is now realized that this type of waste disposal is responsible for many aquifers being significantly polluted with a wide range of xenobiotic and other chemicals.

Landfill gas varies in composition depending on the composition of the waste, its density, temperature and moisture content, the age of the landfill and the extent to which leachate is recycled through the landfill. Typical concentrations of the main components are shown in Table 8.2. The changes in the relative composition of the gases with time are shown in Fig. 8.1. After the utilization of the air within the waste, the onset of anaerobic conditions in phase II eventually leads to methanogenesis (in phases III and IV). This can occur after a few weeks or months and can continue for

Table 8.2 Typical composition of landfill gas (Lisk, 1991)

Component	Percentage by dry volume
Methane	47.4
CO_2	47.0
Oxygen	0.8
Nitrogen	3.7
Hydrogen	0.1
H_2S	0.01
CO	0.1
Aliphatic hydrocarbons	0.1
Aromatic hydrocarbons	0.2
Trace compounds	0.5
Ethene	0.002

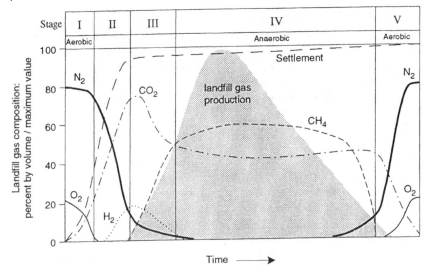

Figure 8.1 Changes in the composition of landfill gas with time (DOE, 1995).

periods of up to 30 or even 70 years. The combustible methane content of the landfill gas renders this a potentially valuable energy resource which can be exploited for either electricity generation (using modified diesel engines) or as a fuel in kilns/furnaces, such as in the production of bricks and cement.

The leachate from landfills will vary widely according to the nature of the waste deposited in it and the stage in the decomposition of the putrescible material. Hazardous waste which has been co-disposed with municipal solid waste could contain significant concentrations of highly toxic chemicals; the landfills licensed for this disposal should have appropriate leachate control. Table 8.3 gives data for the range of concentrations likely to be encountered in landfills.

From Table 8.3, it can be seen that a wide range of concentrations can be found in landfill leachates. The maximum values indicate that these leachates can have very severe polluting effects if they enter surface or groundwaters, especially with regard to biochemical and chemical oxygen demand (BOD and COD). The potentially high concentrations of Cl^- can have important implications for the speciation and mobility/bioavailability of several heavy metals. The Cl^- can combine with Cd and other metal cations to form chloride complexes such as $CdCl^+$, which is relatively stable and less prone to adsorption on mineral surfaces than Cd^{2+}.

Table 8.4 gives an indication of trends in the composition of landfill leachates with age of the landfill. Almost all the constituents, except $NH_3 - N$ are more abundant in the leachates of young landfills where active decomposition of organic material in the waste is taking place.

Table 8.3 The composition of leachates from landfills (based on 15 landfills in the UK)

Constituent	Concentration range (mg/l)
pH	6.2–7.4
BOD	66–11600
COD	<2–8000
TOC	21–4400
Ammonia N	5–730
Nitrate N	<0.2–4.9
Organic N	ND–155
$H_2PO_4^-$	<0.02–3.4
Cl^-	70–2777
SO_4^{2-}	55–456
Na	43–2500
Mg	12–480
K	2–650
Ca	165–1150
Cd	<0.005–0.01
Cr	<0.05–0.14
Cu	<0.01–0.15
Ni	<0.05–0.16
Pb	<0.05–0.22
Zn	<0.05–0.95

From Robinson *et al.* (1982) in Lisk (1991).
BOD, biological oxygen demand; COD, chemical oxygen demand; TOC, total organic carbon.

Table 8.4 Variation with age in the typical concentrations of common constituents of landfill leachates (Lisk, 1991)

Constituent/property	Age of landfill		
	Young	Medium	Old
pH	5.7–8.0	6.4–8.0	6.6–8.3
BOD (g/l)	7.5–17.0	0.37–1.1	0.07–0.26
COD (g/l)	10.0–48.0	1.2–22.0	0.67–1.9
$N(NH_3)$ (g/l)	0.04–1.0	0.03–3.0	0.01–0.9
Cd (mg/l)	0.02–0.10	0.04–0.08	0.01–0.14
Cu (mg/l)	0.08–0.30	0.02–0.11	0.03–0.12
Pb (mg/l)	0.05–0.92	0.04–0.08	0.03–0.12
Zn (mg/l)	0.53–34.2	0.18–0.22	0.19–0.37

Landfilled household waste tends to be more acid than many other types of waste because it contains larger amounts of decomposable organic material. Various acidic organic compounds are produced during the decomposition process.

In modern landfills complying with recent regulations, the leachate accumulating in a sump at the base of the landfill is either recycled through the landfill, where the moisture and nutrients stimulate microbial activity which leads to greater rates of methanogenesis and decomposition of hazardous organic pollutants, or the leachate is taken away for treatment off-site.

8.3.2   Incineration

Incineration involves passing the waste through a chamber at a high temperature (preferably around 1200°C) with an adequate supply of oxygen to oxidize all organic material (p. 306). However, at present, many municipal waste incinerators do not satisfy the new emission regulations mainly because they do not reach a sufficiently high temperature and do not have flue gas cleaning facilities (electrofiltration, mechanical filtration). From 1 December 1996, incineration plants in the UK dealing with municipal solid waste (MSW) must conform to the new HMIP Standards (IPR5/3) which are given in Table 8.5, together with the EC Directive (89/369EC) applying to new incineration plants. Some incinerators are constructed to allow the recovery of some of the heat of combustion, which is either used for generating steam and electricity or for neighbourhood heating.

Although incineration is regarded as a waste disposal method, it is actually only a system of waste reduction; however, it does have the desirable effect of decomposing all organic compounds in the process. An

Table 8.5   Emission standards for incinerators[a]

	EC Directive emissions (ng/m³)		HMIP (IPR5/3) emissions (ng/m³) for > 1 t/h plant[b]
	1–3 t/h plant[b]	> 3 t/h plant[b]	
Total Particulate matter	100	30	30
CO	100	100	100
VOCs (excluding particulates, as total C)	20	20	20
Acid gases			
SO₂	300	300	300
HCl	100	50	30
HF	4	2	2
NOₓ (as NO₂)	–	–	350
Heavy metals			
Cd and Hg	0.2	0.2	0.1 each
Ni and As	1	1	1
Other metals	5	55	1
Dioxins (ng TEQ/m³)	–	–	1

[a] The requirements of combustion conditions are a temperature of at least 850°C for a minimum of 2 seconds in 6% oxygen (or other effective technique for reducing emissions of dioxins) (RCEP, 1993).
[b] Plant size given by its capacity in tonnes per hour incinerated.

ash of inorganic residues is produced which requires disposal in a controlled landfill. However, the disposal of incinerator ash is causing some political debate because if it is classed as a hazardous waste its disposal will be much more expensive that ordinary municipal waste, which, together with the relatively high cost of the incineration operation itself, make the economics of this waste management option look less favourable.

More than 50% of the municipal waste in Japan, Sweden and Switzerland is incinerated, and more than 20% incinerated in Belgium, France, Germany, Italy, Luxembourg and the Netherlands. For the EC countries as a whole, of those incinerators which have been fitted with flue gas treatment, 35% have mechanical filters, 36% have electrofiltration and just 7% have both of these combined (OECD, 1991).

The advantages of incineration are that it can dispose of 99.999% of organic wastes (including chlorinated organic xenobiotics) if properly carried out (high temperature 1200°C, adequate oxygen, etc.), it reduces the volume of waste and energy can be recovered from the process and utilized for electricity generation and combined heat and power. The disadvantages are that it is relatively expensive, there is a danger of highly toxic pollutants (PCDDs and PCDFs) being synthesized and emitted into the atmosphere if conditions are not optimal and, finally, there is an ash produced which needs careful disposal in sanitary landfills.

In the UK the emission limit for VOCs will be set at $50 \, mg/m^3$ in 1997 to meet the requirements of the Environmental Protection Act, 1990.

8.3.3   Composting

Composting involves shredding and separating the putrescible fraction of municipal waste, often mixing it with other organic materials (including sewage sludge) and allowing microbiological decomposition reactions to take place. This involves regular turning of the compost in order to promote aerobic decomposition processes and allowing the compost temperature to rise sufficiently to kill off pathogenic organisms and weed seeds and other undesirable constituents. The finished compost is intended for use as a growing medium for plants, especially as a substitute for peat, which is a scarce resource from an ecosystem at risk of destruction with the accompanying problems of extinction of species. Within Europe, the composting of municipal wastes is carried out to a significant extent in Austria, Belgium, France, Italy, the Netherlands, Portugal and Spain (OECD, 1991).

8.3.4   Recycling

Where materials are recycled (preferably on-site) there will be a saving in waste-disposal costs as well as in inputs to the process. In addition to

Table 8.6 Total waste arisings in England and Wales (House of Commons Env. Comm., 1989; BMA, *Hazardous Waste and Human Health* (1991), by permission of Oxford University Press)

Waste type	Quantity (10^6 t/year)
Liquid industrial effluent	2000
Agriculture wastes	250
Mines and quarries	130
Industrial total	50
hazardous and special	(3.9)
special	(1.5)
Domestic and trade	28
Sewage sludge	24
Power station ash	14
Blast furnace slag	6
Building	3
Medical wastes	0.15
Total	2505.15

materials, energy can also be recovered from the incineration of combustible wastes, or the production of methane from putrescible wastes, which can make a positive contribution to energy budgets. Some examples of recycling in practice are:

- recycled paper is used in many products
- scrap metal is used widely in steel manufacture
- recycled glass is being collected (bottle banks) and used on an increasing scale
- non-ferrous metals including Cu, Pb and Zn are recycled
- old engine oils are treated and reused
- plastics will soon be recycled on an increasing scale after a considerable research and development effort, mainly in the USA.

An example of the waste arisings in a technologically advanced country is provided by the data for England and Wales in Table 8.6.

8.4
Sewage treatment

The treatment of wastewater is of vital importance for preventing the spread of infectious diseases, such as cholera (Section 6.4.1), and protecting river water quality and avoiding noxious odours. In England and Wales, 23×10^6 m^3 of domestic wastewater and 14.1×10^6 m^3 of industrial wastewater are discharged to the sewers every day (Lester, 1990). This sewage is treated by a total of 5000 sewage works around England and Wales, which serve around 44×10^6 people. In addition, the sewage from a further 6×10^6 people living near the coast is discharged into the sea without treatment and up to 2×10^6 people are not connected to the sewage system and use their own septic tank systems (Lester, 1990). The discharge of sewage into the sea around the coastline of Britain has attracted a lot of attention and adverse publicity both concerning the highly polluted state of the North Sea and its marine ecosystem and its possible effects on the health of bathers and surfers using the beaches. As a result of a draft EC Directive in 1989, which was accepted in principle in the UK in 1990, the direct discharge of untreated sewage to the sea will be discontinued by 31 December 1998, but there may be a delay while extra sewage works are built to handle the additional sewage.

In the UK, the original object of sewage treatment was to avoid disease and unpleasant odours from rivers carrying untreated sewage. This is exemplified by the 'Year of the Great Stink' in London in 1858 when a combination of a long dry summer with reduced flows in the River Thames and high rates of sewage discharge into the river (from an increasing number of houses with flush sanitation) caused such a bad odour problem that the business of the Houses of Parliament (which overlook the Thames) had to be adjourned for several days. There had already been three major epidemics of cholera in London before this (1831–2, 1848–9 and 1853–4). Eventually, after further cholera epidemics and the continuing odour problem, a Royal Commission on Sewage Disposal was appointed in 1881. This body was instrumental in putting various remedial measures into action which resulted in marked improvements in the smell and quality of the Thames (Ellis, 1989).

The sewage treatment processes were shown to be very effective in cleaning up wastewaters and so long as the concentrations of the polluting materials are reduced to levels within the capacity of the aquatic ecosystems receiving the purified wastewater, further purification and dilution will occur. With increasing population in many parts of Britain, especially in London and the Thames Valley, it became necessary to re-use treated water further downstream and nowadays in the UK about 30% of all drinking water supplies involve some indirect re-use, which is a much greater proportion than in most other countries (Lester, 1990).

Sewage treatment involves three basic processes (Fish, 1992):

1. the removal of polluting matter from the sewage flow as solids or slurries of solids in water (sludges)
2. the removal of polluting matter from the sewage flow and separated sludges by accelerated natural processes of biochemical breakdown brought about by microorganisms
3. the separation of water from sludges to reduce the volume of sludge for disposal.

These three processes occur during six stages in the treatment of sewage, which are (Fish, 1992):

1. a preliminary screening to remove grit and solids
2. a primary settlement stage in tanks to allow the removal of solids and grease

3. secondary treatment involving the microbial oxidation of organic matter and NH_3 and the further removal of solids
4. a polishing treatment using sand filtration to remove very fine solids
5. tertiary treatment under anaerobic conditions to bring about denitrification and chemical precipitation and remove N and P
6. a sludge treatment stage involving combinations of digestion, thickening, dewatering and drying to prepare the sludge for disposal and also to produce methane (which is utilized to power pumps in the works).

It is important to note that if the sewage is heavily polluted with toxic substances, many of the microorganisms involved in the biochemical processes may be killed and the treatment will be incomplete. This is soon detected and can be coped with, but it causes a lot of extra problems at the sewage works.

Domestic sewage contains approximately 0.1% (1000 mg/l) of impurities, about half of which are dissolved and the rest are in suspension. About 70% of the impurities are organic and include proteins and urea, sugars, starches and cellulose, soap, cooking oil and greases. With sludge production set to rise by 50% over the next decade, there is concern about the organochlorine content. It is believed that dioxins enter the food chain by this route and PCBs have been detected at average levels of 0.29 mg/kg. The inorganics include chloride and metallic salts and road grit where storm water is combined in the sewage. This raw sewage normally has a suspended solids load of approximately 400 mg/l and a BOD of 300 mg/l (Lester, 1990). After full primary and secondary treatment and 6 hours aeration in a diffused air sludge activation plant, the final effluent should contain less than 15 mg/l suspended solids, about 15 mg/l of 5-day BOD and NH_3 nitrogen content of less than 10 mg/l (Fish, 1992) (Section 2.5.3). This is discharged into rivers where it becomes mixed and diluted and could undergo further biochemical purification.

Raw water extracted for public water supply is usually disinfected with chlorine and/or ozone or UV light. There has been concern about the synthesis of organohalides in water as a result of chlorination. Recent studies in the USA have added further weight to the apparent association of a small increase in bladder cancer incidence in long-term consumers of chlorinated water. However, it is acknowledged that the possible slight risk to health is not as great as that for the consumption of non-disinfected water (Neutra and Ostro, 1992). However, there is obviously a need for further careful investigation of this subject since a very large proportion of the populations of technologically advanced countries use chlorine for water disinfection.

The sludge produced from sewage treatment amounted to around 5.5 million (dry) tonnes in the USA in 1992 and the total produced in the original 12 EU countries amounted to 6.3 million tonnes. The sludge produced in the UK was 1.1 million tonnes in the early 1990s but is predicted to rise to 2.2 million tonnes by the year 2006 (McGrath *et al.*, 1994). This sludge needs disposal and the main options for this are: (i) land application, (ii) dumping at sea, (iii) landfilling, and (iv) incineration. The EU has banned the sea disposal of sewage and sludge from the end of 1998 and many other countries are also phasing it out. This implies that the amounts disposed of by the other options will increase, and land application is expected to increase to the greatest extent. In the UK in 1993, 43% of sludge was applied to land and 30% dumped at sea while 22.2% of sludge in the USA is applied to land. In the UK in the 1990s, around 1% of arable land receives sludge in any one year, but around 10% of all this land has received sewage sludge at some time in the past. In most cases, the land receiving sludge has tended to be near to towns (and sewage works). In addition to sludges produced from normal sewage treatment, there is an increasing trend to produce treated sludges, which are considered to have superior properties. Sewage sludge has several beneficial properties for use on agricultural soils. It contains useful amounts of nitrogen and phosphorus which are usually applied in fertilizers and, therefore, the sludge has positive fertilizer value as well as being a source of organic colloids which make a beneficial contribution to soil aggregate stability and soil structure. However, it is the pollutants in sludge which limit its usefulness. Heavy metals are a particular problem because they are concentrated in the sludge and most will accumulate in the soil with continued applications of sludge (Chapter 2 and Section 5.3.3, p. 199). The maximum permissible concentrations of metals in soil after the application of sewage sludge to agricultural land are given in Table 8.7.

In the UK, the Department of the Environment (1993) reported that, compared with 1982/83, the concentrations of most heavy metals in sewage sludge in 1990/91 had shown a marked decrease as a result of waste minimization and the introduction of recycling and clean technology.

Table 8.7 The EC maximum permissible concentrations of heavy metals and other elements in sewage sludge amended soils (EC, 1986; MAFF, 1992)

Element	Maximum concentration (mg/kg dry solids)	Maximum rate of addition over 10 year period (kg/ha)
Cd	3	0.15
Cr	400 (provisional)	15 (provisional)
Cu	100	7.5
Hg	1	0.1
Ni	50	3
Pb	300	15
Zn	300	15
Mo[a]	4	0.2
Se[a]	3	0.15
As[a]	50	0.7
F[a]	500	20

[a] Elements not subject to EC Directive 86/278/EEC.

Table 8.8 Median metal concentration in UK sewage sludges 1982–91 (mg/kg DM) (UK DOE 1993)

	Median (50%ile) metal concentration (mg/kg DH)	
	1982/3	1990/1
Cd	9.0	3.2
Cu	625	473
Cr	124	86
Pb	418	217
Hg	3.0	3.2
Ni	59	37
Zn	1205	889

Table 8.8 gives the median (50%ile) concentration of heavy metals in sewage sludges applied to agricultural land between the two periods.

While several metals showed over a 50% reduction, Zn did not decrease to the same extent and, in the light of the microbial toxicity reported in Section 5.3, there is still major concern about the accumulation of Zn in sewage sludge-amended soils. The Ministry of Agriculture, Fisheries and Food (MAFF) has recommended that the maximum Zn concentration in sludged soils be reduced from 300 mg/kg to 200 mg/kg. One of the reasons why Zn concentrations in sludges are still relatively high is the widespread use of the metal in domestic products such as some types of baby cream, shampoos and cosmetics.

Average application rates for sludges are 10 t dry solids/ha, but in general these tend to be in alternate years and so the time taken for Zn or other limiting metal concentrations to reach the maximum permitted are likely to be many decades. However, although this will be a relatively long time, it will inevitably be reached unless either sewage sludge is disposed of in some other way or the metal concentrations decrease even further.

Recent US environment protection regulations for the maximum concentrations of metals in soils amended with sewage sludge (now officially called 'Biosolids' in the US) give markedly different values to those in force within the EU (US EPA, 1992). These are indicated in Table 8.9.

From Table 8.9, it can be seen that the maximum concentration of metals allowable in sludged soils in the USA is many times more than that in EU regulations, with the exception of Pb. The new US values were based on probabilistic risk assessment using a series of 14 exposure pathways (such as sludge–soil–plant–human (No. 1) and exposure through fish and drinking water (No. 12) and considering the most exposed individuals. Although the limiting concentrations are high, it is not expected that they will be reached; on the contrary, it is expected that sludges will contain lower metal levels and so there will not be a major accumulation in treated agricultural soils.

Table 8.9 Comparison between the maximum allowable concentrations of metals in soils treated with sewage sludge in the EU and the USA (From McGrath *et al.*, 1994)

	Maximum allowable concentrations (mg/kg)						
	Cd	Cu	Cr	Hg	Ni	Pb	Zn
EU[a]	1–3	50–140	100–150	1–1.5	30–75	50–300	150–300
USA	20	750	1500	8.0	210	150	1400

[a] EU values: lower, recommended; higher, mandatory.

These regulations are based solely on the exposure of humans and do not cover ecotoxicological risks. In contrast, the EU limits are based on the precautionary principle and individual EU member states are allowed to use lower limits but none can have higher than the mandatory limit. The philosophy behind the EU values is that if soils receive cumulative amounts of sewage sludge which take the metal concentrations to these maximum concentrations, the fertility of the soil will still have been protected. However, recent research has shown that toxic effects on some species of soil microorganisms can occur below the maximum value of 300 mg/kg Zn (McGrath *et al.*, 1994). The lower maximum limit for Pb in the US regulations is because of the risks to people (especially children ingesting soil, either intentionally or accidentally).

Sewage sludges can also contain a wide range of organic micropollutants and of these PAHs (Section 6.2.5) and dioxin-like compounds (p. 303) constitute the greatest health hazard. PCDDs and PCDFs occur in sludge from industrial/rural catchments in the range 29–67 TEQ/kg dry weight but, surprisingly, unusually high concentrations, in the range 7–192 TEQ/kg, arise from domestic wastewater (Jackson and Edulgee, 1994).

PCDD/F levels decline, as a result of vapour transport and surface photolysis, with a typical half-life of 10–12 years. On this basis, levels in soils will be increasing and higher than the original background after ten annual applications. As a result of uptake by plants and by grazing animals, these compounds enter the food chain (MAFF, 1992). In the UK, mean levels are 9×10^{-2} ng TEQ/kg in root vegetables and 8×10^{-2} ng TEQ/kg in milk.

Control legislation has been enacted by Germany, which has set a limit for PCDD and PCDF in sewage sludge applied to agricultural land at 100 ng TEQ/kg dry weight, with a further restriction to 5 t/ha over 3 years.

**8.5
Hazardous wastes**

8.5.1   The nature and amount of hazardous waste produced

Hazardous wastes consist of individual waste materials and combinations of wastes that are presently, or potentially, hazardous to humans and other living organisms by means of their physical or chemical characteristics, the process by which they are produced, or their effect on human health or the environment (Manahan, 1984; BMA, 1991). The criteria used in the classification of hazardous substances include the type of hazard involved (toxicity, explosiveness, flammability and corrosiveness), the generic category of the substances (e.g. pesticides, wood preservatives, solvents, medicines), technological origins (electroplating, oil refining) and the presence of specific substances (e.g. Cd and Pb compounds, PCBs). Radioactive waste substances are normally classified separately owing to their high toxicity and long life. A brief list of some of the types of substance classified as hazardous includes explosives, compressed gases, flammable liquids, flammable solids, oxidizing materials, corrosive materials, poisonous materials, aetiologic agents (e.g. bacteria and viruses causing infectious diseases) and radioactive materials (Manahan, 1991).

Hazardous wastes can cause considerable pollution of air, water and soil even before they are officially disposed of. This can occur during transit, such as when ships carrying hazardous wastes have sunk or have lost deck cargoes of dangerous chemicals overboard, and accidents involving lorries or trains carrying hazardous wastes. Widespread pollution has also occurred during the temporary storage of wastes in ponds and open tanks where they can volatilize in the air, or leak into surface water, the soil and groundwater. However, in many cases these instances of pollution have not been covered by disposal legislation because that only applies when they are formally disposed of. New legislation in some countries now places an obligation on the generator of the waste to be responsible for it from its generation to its final legitimate disposal.

World total production of hazardous and special wastes $= 338 \times 10^6$ t
   (of which 275×10^6 t, or 81%, generated in the USA)
Total production of hazardous and special wastes in OECD countries $=$
   303×10^6 t
(OECD, 1991).

Of the hazardous waste produced in the USA, 79% came from the chemical industries, petroleum refining produced 7%, the metal industries 2%, and 12% came from miscellaneous other sources.

8.5.2 Hazardous waste management

Source reduction of hazardous wastes. Where possible, the industrial process producing the waste should be modified so that as far as possible the hazardous by-products are either avoided, re-used, or at least minimized. The last approach is referred to as source reduction and has been shown to be economically worthwhile by the 3M Corporation (Minnesota, Mining and Manufacturing Corp.) in the USA. The company launched a 'pollution prevention pays' programme in 1975, which had saved the company $192 million within 10 years and reduced its effluent discharge by 3.7 billion litres, eliminated 10 000 t water pollutants, 140 000 t sludge and 90 000 t air pollutants. However, source reduction is not being adopted as quickly as was expected, and the chemical and allied products industry, which accounts for roughly half of the hazardous waste generated in the USA, has not yet adopted this approach on a large scale. One of the main obstacles to this is the reluctance of competing companies to share information (World Resources Institute, 1992).

The procedures used to treat the hazardous wastes that are produced include:

physical methods (Manahan,1991)
- phase separation (filtration/sedimentation)
- phase transition (distillation, evaporation, physical precipitation)
- phase transfer (extraction, adsorption)
- membrane separation (reverse osmosis, hyper- and ultrafiltration)

chemical methods (Manahan, 1991)
- acid/base neutralization
- chemical extraction and leaching
- chemical precipitation
- oxidation/reduction
- ion exchange
- electrolysis
- hydrolysis
- photolysis

thermal treatment (Section 8.5.3)
- incineration
- wet oxidation
- plasma arc
- molten salt
- superheated water

biotechnological methods (Section 8.5.3 and Chapter 2)
ocean dumping
perpetual storage
- landfill
- underground injection
- salt formations
- arid region, unsaturated zone.

Ocean dumping basically relies on dilution to minimize the hazard from waste substances but in some cases localized effects occur because of poor mixing as a result of stratification, winds, tides and currents. The worst-affected marine ecosystems are those of estuaries (pp. 38–9 and 202–3) and land-locked seas with restricted circulation, with polluting industries either discharging into rivers that drain into the seas or discharging directly into the sea. Examples of extensively polluted seas include the Irish Sea, the Baltic, parts of the North Sea, the Mediterranean, the Black Sea and others. The Irish Sea is the most radioactively polluted sea in the world because of discharges from the Sellafield nuclear reprocessing plant. This sea is also extensively polluted with organic micropollutants, including PCBs, and metals from industrial areas, such as Liverpool. In 1969, a large number of dead sea birds were found in the Irish Sea and it is thought that their deaths were caused, at least in part, by pollutants. Liquid hazardous wastes dumped in sealed containers offshore will leak out eventually and will be dispersed and diluted in due course. The extent to which pollutants accumulate in food chains in the open oceans is not fully understood, although the ubiquitous occurrence of substances such as DDT in marine fauna indicates that this is likely to be significant. The London Dumping Convention of 1972 was intended to reduce the disposal of hazardous substances in the sea and it contained annexes that listed substances which should not be dumped at sea unless it could be demonstrated that they were only present in wastes in trace quantities or would rapidly be rendered harmless in the sea (Tolba *et al.*, 1992).

During 1986–7, 579 000 t of waste (from England and Wales) were dumped at sea. About half of this amount was hazardous waste and so marine dumping is the second most important means of disposal for hazardous waste, after landfilling in the UK. In comparison, during 1988, only 5500 t of hazardous waste was incinerated at sea (BMA, 1991). It is assumed that the plume of incinerator fumes reacted with the seawater and hence did not require elaborate clean-up before emission from the incinerator.

In England and Wales, in 1986–7, 83% of hazardous waste was disposed of by landfill, usually co-disposed with municipal wastes. Around 7% of these hazardous wastes were pretreated chemically or physically to neutralize or stabilize them and so reduce their hazardous properties before they were placed in landfills.

Only 2% of hazardous wastes produced in England and Wales were disposed of by incineration on land (BMA, 1991). However, some of the most highly toxic wastes, including those containing chlorinated hydrocarbons, were rendered harmless by this route. As discussed in Section 6.4.13, the products of incomplete combustion of various chlorinated organic waste could include TCDD (dioxins) and PCDFs (furans).

8.5.3 New technologies for waste disposal

There are a number of new technologies being developed for waste disposal (World Resources Institute, 1992).

Biotechnology. This involves using appropriate microorganisms to decompose organic wastes which are too hazardous to treat by other means. This technique is more appropriate for the treatment of contaminated land, or stripped contaminated topsoil, than hazardous wastes from industrial processes. Although research has been conducted into the use of genetically engineered organisms for this, most of the working techniques involve the use of naturally adapted organisms isolated from sites which have been polluted for a relatively long time. The advantages of this method are that it is relatively inexpensive and does not require expensive specialized equipment. However, it is a relatively slow process which may not completely destroy all the hazardous organic compounds.

Plasma arc destruction. This technique creates temperatures of up to 45 000°C and a plasma state. It is particularly useful for destroying inert organic pollutants, such as PCBs, and materials containing them. The main advantage of the method is the almost complete destruction of refractory hazardous organic substances but it is likely to be expensive.

Molten salt destruction. Hazardous chemicals, such as a DDT, chemical warfare agents, corrosive solvents and acids can be destroyed by treatment in a hot bath of molten salt at 1650°C. Other methods involve a molten mixture of Na_2CO_3 and Na_2SO_4 at 900°C, which has brought about the 99.999% destruction of hexachlorobenzene. This method does not require so much energy as some forms of incineration but is only around 99% effective over a range of compounds.

Superheated water. Water heated to 370°C at 220 bars pressure can dissolve normally insoluble organic chemicals and if oxygen is added to this water the organic pollutants are oxidized to CO_2 and H_2O. Inorganic compounds in the water combine to form salts. This method competes favourably with incineration for the destruction of most organic pollutants.

8.6
Long-term pollution problems of abandoned landfills containing hazardous wastes

Many countries have abandoned landfills or areas of derelict land either where hazardous wastes have caused toxicity or other problems already, or where the risk of potential toxicity problems is so great that it cannot be tolerated and the site will need cleaning-up urgently. Figure 8.2 shows the condition of the area around an abandoned Cu mine.

8.6.1 Love Canal, New York, USA

The incident which drew attention to the problem and established hazardous wastes as a political problem was the Love Canal waste tip near Niagara Falls, in the State of New York, USA. This site, a section of an abandoned excavation for a canal, had been used from 1930 to 1952 for the

Figure 8.2   Abandoned Cu mine at Parys Mountain, Anglesey, North Wales (photo.: B. J. Alloway).

disposal of wastes from a nearby chemical works. It had received about 20 000 t waste, comprising at least 248 different chemicals (including intermediates from the manufacture of chlorinated organic pesticides). Following its closure, the land was taken over in 1953 for building a school and a housing estate. Although various complaints had been made about irritation of children's eyes and respiratory tract, heavy rain and snow melt in the winter of 1977–8 caused groundwater containing chemicals to rise into the basements of houses built near to the tip and some buried drums to break through the covering soil. More severe health effects were reported and the USEPA carried out a series of investigations. Analysis of the groundwater revealed 82 different chemicals, 27 of which were on the EPA's Priority Pollutant List and 11 of these were known carcinogens. Several hazardous VOCs including benzene, toluene, chloroform, trichloro-ethylene, tetrachloroethylene, hexane and xylene were detected in the houses around the site. The site was declared a disaster area and 239 families were evacuated immediately followed by others later. This problem initiated surveys of other sites with serious pollution problems and it is interesting to note that Love Canal came only 25th in order of hazard. The EPA has documented at least 2100 sites used for the disposal of industrial wastes. The Love Canal incident prompted the CERCLA (Comprehensive Environmental Response, Compensation and Liability Act) of 1980, the 'Superfund' legislation. This is intended for the identification of severely polluted sites, their evaluation, monitoring and clean-up. The Love Canal site has cost many millions of dollars so far and has not yet been cleaned up (Manahan, 1984; 1991; BMA, 1991) (Table 6.32, p. 299).

8.6.2   Lekkerkirk, near Rotterdam, the Netherlands

A similar problem to Love Canal was discovered in a new village called Lekkerkirk, near Rotterdam in the Netherlands. Many of the houses in this development were built adjacent to a river in 1970–1 on land raised by a layer of < 3.5 m of household and demolition waste covered with 0.7 m of sand. By 1980, it was realized that rising groundwater had carried pollutants upwards from the underlying wastes into the foundations of the houses. This caused deterioration of plastic drinking-water pipes, contamination of the water, noxious odours inside the houses and toxicity symptoms in the garden plants. As a result of investigations, 250 houses had to be abandoned while $87\,000\,m^3$ of contaminated fill containing 1600 drums of chemicals comprising mainly paint solvents and resins containing toluene, lower boiling point solvents and Cd, Sb, Hg, Pb and Zn were removed and transported by barges to Rotterdam for destruction by incineration. The cost of cleaning the site up in 1981 was the equivalent of £156 million (Manahan, 1984; Finnecy and Pearce, 1986; BMA, 1991).

8.6.3   The Chemstar fire at Carrbrook, Cheshire

The Chemstar fire and its sequel established that one cannot depend upon natural dissipation to reduce solvent pollution of soil to a safe level (Craig and Grzonka, 1994).

Carrbrook, near Stalybridge, is a village built next to a former Victorian textile works. After closure in 1970, the site was rented by Chemstar Ltd, who undertook distillation to recover solvents from industrial residues. Hundreds of tonnes of these feedstocks and recovered solvents were stored on-site in metal drums close to neighbouring homes. In September 1981, an escape of hexane was ignited by the burner of a steam boiler; the fire which followed killed one employee and engulfed most of the buildings and storage on the site. The village was abandoned in the short term and in the long term the factory site was cleared of wreckage and abandoned.

A survey two years later by the Greater Manchester Scientific Services Laboratory identified 400 different chemical contaminants at total concentrations up to 5000 ppm and to a depth of 4 m. Remedial action was discussed by the local council, but none was taken since it was felt that the hazard from solvents would decline in time to safe levels. In 1988 some local residents became aware of the Love Canal incident and drew a parallel with their own experience, which included deaths of pet animals. The toxicity towards animals was subsequently confirmed by the deaths of two large batches of pheasant chicks on adjacent land. A test group of guinea pigs maintained close to the site suffered wasting and early death, while a control group maintained at a higher level 100 m off-site remained healthy.

When the site was re-evaluated during 1988–9, contamination was found to be higher than suggested by the 1983 surveys. Surface soil contained up to a maximum of 71 675 mg/kg of mixed solvents including benzene, trichloroethane and tetrachlorethylene. At a depth of 1 m, overall solvent pollution was 5436 mg/kg with a groundwater content up to 58 240 mg/kg. Benzene was found in soil off-site at 25 mg/kg with PCBs up to 1160 mg/kg and PCDD/PCDF up to TEQ 168 μg/kg, compared with the median soil level in the UK of TEQ 335 ng/kg (Creaser, 1989).

A remedial programme was finally carried out in 1993 by Tameside MBC, who reclaimed the land as a public open space after removal of 10 000 m³ of soil to a licensed landfill with the incineration of 80 m³ of PCB-containing soil. Clean soil was imported and a barrier to groundwater was installed with appropriate monitoring boreholes.

8.6.4   Abandoned waste tips

An OECD environmental study (1991) reported the following estimated numbers of abandoned waste tips in Europe, although not all will have received hazardous wastes

Austria	—around 1000 sites
Netherlands	—5000 sites identified with 350 requiring immediate remediation
former West Germany	—21 000 abandoned landfills with 2000 requiring clean-up
Denmark	—380 sites being cleaned up in 1987
France	—66 abandoned sites ('points noirs') posing grave risks to human health.

8.7
Tanker accidents and oil spillages at sea

Although not strictly waste disposal, oil spillages at sea are important pollution events and the oil slicks produced are quite spectacular and so attract the attention of the world news media. However, it has been estimated that less than 4% of the total amount of oil discharged into the sea is the result of tanker accidents, while 22.5% is from operational discharges. The 1982 estimate for oil entering the sea was 3.1×10^6 t, of which up to 20% was from natural sources outside human control and 45% enters from land via rivers and from coastal refineries etc. Transportation was responsible for up to 40%, with oil tankers accounting for 20–25% of the total (Gourlay, 1988). Nevertheless, spillages associated with tanker accidents do cause a lot of ecological and economic damage, because of the concentrated effect of the slick, which is often near to a shoreline which affects both commercial fisheries and tourism as well as the marine ecosystem.

The largest spills to date have been from the *Atlantic Express*, which released 276 000 t and affected Tobago in 1979, *Castello Belver* (256 000 t), which affected South Africa in 1983, and *Amoco Cadiz* (223 000 t), which affected Brittany, France, in 1978 (Tolba *et al.*, 1992). In comparison with these, the *Exxon Valdeez* spillage of 45 000 t in Alaska and more recently the *Brear* (85 000 t) off the Shetland Isles, Scotland, appear much smaller although their local ecological and economic effects were considerable. It is interesting to note that the number of oil spills greater than 6800 t actually decreased by 74% between 1974 and 1986 (Tolba *et al.*, 1992). The intentional discharge of oil into the Persian Gulf from Kuwait by Iraqi occupying forces also caused similar ecological damage. These same forces also set fire to many oil wells and these took quite a long time to extinguish and prevent the loss of oil. The air pollution, mainly smoke from the oil-well fires, caused the pollution of a large area of land with fallout from the smoke (containing PAHs etc.) and also affected weather patterns.

The spread of oil slicks produced by spillages into the sea depends on the type of oil, how dense it is, how much there is, the tidal, wave and wind conditions, and the rate at which the volatile components evaporate and the denser residue breaks up to form floating lumps. High temperatures and rough seas promote this process, and the slick from the *Brear* in the Shetlands was certainly strongly affected by the gale force winds which persisted for several days after the accident. Some aromatic hydrocarbons dissolve in the water under a slick and increase the ecotoxic effects of the slick. A lot of the oil forms an emulsion with seawater, known as a 'mousse', which eventually washes up on beaches as tar-balls. The hydrocarbons are subject to oxidation catalysed by UV light, and microbial decomposition will also occur. The ecosystems of the sea floor and the inter-tidal zone will all be affected by the oil, although it is often claimed that the dispersants which have been used to break up the slicks have sometimes caused a lot of ecotoxicity also. Sea bird and mammal populations are also badly affected. However, some studies on ecosystems affected by major oil spills have indicated that recovery has been more rapid than was expected.

As stated above, the majority of the oil entering the sea from tankers is from operational discharges, such as when the tanks and pipes are flushed out with sea water before loading with fresh crude oil. This is more dispersed than the slicks from spillages and has less local impact, although floating oil mousse does get washed up on beaches all over the world, especially those near main shipping routes.

8.8 Other multipollutant situations

For the world as a whole, it is possible to generalize and say that the majority of pollution situations involve more than one pollutant although in some cases one or a few compounds may predominate (Chapter 2). The multiple pollutant situation has several implications: in some cases some of the pollutants may have additive or even synergistic effects and in others they may tend to counteract each other's bioaccumulation or harmful effects. Waste disposal is an obvious source of a wide range of pollutants, but there are many other multipollutant sources. Derelict land, defined in the UK by the Department of the Environment as 'land so damaged by industrial or other development that it is incapable of beneficial use without treatment', is in almost all cases a multiple pollutant problem. For example, an old gas works site will be polluted with tars (containing PAHs, benzene, xylene and naphthalene), HCN, phenols, As, Pb, Cu, cyanides, sulphates and sulphides (Section 6.4). Scrapyards are polluted with a range of metals, PCBs, PAHs, hydrocarbons and the products of the partial combustion of plastics.

Even a normally occurring natural hazard such as flooding is a pollution event. Sewers overflow, drums and tanks of various substances in works, farms or even houses leak or burst, and so the water carries a mixture of pollutants (although often dilute) which are deposited on the land that the floodwater covers.

8.8.1   Pollution from warfare and military training

Battlefields and land used for military training, munition and equipment stores will have been contaminated with a wide range of pollutants, including various hydrocarbons from fuel spillages and leaks, PAHs from fires, explosives (such as TNT), mines, bullets, smoke-generating compounds, de-icer and, possibly, chemical warfare materials (CWMs), such as nerve gases, blister agents and choking agents, herbicides (e.g. 'Agent Orange' – Section 6.4.11) and many others. As a result of the end of the 'Cold War' and communism in Central and Eastern Europe, many former military bases and training grounds have been closed down and abandoned and will need to be carefully managed and ultimately cleaned-up. Recent wars and skirmishes around the world will have left a large amount of modern warfare materials to pollute the environment; crashed aircraft will have caused localized pollution. The battlefields of the First and Second World Wars, Korea, Vietnam, Afghanistan, Falklands and Kuwait will still be polluted with persistent pollutants, especially heavy metals which can persist in the soil for hundreds, even thousands of years, and also various persistent organic pollutants.

In Germany, it is estimated that around 1×10^6 ha of land has been used for military training during the last 50 years and this comprises 2.8% of the area of the country (both the former West and East German states). Unlike cases of industrial pollution where the polluter is made to pay for the clean-up of sites where possible, the cost of remediation of former military or warfare sites usually has to be paid for by national governments. In Germany, 256 000 ha were used by the former Soviet forces, 240 000 ha by the East German forces, 200 000 ha by the US, UK, French and Canadian forces and 253 000 ha by the West German forces. It has been found that the most ubiquitous pollutant on military training land is hydrocarbon fuels; air bases are usually the worst affected. Scrap metal is the second most common pollutant, but residential wastes and chemicals are also an important pollution problem. Interestingly, explosives formed the smallest proportion of problem sites but were the most dangerous (Burkhardt, 1995). Hungary has 171 sites of former Soviet forces bases and the most serious pollution problems arise from hydrocarbon fuel leakage at six military air bases.

Aviation fuel (kerosene) spillages/leakages result in a range of hydrocarbons polluting soils and moving down into the groundwater. Although most of the constituents of fuels are hydrophobic, some aromatic hydrocarbons are water soluble and could affect the composition of groundwater used for drinking. Investigations after a large spillage of kerosene (Aviation Turbo fuel TR4 containing 63.5% paraffins, 21% naphthenes, 14% alkyl benzenes, 1% indanes and tetralins and 0.5% naphthalenes) at an airport close to Entzheim near Strasbourg in 1971 showed that light aromatic compounds in the fuel, including benzene, toluene, xylene, 1,2,4- and 1,2,3-trimethylbenzene dissolved in the groundwater (Baradat et al., 1981).

Undiscovered armaments containing conventional high explosives such as TNT cause serious fatal risk from explosions and also water pollution problems. It is estimated that around 100 million pieces of ordnance (primarily land mines) are buried in 64 countries, and unexploded ordnance incidents cause more than 10 000 injuries each year (Moore, 1995). Chemical warfare materials can include nerve gases such as organophosphorus compounds (Sarin and Soman), vesicants/blister agents (Lewisite) and choking agents (phosgene) (Moore, 1995). These materials were manufactured, tested, stored and dumped in many countries and can be a serious risk and pollution hazard both at military bases and at the manufacturing and storage sites.

A former target shooting area for the air force of the Soviet army in East Germany was found to have been polluted with a range of nitro-aromatic compounds from the explosives as well as with the bullets and shells themselves. The most abundant aromatic pollutant was found to be 1,3-dinitrobenzene (Richte and Franke, 1995). With regard to pollution from explosives, the manufacture of TNT produces $40 \, m^3$ of polluted wastewater containing various carcinogenic compounds per tonne of TNT produced. TNT is a serious pollutant problem in the environment becauses it has a low volatility but a relatively high solubility (140 mg/l at 25°C). It contains three nitro groups, which is rare in nature and so microorganisms in the soil will not have become adapted for its decomposition. It is also a highly oxidized compound and so resists oxidative attack. TNT undergoes nine reductive steps to become converted to triaminotoluene (TAT), which is more readily decomposed (Daun et al., 1995).

In Germany in the Second World War, explosives production totalled around 1 600 000 t, of which around 50% was TNT. Therefore, the manufacture of this material, its use in munitions production, storage and detonation will have resulted in large-scale pollution problems.

Warfare will have involved contamination by explosives, metals (in shells, bullets and wrecked machines/vehicles), leakages of chemicals and hydrocarbons (especially from bombing industrial areas) and PAHs from burning materials. After the Second World War, many urban areas which had suffered destruction by bombing were rebuilt. Although unexploded bombs are still occasionally found, little mention seems to be made about the possible problems of soil pollution, which may have persisted for many years after the war. After each major war, or anticipated war emergency, there is usually widescale dumping of unused war materials, which are likely to be hazardous to store and difficult to dismantle. In Britain after the Second World War, large amounts of munitions (1.17 million tonnes) were dumped in deep channels at various locations off the coastline. Some 50 years later it is becoming increasingly common for some of the dumped items to be washed up on beaches, where they pose a severe danger to people who might touch them. In October, 1995, 4500 incendiary bombs containing phosphorus, benzene and cellulose (designed to ignite on contact with air) were washed up on beaches along part of the West Coast of Scotland. Although not officially confirmed, it is expected that these munitions were originally dumped in an underwater trench 250 m deep and 5 km wide about 10 km off the Scottish coast. This dump was probably disturbed by the laying of a submarine gas pipeline in a shallow trench on the sea floor which crossed the area (Edwards, 1995). This case illustrates the problems likely to be encountered in many situations where munitions have been dumped. This is another (literal) case of a 'chemical time bomb'.

8.9
Chemical time bombs

The concept of chemical time bombs proposed by Stigliani (1991) and others is a useful way of focusing attention on situations where hazardous chemicals may suddenly be released into the environment. As listed by Tolba *et al.* (1992) these can include rapid acidification of soils and lakes as a result of depletion of the buffering capacities of forest soils, leaching of phosphates from agricultural soils into aquatic ecosystems resulting from saturation of sorptive mechanisms, leaching and plant uptake of metals from polluted agricultural soils through acidification or reduced sorption capacities, metals in coastal waters and estuaries from desorption from sediments, and the release of sulphuric acid and metals from wetlands or drying out caused by drainage or climatic change. In addition to these general types of source, individual sources of pollutants, such as hazardous wastes in containers dumped in the sea or in landfills, may also cause a time bomb-like release of pollutants, as exemplified by Love Canal (Section 7.6.1) and Minamata (p. 214).

Changes in land use, such as set-aside and afforestation on former arable farm land can have marked effects on the mobility and bioavailability of heavy metal pollutants in soils. Ironically, in some cases the land-use change may have been carried out in order to improve the environmental amenities in the vicinity of major urban areas. There is an increasing demand for woodland and nature reserves near to urban centres and in some European countries this is often seen as a higher priority use of scarce land resources than agriculture. Trees have the effect of acidifying soils as a result of the discontinuation of liming, the formation of acid exudates and leaf decomposition products, and the entrapment/scavenging of acidic atmospheric pollutants such as SO_2, NO_2 and $(NH_4)_2SO_4$. The increased leaching of heavy metals such as Cd in acidic soils will eventually lead to increased concentrations in the groundwater, which can be of major significance to human health where the aquifers are a source of drinking water.

References

Baradat, Y., Lemlin, J. S., Sibra, P. and Somerville, H. (1981) *CONCAWE, Report No. 8/81*. Oil Companies European Organization for Environmental Health and Protection. The Hague.

BMA (British Medical Association) (1991) *Hazardous Waste and Human Health*. Oxford University Press, Oxford.

Burkhardt, (1995) Oral Communication at *Fifth International Conference on Contaminated Soil*, Maastricht, October 1995.

EC (Council of the European Communities) (1986) Directive 86/278/EEC on the *Protection of the Environment and in Particular of the Soil, When Sewage Sludge is Used*. EEC, Brussels.

Craig, T. and Grzonka, R. (1994) A case study of land contamination by solvents. PCBs and dioxins. *Land Contamination and Reclamation*, **2**, 19–25.

Creaser, C. S. (1989) Survey of background levels of PCDDs and PCDFs in urban British soils. *Chemosphere*, **21**, 931.

Daun, G., Lenke, H., Desiere F., Stolpmann, H., Warrelmann, J., Reuss, M. and Knackmuss, H. J. (1995) In *Contaminated Soil '95* (eds. W. J. van den Brink, R. Bosman and F. Arendt), pp. 337–346. Kluwer Academic, Dordrecht.

Department of the Environment (1993) *UK Sewage Sludge Survey*. Final Report. Consultants in Environmental Sciences Ltd, Gateshead.

Department of the Environment (1995) Waste Management Paper No. 266: *Landfilling Wastes*. HMSO.

Edwards, R. (1995) *New Scientist*, 18 November, 16–17.

Finnecy, E. E. and Pearce, K. W. (1986) In *Understanding Our Environment* (ed. R. E. Hester). Royal Society of Chemistry, London.

Fish, H. (1992) In *Understanding Our Environment* (2nd edn), Ch. 3. (ed. R. M. Harrison). Royal Society of Chemistry, Cambridge.

Gourlay, K. A. (1988) *Poisoners of the Seas*. Zed Books, London.

House of Commons Environmental Committee (1989) 1st Report, *Contaminated Land*. HMSO, London.

Jackson, A. P. and Edulgee, G. H. (1994) An assessment of the risks associated with PCDDs and PCDFs following the application of sewage sludge to agricultural land in the UK. *Chemosphere*, **29**, 2523–2543.

Lester, J. E. (1990) In *Pollution: Causes, Effects and Control* (2nd edn), (ed. R. M. Harrison). Royal Society of Chemistry, Cambridge.

Lisk, D. J. (1991) Environmental effects of landfills. *Science of the Total Environment*, **100**, 415–468.

MAFF (Ministry of Agriculture, Fisheries and Food) (1992). *Dioxins in Food*, Food Surveillance Paper No 31. HMSO, London,

MAFF (Ministry of Agriculture, Fisheries and Food and Welsh Office Agriculture Department) (1992) *Code of Good Agricultural Practice for the Protection of Soil*, Draft Consultation Document. MAFF, London.

Manahan, S. E. (1984) *Environmental Chemistry* (4th edn). Brooks/Cole (Wadsworth), Monterey, CA.

Manahan, S. E. (1991) *Environmental Chemistry* (5th edn) Lewis Publishers, Chelsea, MI.

McGrath, S. J., Chang, A. C., Page, A. L. and Witter, E. (1994) *Envir. Rev.*, **2**, pp. 108–118.

Moore, T. (1995) *Environ. Eng. World*, May–June, 28–33.

Neutra, R. R. and Ostro. B. (1992) *Sci. Total Environ.*, **127**, 91.

OECD (Organization for Economic Co-operation and Development) (1991) *The state of the Environment*. OECD, Paris.

Richter, M. and Franke, C. (1995) Distribution and mobilisation of introaromatic compounds in a former military shooting area. In *Contaminated Soil '95* (eds W. J. van den Brink, R. Bosman and F. Avendt). Kluwer Academic Publishers, Dordrecht.

Robinson, H. D., Barber, C. and Morris, P. J. (1982) *Water Pollution Control*, **54**, 465–478.

Royal Commission on Environmental Pollution (1993) Report No. 17. *Incineration of Waste*. HMSO, London.

Stigliani, W. M. (eds) (1991) *Chemical Time Bombs: Definitions, Concepts and Examples*. IIASA, Laxenburg, Austria.

Tolba, M. K., El-Kholy Osama, A., El-Hinnawi, E., Holdgate, M. W., McMichael, D. F. and Munn, R. E. (eds) (1992) *The World Environment, 1972–1992*. UNEP, Chapman and Hall, London.

US EPA (1992) 40 CFR Parts 257, 403 and 503 [FRL-4203-3] *Standards for the Use and Disposal of Sewage Sludge*. United States Environmental Protection Agency, Washington DC.

World Resources Institute (1992) *World Resources 1992–93*. Oxford University Press, Oxford.

Table of units Appendix and conversions

SI prefixes

10	deca (da)	10^{-1}	deci (d)
10^2	hecto (h)	10^{-2}	centi (c)
10^3	kilo (k)	10^{-3}	milli (m)
10^6	mega (M)	10^{-6}	micro (μ)
10^9	giga (G)	10^{-9}	nano (n)
10^{12}	tera (T)	10^{-12}	pico (p)
10^{15}	peta (P)	10^{-15}	femto (f)

SI units

length	metre (m)
mass	kilogram (kg)
time	seconds (s)
temperature	kelvin (K)
force	newton (N)
pressure	pascal (P)
energy	joule (J)
power	watt (W)
heat	joule (J)

In the following table the larger the unit the smaller it appears in magnitude relative to the left hand column. Equivalence of a pair of units is given by their ratio, for example:

$$\text{cm}^3/\text{litre} = 1/10^{-3} = 1000\,\text{lb/kg} = 2.2 \times 10^{-3}/10^{-3} = 2.2$$

Volume

cm^3	litre	m^3	ft^3
1	10^{-3}	10^{-6}	3.53×10^{-5}

Mass

g	kg	lb	t (metric)
1	10^{-3}	2.2×10^{-3}	10^{-6}

Energy

erg	joule	kW	MeV
1	10^{-7}	2.78×10^{-14}	6.24×10^5

Pressure

Pa	dyne/cm^2	p.s.i.	atmospheres
1	10	1.45×10^{-4}	9.87×10^{-6}

Area

cm^2	ft^2	acre	hectare
1	1.08×10^{-3}	2.47×10^{-8}	10^{-8}

Index

AA spectroscopy, *see* Atomic
   Absorption Spectroscopy
Accurate mass
   determination 103
Acetaldehyde 332
   from cigarettes 337
Acetic acid 181
   solution pH 181
Acetone
   from cigarettes 337
   in paint works 331
   USA production 271
Acid rain 174, 184–6
Activity coefficient 53
   of analytes 92–3
Adenine 70
Adhesives 272–3
   emissions indoors 333
Adiabatic lapse rate 28
Advanced gas-cooled reactor
   (AGR) 227–8
AEA Technology
   pollution monitoring 187
Aerosols 274
   analysis by GLC 104
   components in Los Angeles
   air 251
Aflatoxin 68, 76
Agent Orange 298–301, 304
Agricultural land
   monitoring 146–8
   sludge treatment 363–5
Agriculture
   Centre for Strategy, Reading 174
   greenhouse warming 172–3

as source of metal pollution 197
Air
   clean, composition 165
   indoors 251
   organochlorines in 295
Air pollution 7–9
   analysis by GLC 99–100
   and buildings 187
   comparative data for cities 263
   direct sampling 107–9
   excursions 164
   monitoring 141–3
   by nitrosamines 338
   note on units 186
   from oil fires 373
   at sea 285
   transport of 27–34
   see also Indoor pollution
Alaska
   CO_2 level 165
Aldicarb 320
Aldrin
   and death of birds 290–1
   in drinking water 82
   in Lake Kariba 289
   synthesis 286, 289
Algae
   in food chains 289
Algal blooms 281
Alkali Act (1862) 7
Alkanes, *see* Hydrocarbons
Allergens 344
Allethrin 310
Allotments
   pollution input 148

Alpha particle 221, 346
Aluminium
   levels in drinking water 217
   principal features 217
   toxicity 183, 184
Alzheimer's disease 218
Ames' test 70
Aminoacids 312, 317
Ammonia
   from landfill leachate 356–7
Ammonium nitrate 183
Amoco Cadiz 373
Amsterdam
   vehicle usage 177
Anaemia
   lead induced 216
Analysis
   of carbon dioxide 111
   of carbon monoxide 111
   of nicotine 338
   of NO_x 110
   of oxygen 109–10
   of ozone 110–11
   of PAH 118
   of sulphur dioxide 111
Analytical quality assurance
   137–9
Ancient civilizations 191
Animals
   monitoring 150
Antarctic
   carbon dioxide levels 165
   icebergs 172
   ozone layer 160, 161
   penguins 26

Anthracene
  structure 268
Anthracite 253
Aphids 314, 319
Aquifers
  chemical dispersion in 41–5
  leachate pollution 356
Arctic
  brown snow 61
  seal 295
Armaments, *see* Warfare agents
Aroclors 293
Aromatic compounds
  metabolism 271
  soil standards 78–80
  see also Hydrocarbons
Arsenic
  LD_{50} in mammals 208
  principal features 211–12
  in soil and water 209
Asbestos 282
  analysis 341
  from factories 342, 343
  indoors 340–3
  in the iron industry 339
  in mercury cells 282
  in offices 341
  types 341
Asbestosis 242
Asthma 250
  causes indoors 343–4
  link with formaldehyde 331
Athens
  air pollution 263–5
Atlantic Express 372
Atomic absorption
    spectroscopy 123–37
  basic instrument 126
  basic theory 124–5
  of cadmium 130, 139
  calibration curve 126
  detection limits 133
  double beam 127
  graphite furnace 130
  interferences 127–9, 9, 133
  of lithium 124
  sample preparation 129
  small samples 130
  of sodium in concrete 128–9

variation with temperature
    125
  see also Atomic emission
    spectroscopy
Atomic emission spectroscopy
    131–6
  background correction 135–7
  basic theory 124–5
  of boron 136
  of copper 136
  derivation of intensity 135–6
  detection limits 133
  of metals in rainwater 135–6
  plasma source 131–2
Atomic Energy Commission
    234–6
Atrazine 40, 320–2
Aviation fuel 43
  pollution from 160

Bacteria
  action on waste 355
  and coalification 253
  in effluent 281
  intestinal 337
  methylating 206
  oxidizing 60–1, 195
  in soil 183
  in streams 143–4
Barium 221
Batteries
  metal content 199
Battlefields
  pollution 374–6
Becquerel, H. 220
Benzaldehyde 267
Benzene
  in air 262, 266
  cancer induction 227
  from cigarettes 337
  for detergent synthesis 279
  indoors 330
  in paint works 331
  in petrol 277
  toxicity 227
Benzo-[a]-pyrene 116
  carcinogenicity 117
  from cigarettes 336, 337
  and skin cancer 267

Benzthiazole
  from an industrial plume 107–8
Benzyl radical 267
Beryllium
  mammalian LD_{50} 208
  principal features 218
  soil criteria 211
BHA 72
BHT 72
Binghampton
  PCB pollution 296
Bioavailability
  of DDD 289
  of metals to plants 206
Biochemical action
  of microorganisms in water 45–7
Biomagnification 289
Biomonitoring 150–2
Bioresmethrin 310
Birmingham
  solvent contamination 44
Bisphenol A 72
Bitumen
  use and runoff 269–70
Black Forest 184
Blood
  lead levels 344–6
Blue-baby syndrome 45
Blue-green algae 144
BOD
  cost/control 6
  measurement 46
  of landfill leachate 356–7
  in polluted streams 48
  of sewage 362
Boiling water reactor (BWR) 227
Bombay
  air pollution 263
Bordeaux Mixture 213
Boric acid
  as moderator 227
Botulism 76
Brazil
  asbestos hazard 343
  indoor pollution 330
Brear 373
Breeder reactors (LMFBR) 229–30
Brevitoxin
  and death of dolphins 295

Brick kilns
  fluorine from 219
  sulphur and molybdenum
    from 220
Brookings Institute 173
Brown snow 61
Buffer action 181
Buffer capacity 183
Butadiene
  indoors 330
  USA production 271

Cabbage root fly
  DDT resistance 289
Cacodylic acid 300
Cadaverine 323
Cadmium
  in diet 199
  from landfill 356–7
  mammalian LD_{50} 208
  principal features 212–13
  production 191
  from smelters 198
Caesium 137 230
  from Chernobyl 231
Calcium
  sequestration 281
  sulphate 188–9
Calcutta
  air pollution 263
Calder Hall 231
Californian seal 295
Camacari Petroleum Complex
  risk to workers 330
Camelford
  water pollution 217
Canadian standards
  for metals in drinking
    water 204
  for metals in soils 211
  for soils 80
  for water 82
Cancer
  from asbestos 341, 342
  benzene and 277
  of bladder 362
  in coke oven workers 269
  in gas workers 268
  of lung 339

  annual death rate in USA 347
  neoplasms 242
  from nitrosamines 338
  of skin 267
  from vinyl chloride 277, 293
CANDU reactor 228–9
Capillary columns for GLC
  advantages 98
Carbamate pesticides 318–22
  mechanism of action 313
Carbaryl 319
  synthesis from phosgene 319
Carbon
  for odour control 324
Carbon dioxide 165–74
  critical concentration 84
  emissions/capita 168
  emissions by fuel type 167
  in ice 166
  increase in 165–7
  indoors 336
  IR spectrum 111, 168–9
  in landfill gas 355
  predicted levels 171
  sources 165
  water solubility 166
Carbon monoxide 247
  in chimney effluent 112–13
  from cigarettes 336
  critical concentration 84
  from incineration 336
  indoors 335–6
  toxicity 336
  from vehicles 256
Carbon tetrachloride 283
  cancer from 278
  production 276
Carbonyl compounds
  exposure limits 332
Carcinogenesis
  definition 69
Casco Bay, Maine
  sediments 285
Castello Belver 372
Catalytic conversion 177–8, 257–8
Cation exchange capacity 55
Cellulose 252
Central Electricity Generating
  Board 182

Certified reference materials
  (CRMs) 138, 139
CFCs, *see* Chlorofluorocarbons
Chalcopyrite
  oxidation 195
Chemical ionization
  for GC/MS 104–5, 107
Chemical time bombs
  sudden release of pollutants
    376–7
CHEMSTAR fire 371–2
Chernobyl
  nuclear accident 236–7
  spread of caesium 137, 231
Chezy equation 36
Chimney plumes 29–31
China
  population pressure 167,
    168
  soil transport 61
Chloral 283
Chlor-alkali process 214, 282
Chlorine
  addition to ethylene 292
  in coal 179
  by electrolysis 282
  production 282–3
  radical reactions 161
  in sewage treatment 362
Chlorofluorocarbons
  analysis by GLC 104
  control 162
  IR spectra 168
  for metal cleaning 275
  nomenclature 161
  reactions in the stratosphere
    160–1
  release to air 160
  from spray cans 274
Chloroform 145
  in drinking water 82, 277
  reaction in incinerators 306
  toxicology 277–8
  WHO guideline 278
Chlorophenols
  HPLC analysis 121–3
Chlorpyrifos 314–16
Cholera 282, 360, 361
Cholinesterase 312–13, 318

Chromatography
  column 87
  see also TLC, GLC; HPLC
Chromium
  in coal ash 198
  principal features 214
  production 191
Chrysanthemum sp. 308
Chrysene 116
  carcinogenicity 117
Ciba-Geigy Company
Cigarettes 336–40
  effect of nicotine 311–12
  production 270
  pyrolysis products 340
  and radon exposure 347
Cinerolone 308
Clay minerals 53, 205
Clean air
  composition 165
Clean Air Act (1956) 7, 181, 248, 249
Clear Lake 289–90
Coal
  beryllium in 218
  composition 253
  consumption 179
  fluorine content 220
  formation 252–3
  sulphur content 179
  tar industry 267
Coalite Chemical Company 299
Coatings 273–4
Cobalt 60 223
Cockroach 344
COD
  of landfill leachate 356–7
Cold fusion 239–40
Concrete
  sodium in 128
Contaminated land
  pollutants in 77–81
Control zones
  at Séveso 302–3
Cooking
  air pollution from 335
Copenhagen
  International Meeting 162
Copper
  in landfill leachate 357

principal features 213–14
production 191
sulphide, solubility 53
Coprecipitation
  of metals 57–8
Curie, M. and J. 220, 223
Cyanide
  on derelict sites 79
Cyclohexane
  USA production 271
Cytosine 70
Czech
  lignite 8
  sulphur in coal 179

Dalapon 288
Darcy's law 40
DDD
  structure 284
  in water body 289–90
DDT 309
  in brown snow 61
  distribution coefficient and
    solubility 59
  HPLC analysis 121
  in Lake Kariba 289
  resistance to 292
  synthesis 283–4
  in Third World 11
  toxicity and LD_{50} in the rat 288
Dealkylation 47
Death
  statistical risk of 76
Dehalogenation 48
Delhi
  air pollution 263, 265
Deltamethrin 310
Denmark
  aquifers 44
Dental alloys 199
Derelict land
  monitoring 148–9
Derris elliptica 311
Desertification 172
Desorption
  of air samples 99
Detection limits
  for metals 133
Detergents 279–81

builders 280
displacing soap 279
pollution by 281
synthesis 279
types 279–80
Deuterium
  fusion reaction 239
Diazinon 314, 317
Diazotization 117
Dibenzofurans 296–9
Dichloroacetaldehyde 317
Dichloroethane 292
Dichlorophenoxyacetic acid
    287, 288
  aqueous solubility 59
  distribution coefficient 59
Dichlorvos 314–16
Dieldrin
  and death of birds 290–1
  in drinking water 82
  HPLC analysis 121
  resistance to 292
  synthesis 286
Diels–Alder reaction 285
Diesel engines 178–9
  particles 248
  smoke 198
Dinitro-*o*-cresol (DNOC) 288
Dioxins 297–305
  detection limit 302
Direct air sampling 107–9
  in EC countries 264
DNA
  binding PAH 269
  structure 70
DOE (UK)
  definition of derelict land 374
  on nuclear plant 231
  red list of pollutants 24
  standards for water 82
  trigger concentration for
    pollutants 79
Doll, Sir Richard 267
Dolphin
  deaths 295
Dose response
  definition 66
Drax power station 188–9
Drinking water

carbon tetrachloride in 278
chloroform in 82
disinfection 362
monitoring 146
pollutants in 81–3
see also Water
Dry cleaning 275–6

Earth's crust
composition 192–3
Eastern Europe
wastes 354
EC
abandoned waste tips 372
Common Agriculture Policy 45
dangerous substances 22–4
directives
on sewage 360–1
on sludge amended soils 363–5
energy use 255
pollution controls
for cadmium 199
for NO$_x$ 187–8
for ozone 164
for SO$_2$ 187–8, 262
standards
for drinking water 204
for emissions from cars 258
for incinerators 358
Ecotoxicology 75
Eddy diffusion 97
Effective stack height 32
El Niño 171
Electrochemical cell
zinc/copper 49
Electrode potential 49
Electron capture detector (ECD)
for GLC 94–5
Electron volt 124, 222
Electronics
metal components 199
Emission
from furnishings 332
from incinerators
of CO 112–13, 336
of PCDD/PCDF 306–7
of THC 112–3
standards 358
Emphysema 218

Enantiomers
of pyrethrins 308–9
Endosulfan 286–7
Energy
critical for fission 224
demand in Third World 167
major users 255
of the sun 239
world use 12
yield from fission 225
Environmental monitoring
of air 141–2
biological indicators 150–2
objectives 140
of particles 141–2
of rivers 143–5
of water 143–6
Environmental Protection Act 86
EPA
asbestos sampling 341
classification of hazardous
substances 20–2
index of standards 251
Priority Pollutants 370
standards for
air quality 261–2
for NO$_x$ 188
for ozone 164
for particles 252
for SO$_2$ 188
see also USA
Epoxidation
of Aldrin 286
of aromatics 271
of PAH 269
of vinyl chloride 293
Errors of measurement 138
Estuaries
dispersion of pollutants 38–9
pollution of 367–8
sediment inclusions 39
Ethanol
as fuel 257
human toxicity 300
Ethylene
for detergents 281
USA production 271
Ethylene glycol
USA production 271

EURONEST 330
European Community, see EC
European Community Bureau 138
Exceedences 261
Excited state
of lithium 124
temperature variation 125
Exxon Valdez 373

Fast atom bombardment (FAB)
for GC/MS 105
Ferrihydrite
co-precipitation by 57
Fertilizers
metal content 197
Fick's Law 42
Fish
chromium poisoning 214
depletion of stocks 183
LC$_{50}$ values
for Aldrin 286
for Dieldrin 286
of organophosphorus
compounds 315
for pyrethrins 311
pyrethroid toxicity 310–11
Flame ionization detector (FID)
93–4
Flavonols
in tea 339, 340
Fluidized bed combustion 189–90
Fluorine
from kilns 219
in plants 218
principal features 219
in water 220
Fluorosis 219
Food chain 147
in Clear Lake 289
in Lake Kariba 289–90
in oceans 368
sewage sludge input 362
strontium 90 in 230
Foodstuffs
Aldicarb in melons 320
nitrite in 337
PAH in 270
PCDD/PCDF in 365
Forestry Commission 184

Forests
    damage in Europe 184–6
    early data 186
    and global warming 173
Formaldehyde
    from cigarettes 337
    formation 259
    indoors 331–2
Fossil fuel
    additives 197
    emissions on combustion 247
    metal content 197
    usage 170, 345
France
    study on acid rain 184
    targets for NO_x and SO_2 187
Free radicals, see Radical
        reactions
Freundlich isotherm 205
Fungicides
    metal content 197
Furnishings 332, 333

Gamma radiation 222, 223
Gangue minerals 197
Garden soil
    pollutant input 148
Gas liquid chromatography, see
        GLC
Gas works
    abandoned sites 374
    as pollution sources 19
Gases
    dangerous 84
    sampling 104
Gaussian model
    form of GLC peaks 96
    for plumes 31–3
Geneva
    World Climate Conference 170
Germany
    emissions of NO_x and SO_2 176
    forest damage 185–6
    land for military training
        374–5
GLC 90–104
    activity coefficients 92–3
    capillary columns 98
    of chlorofluorocarbons 104

column components 91
compared to HPLC 117, 119
detectors 93–5, 99–103
optimum conditions 97–8
of PAH 338
principal parameters 95–6
temperature programmed 96–7
Glue sniffing, see Solvent abuse
Glutathione 271
Glycerol
    ionization 105
Glyphosate 314–17
Great Dun Fell 164
Great Lakes 38
    death of birds 295
Green River Basin 8
Greenhouse effect 167–74
    computer models 170
    consequences 171–2
    control 173–4
    methane balance 257
    monitoring 174
Gross domestic product (GDP)
    per capita 12
Groundwater
    at air bases 374–5
    dispersion of pollutants 39–41
    at Love Canal 369–370
    sampling 145
    see also Water
Guanine 343–4

Hahn, O. 224
Half-life
    of atrazine 320
    of plutonium 224
    of potassium 223
    of pyrethrins 309
    of radionuclides 221, 222
    of sulphur 223
Hardness
    of water 280, 281
Harrisburg
    nuclear accident 234–6
Hazardous waste 365–8
    management methods 367
    world production 366
Heavy metals 190–217
    in aerosols 200–1

and agriculture 197
analysis 211
in the atmosphere 194
biochemical properties 191
defined 190
in drinking water 203–4
effect on environmental media
    199–200
from electronics industry 199
at former landfill 371
from fossil fuels 197
in human diet 208
from incinerators 358
inhalation 201
in marine invertebrates 202–3
methylation of 206
mobilization 281
production 191
in rocks 193–4
sources 192–9
toxicity 207, 208
uses 190–1
Heptachlor
    HPLC analysis 121
Herbicides
    in fish 299
    organochlorine 287–8
    toxic components 297–305
Herring gull 295
HEPT (GLC) 96–8
Hexachlorobenzene
    destruction 369
    in drinking water 82
    formation from $CHCl_3$ 306
Hexachlorocyclohexane
    in brown snow 61
    see also Lindane
Hexachlorocyclopentadiene
    285–6
n-Hexane
    fire at CHEMSTAR 371
    toxicology 276–7
High pressure liquid
        chromatography,
    see HPLC
HMIP (UK) 113
    emission standards for
        incinerators 358
Hofmann La Roche Company 303

Homeostasis 192, 207
House dust mite 343–4
House of Commons Select
    Committee on Energy 231
HPLC 113–21
    of amines 117–19
    compared to GLC 117, 119
    detection of derivatives 117,
        119, 123
    detectors 115–16
    linked to mass spectrometer 107
    of PAH 118
    of phenols 117, 119
    of polluted air 117
    of polluted water 120–3
    solvents 115
    trace enrichment 120, 121
    typical components 114
Human diet 207, 208
Human food chain 212, 213, 215
Human sperm 5
Humus 54–5
Hydrides
    of metals 211
Hydrocarbons
    at air bases 375
    in cities 260–6
    from furnishings 333
    indoors 332
    reactions 259, 266
    separation by GLC 97, 100
    urban pollution by 262–7
    from vehicles 256
Hydrofluoric acid 149, 250,
    219–20
Hydrogen chloride 292
Hydrogen sulphide
    critical concentration 84
    in petroleum 253
Hydrolysis
    of herbicides 47
    of pesticides 315, 318
Hydroxyl radical 34–5, 180
    and alkanes 258–60, 266
    and aromatic hydrocarbons 339
Hydroxylation 47

Ice cores 166–7, 170
ICMESA factory 299, 301–2

ICP–OES, see Atomic emission
    spectroscopy
Igneous rocks
    as sources of metals 193
Immune system
    defects 71
Incineration
    carbon monoxide from
        111–13, 336
    of clinical waste 111–13
    emission standards 358
    of municipal waste 306, 358
    at sea 306, 368
    of sewage 324
India
    energy demand 168
Indian Ocean 174
    organochlorines in air 285,
        295
Indolylacetic acid 287
Indoor pollutants
    asbestos 340–3
    carbon monoxide 335–6
    mutagens 339–40
    oxides of nitrogen 334–5
    ozone 333–4
    PAH 338–9
    volatile organic compounds 329
    see also Indoor pollution
Indoor pollution 329–48
    sources 26
    see also Indoor pollutants
Infrared radiation
    non-dispersive gas analyser 111
    trapping by greenhouse gases
        168
Infrared spectra
    of carbon dioxide 111, 168–9
    of carbon monoxide 111
    of sulphur dioxide 111
    of water 169
Inks 273–4
Inorganic pollutants
    standards for soils 78
    ozone 157–64
Insecticidal strips 316
Insecticides
    banned in the USA 321
    toxicity 288–90

International Agency for Research
    on Cancer 269
International Atomic Energy
    Agency (Vienna) 138
    scale of risk 232
International Labour Office 343
Inversions
    atmospheric 28–9, 31, 261
Iodine
    in milk 234
    radioactive 231
    from Three Mile Island 236
Iodofenphos 314–16
Ion exchange
    in soils 54–6, 205
Ionization potential
    of lithium 124
    of potassium 125
Iraq
    mercury poisoning 215
Irish Sea
    pollution 230, 368
Isomers
    of DDT 283–4
    of chrysanthemic acid 308–9
    of Lindane 284
    of PCBs 293–4
Itai-itai disease 213
Italy
    production of NO_x and
        SO_2 187

Japan
    adhesive production 272
    energy use 255
    waste production 354
Jintsu Valley
    cadmium poisoning 213
Joint European Torus (JET) 239
Jordan
    soil pollution 295

Kapenta 289, 290
Kennaway, Sir Ernest 267
Kerogen 253
Kidney damage
    by lead 215
    by organochlorides 277
Kiev 236

Kuwait
  oil pollution 373

Labio attivelis 290
Lake Baikal 38
Lake Kariba
  biomagnification of DDT
    289, 290
  food chain relationships 289, 290
Lakes
  acidification 182
  detergents in 281
  dispersion of pollutants in 37–8
  pollution by chlorophenols
    121–3
  sampling 144
Lamber–Beer law 116, 125
Land
  sources of contamination 17–18
Landfill 354–8
  abandoned sites 369–72
  composition of gas 355–6
  leachates 42–4, 356–7
  methane from 169
Lead
  EC level for blood 216
  on foliage 201
  indoors 344–6
  in landfill leachate 357
  leachates 42–4
  in particles 197, 201
  principal features 215–16
  production 191
  in water 146
Leaves
  damage by PAN 259
  heavy metals in 208
Leeds
  asbestos pollution 342
Lekkerkirk
  polluted site 371
Lichen
  as pollution indicators 184
Lignin
  chlorinated 307
  structure 252
Lignite
  composition 253
  Czech 8

Limits to Growth 9
Limonene
  indoors 333
Lindane
  control levels 285
  cross resistance 289
  distribution coefficient 59
  isomers 284
  synthesis from benzene 285
  see also Hexachlorocyclohexane
Lithium
  energy levels 124
Liver
  damage by organochlorines 277,
    278
Locusts 317, 344
London
  air pollution 262–3
  smog 181, 249
  vehicle usage 177
Longitudinal diffusion 97
Los Angeles
  aerosol components 251, 261
  air pollution 28, 263
  indoor pollution 331
  vehicle usage 177
Love Canal 299, 369–70
Lullington Heath 164
Lung disease 218, 341–3, 339
Luxembourg
  forest damage 185–6
Lymphomas 277

Madrid
  air pollution 263–5
MAFF
  maximum levels of metals in
    sludge amended soils 363
Malaria 292
  pesticidal control 316
Malathion 314–15
Manning equation 36
Mars 170
Mass spectra 99–107
  accurate mass 102, 103–4
  detection limits 133
  GLC detection
  of organochlorines 101–2
  of Propoxur 106–7

see also MS/MS
Mass transfer 98
Maximum acceptable risk 232
Medicinal preparations
  metals in 199
Melbourne
  vehicle usage 177
Mercury 283
  analysis 123, 133
  poisoning in Iraq 215
  principal features 214–15
  slimicides 215
  in soils 215
Metal cleaning 275
Metal ions
  cation exchange 54, 205
  co-precipitation 201
  in drinking water 204
Metallothionein proteins 192, 207
Metals
  AA spectroscopy 123–37
  particles 343
  selective absorption 205
  trigger concentrations 210
Methaemaglobinaemia 45
Methane 258–9
  chemical ionization 104
  and greenhouse effect 257
  IR trapping 168
  from landfill 355–6
  reactions 258
Methanogenesis 355–6
Methanol
  as fuel 257
  USA production 271
Methoxychlor
  HPLC analysis 121
  structure 284
Methyl bromide 163
Methyl radical 258
Methylation
  of heavy metals 206, 214
Meuse Valley 181
Mexico City
  air pollution 263–6
Mice
  cancer from $CHCl_3$ 277
  teratogenicity of TCDD 300
  toxicity of PAH 268

Microorganisms
  anaerobic 54
  detoxification by 14
  methylating 55
  oxidizing 55, 57, 60
  for waste disposal 368–9
  in water 45–6
  zinc toxicity 365
Milan 303
Minamata Bay
  mercury poisoning 214
Mineral fibres 240–2
Mining
  of asbestos 343
  and pollution 195–7
Minnesota Mining and
  Manufacturing Corp. 366
Mixed function oxidase 315
Mobile monitoring
  of incinerator gases 108–9
  of PM$_{10}$ emissions 141
  of urban air pollutants 264–5
Molecular extinction coefficient
  116
Molecular ion 101, 104, 107
Molten salts
  for waste disposal 369
Molybdenum
  in livestock 220
Monocrotophos 314–17
Monsanto Company 293
Montreal Protocol 162, 272, 275
Moscow
  vehicle usage 177
Motorway
  benzene level 277
MS/MS 105–7
  of Propoxur 106
  reverse geometry spectrometer
    105
  see also Mass spectra
Multiplication factor
  for neutrons 224–5
Multipollutant situations 373–4
Munich
  vehicle usage 177
Municipal incinerators 306, 358
Mussels
  metals in 151, 203

Mustard gas 318
Muon 240
Mutagens
  definition of mutagenesis 69
  indoor sources 338
  on particles 340
  in tea 339

NAAQS (EPA)
  table of standards 261–2
  and carbon monoxide 336
  and ozone 333
Naphthalene 268
NASA 170
National Measurement
  Accreditation Service 139
National Resources Defense
  Council (USA) 274
National Radiological
  Protection Board (UK) 348
National Rivers Authority 81,
  278
Natural gas
  composition 257
  as fuel 257
Neptunium 224
Nernst equation 50
Nerve poisons 288, 309
  mechanism of action 312–13
Netherlands
  air pollution at Bilthoven
    99–100
  aquifers 44
  forest damage 185–6
  pollution at Lekkerkirk 371
  standards for soils 77–8, 209
  standards for water 209
New Jersey
  indoor air pollution 332
New York
  air pollution 263
  radon risk 346
  vehicle usage 177
New York State
  lake pollution 182
Nicotine 311
  from cigarettes 336–8
Nissan Company 275
Nitrate

  in soil and water 45
Nitric acid 149
  indoors 334, 335
Nitric oxide
  reactions
    with ammonia 190
    with carbon 190
    with ozone 334
    with SO$_2$ 180
  toxicity
Nitriloacetic acid 281
Nitrite
  in cigarettes 337
Nitrobenzene
  UV absorption 116
Nitrocompounds 269
  from explosives 375
Nitrogen fixation
  by vehicles 163, 174
  during coal burning 189
Nitrogen oxides
  in chimney effluent 112–13
  determination 110
  see also Oxides of nitrogen
Nitrogen peroxide
  indoors 334–5
  as ozone source 163
  PAN formation 259
Nitropyrene 251
Nitrosamines
  from cigarettes 336–8
  from nitrate ions 44–5
Nitrous acid
  indoors 334, 335, 337
Nitrous oxide
  in air 174
  from fertilizers 170
  IR spectrum 168
  reaction with oxygen 175
  sources 174
Non-ferrous metals
  minor metals, pollution by
    193, 198
Norway
  heavy metal deposition 198, 201
Nuclear accidents 223–7
Nuclear fission
  criteria for 224
  fission yield 225

Nuclear fission *contd.*
 products 230–1
 see also Nuclear power
Nuclear power
 commercial viability 238
 fuel preparation 226
 future of 231
 generation 225–30
 reactor sites
  in the UK 233
  in the USA 232
 reactor types 226–30
 scale of risk 232
 social aspects 237

Ocean dumping 367
 of munitions 376
Octanol–water
 partition coefficient 58
Octylphenol 72, 74
Odd oxygen 158
Odours 322–4
 indoors 371
 threshold levels 323
OECD countries
 industrial waste 354
 total hazardous waste 366
Oestrogenic chemicals 5
 activity 73–4
 and male fertility 73
 in rivers 73
 structures 72
Office buildings
 indoor buildings 334
Oil
 consumption worldwide 179
 tanker accidents 372–3
Ore minerals 196
Organic compounds
 biodegradation
  controlling factors 59
  in natural waters 52–3
  of organochlorines 60
  type reactions 47
 USA production 271
 see also Organic solvents
Organic pollutants
 adsorption and decom-
  position on soil 58–61

WHO guidelines for water 82–3
 see also Organic solvents;
  Organochlorines
Organic solvents
 fire at CHEMSTAR 371
 in paintworks 331
 polluting aquifers 44–5, 61
 toxicology 276
Organochlorines 277–8,
  282–306
 Agent Orange 298–301
 Binghampton incident 296
 carcinogenicity 303–5
 degradation in soil 60
 dioxin incidents 299
 in dolphins 295
 indoors 331
 PCB incidents 295
 production 282–3
 risk assessment 304–5
 at Séveso 301–3
 in sewage sludge 362, 365
 soil standards 78–80
 USEPA 303–5
Organophosphorus pesticides
  312–17
 toxicity data 315
Otter
 decline of 295
Overpotential 50–1
Oxford University
 building damage 187
Oxidation 46–52, 315
 of hydrocarbons 332
 microbial 60
 in superheated water 369
Oxides of nitrogen
 and acid rain 174
 in cities 260–6
 emissions
  in Germany 176
  by source 175
  from vehicles 176, 256
 indoors 334–5
 and ozone 174
 in the UK 262–4
 reaction with radicals 175
 see also Nitric oxide;
  Nitrogen peroxide;

Nitrous oxide
Oxides of sulphur, *see* Sulphur
  dioxide; Sulphur trioxide
Ozone 157–64
 air quality standards 164
 in cities 263–5
 depletion 159–62
  by CFCs 161, 274
  by solvents 273
 diurnal levels 163–4
 formation 157
 indoors 333–4
 IR spectrum 168
 reaction with NO 160, 334
 properties 157
 stratospheric 158–62
 TLV 164
 tropospheric 162–4
 in water 164

P-450 cytochrome 271, 305, 339
PAH 12, 267–71
 determination by HPLC 116,
  118
 in food and water 270
 indoors 338–9
 oxidation 61
 in particles 251
 from petrol 269
 in sewage sludge 365
 soil standards 78–80
PAN
 formation 35, 259–60
Paraquat
 adsorption 54
Parathion 314–15
Paris
 vehicle usage 177
Particles 248
 from cigarettes 337
 in cities 263–5
 coal dust 241
 filtration 109, 248–9
 indoors 329, 343
 silica 241
 soot 247
 see also PM_{10}
PCB 293–7
 dioxin-like compounds 298, 303

incidents 295
in the sea 368
synthesis 294
Yusho incident 294
PCDD 297–305
attack by microorganisms 307
carcinogenicity 303–4
cost of control 305
incidents 299
from incinerators 306–7
levels in serum 301
risk assessment 304–5
Séveso incident 301–3
teratogenicity 300
toxic effects 303
PCDF
in humans 296
from PCBs 296–7
Peat
composition 253
Pentachloroanisole
in brown snow 61
Pentachlorophenol
in drinking water 83
in lake water 121–3
Perchloric acid 149
Peregrine falcon
in the UK 291
Permethrin 310
Peroxy radical 259, 261
catalysis by 292
in odour treatment 324
ozone generation 266
Perylene
structure and activity 268
Pesticides 25
in aquifers 61
banned in USA 321
carbamate 318–20
control of 291–2
growth in use 86
natural 307–12
organophosphorus 312–18
pollution by 370
separation by HPLC 121
soil standards 78
TLC of 89, 90
WHO guidelines for water 82
Petrochemicals

production 282–3
Petrol
benzene content 277
compared with natural gas 257
critical concentration in air 84
leaded/unleaded usage 345–6
see also Petroleum
Petroleum
consumption 253
fractions 255
organic components 254
refining costs and control 6
release 255
see also Petrol
pH
of landfill leachate 357
of stomach 337
of surface waters 183
typical values 182
Phenanthrene 268
Phenoclor 293
Phenols
analysis 117
formation in air 266
in mammals 271
USA production 271
see also Chlorophenols
Phosgene 278, 319, 375
Phosphate
in detergents 280–1
Photochemical reactions 35,
309, 332
Phthalates 73–4
as stationary phase in GLC 93
Physostigmine 318–19
Phytochelatins 192, 207
Phytotoxicity 207
Pica 201
Picloram
in humans 300
Pinene
indoors 333
Pitchblende 220
Pittsburg
SO_2 pollution 181
Planck's equation 124
Plants
fluorine in 220
monitoring 149–50

uptake of metals 206, 209, 218
Plasma arc
for emission spectroscopy
131–2
for waste disposal 369
Plastics
hot melting 273
metal content 199
Plumes
in aquifers 43
chimney 29–31
tracing 223
Plutonium
in breeder reactors 230
fission of 224
release at Cheliabinsk 234
toxicity 231
PM_{10}
in cities 263–5
indoors 329
in the USA 248
in Utah, mortality 250
see also Particles
Pneumoconiosis 242
Pollutants
in contaminated land 77–81
critical concentration of gases
84
dispersion in water 36–45
in drinking water 81–3
effect on mammals 68–9
fate of 62–3
primary and secondary 5–6
reactions in air 34–5
trapping by TENAX 99
types and their transport
18–19
from vehicles 177–9
see also Multipollutant
situations; Pollution
Pollution
by aviation 160
controls 164
definition 5, 147
of estuaries 38–9
of groundwater 39–41
by hydrocarbons 163
indoor 26, 33–4, 329–50
of the Irish Sea 230

Pollution *contd.*
  of lakes 37–8
  in London 181
  long-term problems 369–72
  by metals 190–220
  from mining 195–7
  model pathway 18
  non-point sources 33–4
  by PCBs 294–5
  Pittsburg incident 181
  of rivers 36–7
  of the sea 368
  of soil 146
  from warfare agents 374–6
  of wells 286–7
  see also Indoor pollution;
    Pollutants
Polonium 220
Polybromobiphenyls 295
Polychloroterphenyls 294
Polyvinyl chloride (PVC) 292–3
Population
  and energy demand 167–8
  of Europe 10
  of the Third World 11
  of the world 10
Potassium
  radioactivity 223
Power stations
  emissions
      of NO_x 179
      of SO_2 180
      uranium from 198
  see also Nuclear power
Pre-column
  for HPLC 120
Predatory birds
  death of 290–1
Pressurized water reactor
  (PWR) 226–7
  release of hydrogen from 236
Printing
  metals from 199
Proinsecticides 315
Propylene
  for detergents 279
  USA production 271
Pyrene 268
Pyrethrins 308–9

structure activity relations
    308–9
  synthetic analogues 309–11
  toxicity data 311
Pyrethrolone 308
Pyrimicarb 319, 320
Pyrite
  bacterial oxidation 51
  in coal 188
  reactions 57

Quadrupole
  focusing 107
Quantum theory 124
Quebec
  incineration study 306

RAD 222
Radical reactions 34–5
  of CFCs 161
  inactivation 74
  with PAH 269
  in PAN formation 259
  in polymerization 292–3
Radioactivity
  in buildings 221
  human exposure 236
  protection from 222–3
  toxicity measurement 222
  types of emission 220–2
  units 222, 347
  see also Radionuclides
Radionuclides 220–42
  artificial production 223
  emission energies 22
  pollution from submarines 237
  from radon 346
  see also Radioactivity
Radon
  indoors 346–8
  remedial measures 348
Rats
  cancer from $CHCl_3$ 277
  LD_{50} values
      organophosphorus pesticides
      315
      pyrethrins 311
  teratogenicity 300
  toxicity of PAH 268

Red list (DOE) 22, 24, 322
Red spider mite 314
Redox reactions 46–52, 201
Reduction 46–50
  potential 50–1
REM 222, 236
Response factor
  in HPLC 123
Resistant strains 289
Rhine
  pesticide pollution 36
Risk
  from chemicals on land 77–81
  from radon 346
Rivers
  monitoring of 143–5
  pollutant transport in 36–7
  Pollution Act (1876) 7
RNA 269
Rockwool 342
Roentgen 22
Rotenone 311
Rothamstead Research Station 309
Ruthenium
  radioactive 231

SAAB Company 275
Sampling
  of chimney effluent 108–9
San Francisco
  air pollution 28
Sandoz Company 36
Sarin 312–13, 318
Scripps Institute (La Jolla) 174
Sea levels
  increasing 172
Sea water
  heavy metals in 202
  sewage discharge 360–1
  sludge disposal 363
  waste dumping 368
Seal
  organochlorines in milk 295
  populations 295
Sedimentary rocks
  as source of metals 193
Selenium
  mammalian LD_{50} values
Sellafield (Windscale) 233–4, 368

Septic tanks 44
Séveso 301–3
Sewage treatment 360–5
    amounts of sludge 362–3
    metals in sludge 199, 363–5
    odours 323
    organochlorines in sludge 365
    sea disposal 13
    stages 361–2
Shale oils 8
Shanghai
    air pollution 263
Sheep poisoning 213
Sick office syndrome
    allergens and 344
    asbestos levels 341, 342
    NO_x and ozone levels 334
    see also Indoor pollution
Silent Spring 9
Silicon tetrafluoride 219
Silicosis 242
Silylation
    for GLC 91–2
Simazine 321–2
Skatole 323
Smelters
    arsenic from 198
    at Avonmouth 198
    beryllium from 219
    damage to plants 184
    lead from 215
    in Lower Swansea Valley 198
    at Zlatna, Romania 196
    see also Smoke
Smoke
    cigarette 336–40
    and nicotine risk 336
    toxic components 337
    see also Smelters
Soil
    colloids 56
    composition 53–5
    decomposition of organics in
        58–61
    on derelict land 148–9
    in gardens 148
    heavy metals
        deposition of 191
        maximum levels in 363–5

in soil 205
organic matter in 54–5
PCBs in 295, 321–2, 372
pesticides in 291
physico-chemical properties
    53–5
pollution monitoring 146–9
risk from chemicals 77–81
solvent pollution 371–2
transportation 61
typical profile 62
Solar energy 167–8
Solubility product 53
Solvent abuse 272, 276–8
Solvents, see Organic solvents
Soot
    organochlorines in 296
South America
    use of pesticides 316, 317
Spain
    NO_x and SO_2 production 187
Sphalerite
    oxidation 195
Stability field
    of water 50–1
Steel
    aerosols from 198
    PCDD from 299
Stratosphere 27
Straw burning 249
Strontium 90 230
Structure–activity relations
    308–9
Strychnine
    lethal dose 300
Styrene
    indoors 333
    USA production 271
Sudan
    Endosulfan poisoning 287
Sugar cane 174
·Sulphonate detergents 279–80
Sulphur
    in petrol 178
    radioactive 223
    removal from fuels 188–9
Sulphur dioxide 179–90
    in chimney effluent 112–13
    in cities 263–5

critical concentration 84
effect on the environment 181
emissions in Germany 176
indoors 336
in lakes 181–3
toxicity 84, 181, 249–50
units of measurements 186
Sulphur trioxide 180
Sulphuric acid
    formation in air 180
    in smuts 247
    sulphonation by 279
Sun
    plasma 238
'Superfund' legislation (USA)
    370
Superheated water
    for oxidation of organic
        chemicals 369
Surfactants, see Detergents
Swansea Valley
    smelter 198
Sweden
    lake pollution 182
Switzerland
    forest damage 185–6
    petrol consumption 345
Synergism 373–4
Synthesis
    of Aldrin 286
    of carbamates 320
    of chlorophenols 298, 305
    of dioxin 298
    of Lindane 284–5
    of Parathion 315
    of PCBs 294
    of PCDFs 297
    of PVC 292–3
    of pyrethrins 309–10
    of vinyl chloride 292
Systemic insecticides 316

Tabun 318
Tailings
    from mines 195
Taipei City
    indoor pollution 330
Taiwan
    PCB poisoning 295

Tanker accidents 372–3
Tar oils 150
Target values
  for soils 78–80
TCDD 297–305
  in air 306, 307
  and birth defects 298, 300
  carcinogenicity 303–4
  chlorophenols and 297–8
  distribution coefficient 59
  in humans 300–1
  pollution incidents 299
  risk assessment 304–5
  at Séveso 301–3
  in soil 147
  solubility 59
Tea plant 218, 220
Tenax 99
Teratogenesis
  of benzpyrene
  defined 69, 268
  of Endosulfan 287
Test panels 323
Tetraethylpyrophosphate 312
Thiols
  in sewage 323
Third World
  hazards at work 343
  indoor pollution 251
  pesticide use 284–5, 292
  population 10–11
  respiratory disease 252, 342
  urban pollution 251
Thorium 221
Three Mile Island 234–6
Threshold concentration 323
Thyroid gland
  and iodine 131 231
Tiger fish 290
Tilapia rendalli 290
Times Beach, MS 299
TLC 86, 88–90
Tobacco, see Cigarettes
Tokyo
  street dust components 338
  terrorism 318
Toluene
  at former landfill 371
  indoors 330

metabolism 65–6
oxidation in air 267
in paintworks 331
production in the USA 271
toxicity 277
Total hydrocarbons
  in chimney effluent 112–13
Toxic equivalent factor (TEF)
  298
Toxicity
  of Aldicarb 320
  of Aldrin and Dieldrin 286
  of aromatic hydrocarbons 330
  of asbestos 341–3
  of coal dust 242
  of Endosulfan 287
  of lead 344
  of Lindane 285
  of VOCs 330
Toxicology
  basic principles 64–8
Toxins 76
  natural, lethal doses 66–7
Trace enrichment
  of chlorophenols 121–2
  of pesticides 120–1
Trace metals
  coprecipitation 57
  replacement order on soils 56
Transfer coefficients 205–6
Transformer fires 296–7
Transport
  in cities 262–6
  as pollution source 20
Trees
  damage classification 185
Tributyltin 203
Trichloroethane 272–4
  for dry cleaning 275
  indoors 330
  ozone depletion by 272–3
  water pollution by 278
Trichloroethylene
  mass spectrum 102
Trichlorophenoxyacetic acid
  287–8, 298, 300
Trifluralin
  in brown snow 61
Trigger concentrations 77

of metals 210
Tritium
  fusion reaction 239
Troposphere 27
Tryptophan
  pyrolysis products 340
Tsetse fly 292
Tumbleweed 317

Ultraviolet light
  dealkylation by 322
  detector for HPLC 116
  as disinfectant 362
  and ozone 160
  for synthesis 284
Union Carbide Company 319
United Kingdom
  diesel vehicles 178
  emissions
    of NO_x 179
    of SO_2 179
  forest damage 185
  Forestry Commission 184
  nuclear stations 233
  petrochemicals 255
  radon, risk from 346, 348
  statutory limits
    for carbonyl compounds 332
    for vehicle emissions 248
    metals in drinking water 204
  total waste arising 360
  urban pollution 262
United Nations panel on
  Climate Change 173, 174
Ural mountains
  nuclear accident 234
Uranium 238
  decay series 221
Urban pollution 260–6
  diurnal cycle 260
US Bureau of Standards 138
US Health Effects Research
  Laboratory 74
USA
  chemical production 271, 283
  energy use 255
  GDP 12
  hazardous waste 366
  indoor pollution 333

lead deposition 344
lung cancer death rate 347
maximum levels of metals in
    amended soils 365
municipal waste 354
standards for
    asbestos 342
    metals in drinking water
        204
    water 82
see also EPA

Van Deemter equation 97
Vehicle emissions 177–9,
    256
by class of vehicle 256
control 256–8
hybrid vehicles, reduction
    from 256
in major cities 177
in street dust 338
see also Air Pollution
Ventilation coefficient 29
Venus 170
Vietnam War 298, 300
Vinyl chloride 69, 292–3
mass spectrum 102
toxicology 277
from waste 346
VOCs
emission standards 359
indoors 330–3
see also Hydrocarbons
Volcanoes
metal emissions 191, 195
Vulcan 306

Waldsterben 9
Warfare agents 312, 317–18, 374–6
disposal 317–18
pollution from 374–6
Waste disposal 354–69
amounts dumped at sea 368
landfilling 354
by microorganisms 368
sewage 360–5
see also Wastes
Wastes 13, 353–77
abandoned sites in Europe 372
amounts dumped at sea 368
amounts from industry 354
amounts produced, by type 353
composting 359
in England and Wales 360
from mines 197
municipal yield per capita 354
recycling 359–60
volatiles 346
Water
analysis of polluted 120–3
asbestos in 341, 342
chlorination 145
dispersion of pollutants 36–45
drinking standards
    for metals in 204
    for other inorganic in 204
    PAH in 270
pollution
    monitoring 143–6
    by solvents 276–8
    by triazines 321–2
softening 281
stability field of 50–1

see also Drinking water
Wells 40–1
polluted 286–7, 320
Wheat bulb fly 290
White rot fungus 307
Windscale (Sellafield) 233–4, 368
Wood 257
burning 336
composition 253
for fuel 12, 167
metals in preservatives 197
World Health Organisation (WHO)
guidelines
    for air quality 262
    for drinking water 82, 83
    for indoor air 251
    for metals in water 204
    for NO_x and SO_2 188
    for ozone 164
    for particles 252
    for smoke 250

X-ray diffraction 341
Xenon as moderator 236
Xylenes
indoors 330, 331
production in the USA 271

Yusho incident
PCB poisoning 294–5

Zeolites 281
Zinc
in landfill leachate 357
ores 195–6
principal features 215–17
production 191